MW00736822

# The Basics of Film Processing in Medical Imaging

Arthur G. Haus
Susan M. Jaskulski

Medical Physics Publishing
Madison, Wisconsin

ISBN: 0-944838-78-2 Hardcover 0-944838-79-0 Softcover

*Library of Congress Cataloging-in-Publication Data*

Haus, Arthur G.
Basics of film processing in medical imaging / Arthur G. Haus, Susan M. Jaskulski.
p. cm.
Includes bibliographical references and index.
ISBN 0-944838-78-2 (hardcover). -- ISBN 0-944838-79-0 (softcover)
1. Radiography, Medical--Processing. 2. Radiography--Films.
I. Jaskulski, Susan M., 1953- . II. Title.
[DNLM: 1. X-Ray Film. 2. Diagnostic Imaging. 3. Quality Control.
WN 150 H376b 1997]
RA78.H38 1997
616.07'572--dc21
DNLM/DLC
for Library of Congress 97-18596
CIP

Medical Physics Publishing
4513 Vernon Blvd.
Madison, WI 53705
(608) 262-4021

*This book is dedicated:*

*by Arthur Haus to his wife, Sandy; his children, Tim and Kris (Niles); his grandchildren, Connor and Madison Mae; and his parents, Jean and George.*

*by Susan Jaskulski to her mother, Doria.*

# Contents

# Foreword

The role of medical imaging is visualization of selected parts of the human body, to determine whether they are normal or pathologic. Keeping in mind that microscopic examination of any tissue sample can be accomplished on the order of microns with a concomitant high contrast, the medical imager has to work with images of vastly inferior resolution. In order to accomplish the difficult task of in vivo imaging, one has to understand both the object being imaged and the capabilities and limitations of the imaging process, which functions as a chain of events. When film is used as the primary receptor, the quality of the final image will depend on the quality of the film itself and the processing to which it is subjected. The paradox of imaging is that although processing is the weakest link in the chain, it is also the one that has been the most neglected. Few modalities of medical imaging are as sensitive to variations in exposure and processing as is screen-film mammography, which has served as a catalyst for their optimization. In order to visualize the subtle soft tissue attenuation differences necessary to find breast cancer at an early stage, focus had to be directed to all aspects of the imaging chain, particularly film processing. The opposing demands of improved image quality and reduced radiation dose have necessitated a thorough review of the entire process. Although much progress has been made, this is a dynamic field with ever-improving film and chemistry technology requiring frequent review of the imaging chain.

This book, *The Basics of Film Processing in Medical Imaging*, is a most welcome reference for all of us in the medical imaging profession. The authors, Art Haus and Susan Jaskulski, are indeed superbly qualified to share their expertise with us. All aspects of the complex subject of film processing are dealt with in an exemplary fashion. The authors are to be commended for an excellent job.


László Tabár, M.D.
Associate Professor
Director, Department of Mammography
Falun Central Hospital
Falun, Sweden

# Preface

Film processing has been a much discussed subject in medical imaging since the 1970s, but the conferences and books devoted solely to it are few and far between and, therefore, stand out as significant milestones.

In March of 1977, a three-day conference called the "Second Image Receptor Conference: Radiographic Film Processing" was held in Washington, DC. This conference, with a faculty of recognized experts and sponsored by the Bureau of Radiological Health, Food and Drug Administration, Public Health Service, U.S. Department of Health, Education, and Welfare, was "held to address solutions to problems associated with radiographic film processing resulting in poor image quality and increased patient radiation exposure. It was concluded by the group that significant problems with radiographic processing do exist; solutions to these problems include both improvements in processing equipment performance and the development and promulgation via education and guidelines of facility-based quality assurance programs."

In 1976 and 1977, two books authored by Joel E. Gray were published by the Food and Drug Administration: *Photographic Quality Assurance in Diagnostic Radiology, Nuclear Medicine, and Radiation Therapy, Volumes 1 and 2* (HEW publication no. [FDA]76-8043 and HEW publication no. [FDA]77-8018).

In 1983, *Quality Control in Diagnostic Imaging* was published by Joel E. Gray, with coauthors Norlin T. Winkler, John Stears, and Eugene D. Frank.

In 1986, the American College of Radiology (ACR) introduced the Mammography Accreditation Program (MAP). This program currently requires submission of processor quality control data.

In 1990, the ACR Committee on Quality Assurance in Mammography, chaired by R. Edward Hendrick, developed manuals on quality control in mammography for the radiologist, radiologic technologist, and medical physicist. Revisions have been made to the manuals in 1992, 1994, and 1996. Several of the tests that are performed are related to film processing.

In March of 1992, a symposium called "Film Processing in Medical Imaging: A Practical Update for the '90s" was held in Rochester, NY. This meeting was cosponsored by the Upstate New York Chapter of the American Association of Physicists in

Medicine, the Rochester Institute of Technology (RIT) Center for Imaging Science, and Eastman Kodak Company. The proceedings of this conference were published in 1993 as *Film Processing in Medical Imaging* (Arthur G. Haus, editor).

This book, *The Basics of Film Processing in Medical Imaging*, is intended to meet the ongoing demand from all individuals involved in medical imaging for useful, practical information on film processing, a need that still exists despite the passage of over 20 years since the first conference held by the Food and Drug Administration. It is not an updated version of the published proceedings from the conference held at RIT but a compilation of current information. In addition, this book contains discussions of many of the questions/issues that continue to surface again and again.

*The Basics of Film Processing in Medical Imaging* is organized into seven chapters, including film, chemicals, processors, image quality, quality control, artifacts, and troubleshooting. Most chapters begin with historical or background information to highlight how far processing in medical imaging has already come, even though work certainly remains to be done. Appendices include reciprocity law failure and latent image fading, manual processing, handling and processing of mammography film, the mammographic darkroom, cleaning intensifying screens, mobile van film processing, and the sensitometric technique for the evaluation of processing (STEP) test. References are included at the end of each chapter. Note that the publications listed as references are not intended to be all-inclusive, but those still generally available. There are some underlying themes throughout this book—specifically, how important both following the manufacturers' recommendations and regular maintenance of the processing environment are to achieve properly processed images.

Some might think that medical imaging, poised on the brink of new imaging technologies that will be essentially filmless—such as picture archiving and communication systems (PACS) with soft copy display, and dry-processed films for medical imaging—has no need for information on conventional film processing using liquid forms of chemicals. More research will be needed, however, to understand the implications of using any new technologies in terms of image stability, film handling requirements, and quality control before they move from research facilities to general use by all medical imaging facilities.

The need for knowledge about conventional film processing and about the quality control activities necessary to achieve and maintain high image quality will continue for many years yet to come. This book, therefore, is intended to be a ready reference while that need exists.

ARTHUR G. HAUS
SUSAN M. JASKULSKI

# Acknowledgments

We want to recognize and thank a number of people for their help and support in preparing this book.

A special thanks goes to László Tabár, Associate Professor, Falun Central Hospital, Falun, Sweden, for writing the foreword and being a source of inspiration to both of us.

We also want to recognize Joel Gray, Consultant, Rochester, Minnesota, and R. Edward Hendrick, Associate Professor, University of Colorado, Denver, Colorado, for their leadership and many contributions in quality control. We also thank Joel Gray for reviewing Chapter five on quality control.

We thank the following people from Eastman Kodak Company for providing input and reviewing various chapters:

Robert Dickerson (Chapter one), Franklin Brayer, Richard Cataldi, Peter Cumbo, Alan Fitterman, and Christopher Rau (Chapter two), Kenneth Oemcke (Chapter three), Philip Bunch and Kenneth Huff (Chapter four), Charles Baker (Chapter five), John Widmer (Chapter six), Ronald Lillie (Chapters six and seven), Karen Ursel (Chapter seven), Laura Gronberg (Appendices), and Norman Geil (all chapters).

We also thank Martin Coyne for supporting the writing of this book; Donna Fink for her exceptional efforts in reviewing the entire volume and providing editorial comments; and Debra Roschetzky, Thomas Bonisteel, and Linda Blauers for providing administrative assistance.

We are grateful to E. Lee Kitts, Research Fellow, Sterling Diagnostic Imaging, Inc., Brevard, North Carolina, for providing input on Chapter five on quality control, especially concerning the crossover procedure.

We also thank Carol Mount, Medical Imaging Technical Specialist, The Mayo Clinic, Rochester, Minnesota, for providing the guidelines for diagnosing artifacts on film caused by equipment in Chapter six.

We thank Orhan Suleiman, Chief, Radiation Programs Branch, Division of Mammography Quality and Radiation Programs, Food and Drug Administration, for providing input on Appendix G, the STEP test.

We thank the following companies and organizations for providing figures: Agfa Gevaert NV., The American College of Radiology, Eastman Kodak Company, Konica Medical Corporation, Imation Corporation, Nuclear Associates, Sterling Diagnostic Imaging, Inc., and Radiation Measurements, Inc. (RMI).

A special thanks goes to Elizabeth Seaman, Managing Editor, Medical Physics Publishing, Madison, Wisconsin, for her careful review of the book and for providing excellent editorial comments and suggestions. We are also grateful to Peggy Lescrenier, President, Medical Physics Publishing, for supporting this project.

Finally, we have had the opportunity to work with many individuals who have provided expertise, leadership, support, and education on quality control. We want to thank the following medical physicists, technologists, radiologists, and other professionals who have been particularly valued colleagues:

> Valerie Andolina, Patricia Barnette, Gary Barnes, Lawrence Bassett, Daniel Bednarek, Richard E. Bird, Margaret Botsco, Terese Bogucki, Libby Brateman, Jo Ann Bresch, Stewart Bushong, Patrick Butler, Priscilla Butler, John Coscia, Kenneth Coleman, Richard Cross, Angeline Cullinan, John Cullinan, Peter Dempsey, Debra Deibel, Gerald Dodd, Carl D'Orsi, Kathleen Durrell, G. W. Eklund, Stephen Feig, Donald Frey, Rodney Fuller, Joel Gray, Kathleen Harris, Mary Ann Harvey, Rita Heinlein, R. Edward Hendrick, Maynard High, Russell Holland, Valerie Jackson, Caroline Kimme-Smith, E. Lee Kitts, Daniel Kopans, Stuart Korchin, Shelly Lillé, Michael Linver, Andrew Maidment, Anna Marie Mason, Richard Massoth, John McCrohan, Robert McLelland, Robert Meisch, Ellen Mendelson, Richard Morin, Ingrid Naugle, Edward Nickoloff, Thelma Papini, Connie Petrovich, Douglas Pfeiffer, Etta Pisano, Robert Pizzutiello, Jr., William Poller, Christine Quinn, Raymond Rossi, Lawrence Rothenberg, Connie Rufenbarger, Ellen Shaw de Paredes, Edward Sickles, Robert Smith, Perry Sprawls, Carol Stelling, Keith Strauss, Orhan Suleiman, Daniel Sullivan, László Tabár, Raymond Tanner, Jerry Thomas, Irena Tocino, Robert Uzenoff, Carl Vyborny, Christine Watt, Gini Wentz, Pamela Wilcox-Buchalla, Kathleen Willison, Brian Wing, Martin Yaffe, Wende Logan Young, and Marie Zinninger.

# 1

# Film

- History
- Film Emulsion Characteristics
- Latent Image Formation and Amplification
- Storage

Film used for medical radiographic imaging has evolved over the last century, beginning with its first use in 1896, the year after Roentgen discovered x-rays. Its history is intertwined with important developments in emulsion supports or bases and fluorescent intensifying screens.

## History

The first photographic image was recorded in the early 1800s (Figure 1-1), and Joseph Nicephore Niepce of France is credited with making the first permanent photograph in 1826 by exposing a light-sensitive material in a camera obscura. The camera was constructed by making a small hole in a large box. The image was formed by a lens, reflected by a mirror onto a ground glass, and then hand-traced to make it permanent.

In England in 1851, Frederick Scott Archer used glass with the wet collodion process. Wet collodion (a sticky wet substance) was used as a binder for coating silver salts on glass. A disadvantage of this method was that the collodion had to remain wet during exposure and development.

Gelatin silver bromide dry plates were invented by Richard L. Maddox (England, 1871). Gelatin replaced collodion, making the photographic process easier to manage because the plates were used dry.

The manufacture of dry plates soon began in several countries. The plates were coated by hand until 1879, when George Eastman invented a plate-coating machine (Figure 1-2). Most of the coatings were still on glass, but paper was also used as a support for the sensitive emulsion. In 1889, Eastman introduced flexible transparent film made of cellulose nitrate to support the emulsion (Figure 1-3). This greatly advanced photography and made roll film possible. When Roentgen discovered x-rays in 1895, glass plates, flexible films, and sensitized papers were already available to record the radiographic image, and his original communication indicated the importance of the photographic plate as a means of recording the radiographic image. He demonstrated his discovery with a radiograph of his wife's hand.

Most early radiographs were made on glass plates because it was generally felt that plates were superior to film and because their characteristics were better understood. Emulsion was coated onto only one side of the glass.

The earliest plate made strictly for radiographic purposes probably came from Dr. Carl Schleussner (1896), a German plate manufacturer who, at Roentgen's request, made some plates with a thick silver bromide emulsion (Figure 1-4). The Schleussner rapid photographic plate had a thicker emulsion coating than the usual English and American plates and, therefore, exhibited greater photographic density. These plates became popular in the United States and England.

Intensifying screens were also under development during this time. From January to March 1896, Thomas A. Edison and his associates tested approximately 8500 different materials in their effort to build a new incandescent lamp. During this work, about 1800 substances were found to be fluorescent. Of these substances,

**Figure 1-1** The first known photograph. In 1827, Joseph Nicephore Niepce made a crude photograph of his house in Saint Loup-de Varenness using a camera obscura. The photograph required eight hours of exposure time. (Reprinted with permission from RadioGraphics 1989; 9:1204.)

**Figure 1-2** Photograph of a box of Eastman's gelatine dry plates, 1884. (Reprinted with permission from the George Eastman House.)

**Figure 1-3** In 1889, George Eastman introduced flexible film in rolls, a lightweight, non-breakable substitute for glass. The transparent nitrocellulose roll film was cast on these 200-foot-long tables. (Reprinted with permission from the George Eastman House.)

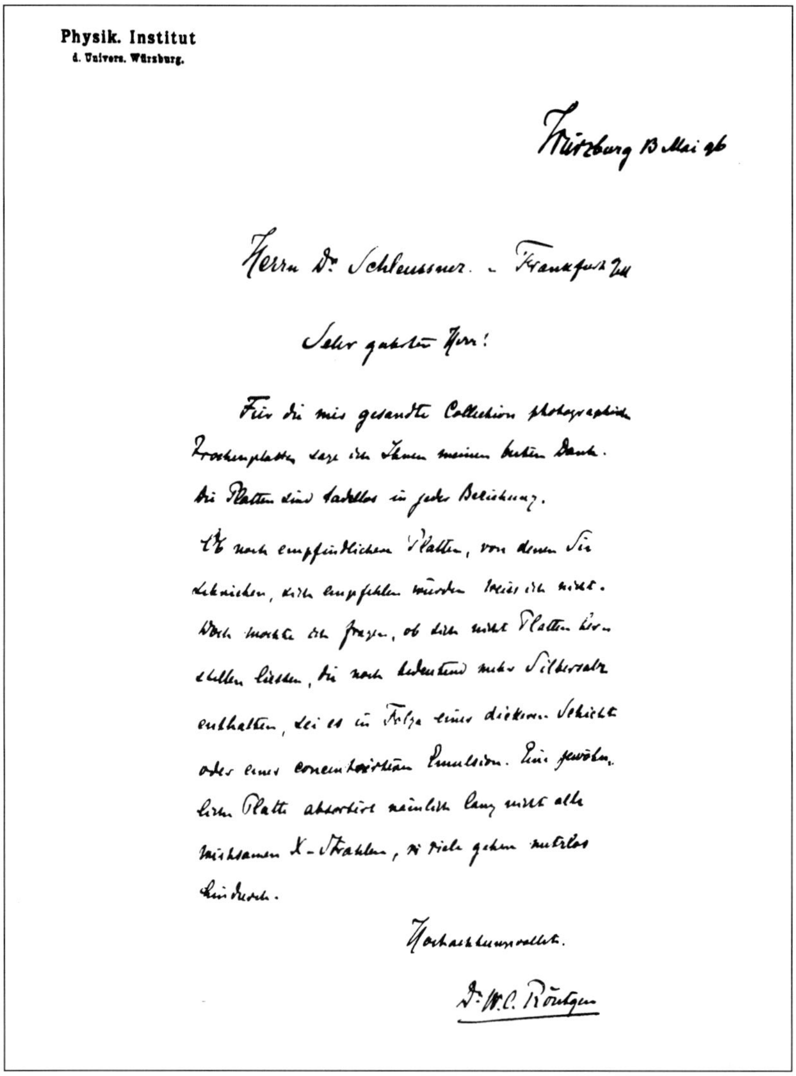

**Physik. Institut**
d. Univers. Würzburg.

Würzburg 13 Mai 96

Herrn Dr Schleussner – Frankfurt a/M

Sehr geehrter Herr!

Für die mir gesandte Collection photographischer Trockenplatten sage ich Ihnen meinen besten Dank. Die Platten sind tadellos in jeder Beziehung. Ob noch empfindlichere Platten, von denen Sie schreiben, sich empfehlen würden weiss ich nicht. Doch möchte ich fragen, ob sich nicht Platten herstellen liessen, die noch bedeutend mehr Silbersalz enthalten, sei es in Folge einer dickeren Schicht oder einer concentrirteren Emulsion. Eine gewöhnliche Platte absorbirt nämlich lange nicht alle wirksamen X-Strahlen, zu viele gehen nutzlos hindurch.

Hochachtungsvollst.

Dr W.C. Röntgen

**Figure 1-4** Letter from Dr. Roentgen to Dr. Schleussner, 1886. The translation of the letter is as follows:

To: Herr Dr. Schleussner in Frankfurt/Main
Most Honorable Sir!
My best thanks for the collection of dry plates that you sent me. The plates are flawless in every respect. Whether still more sensitive plates that you wrote about might be recommendable, I don't know. But I should like to ask whether plates could be manufactured which contain significantly more silver salt, be it a consequence of a thicker layer or of a more concentrated emulsion. For a common plate does by far absorb all effective X-rays, too many pass through uselessly.
With Utmost Respect,
Dr. W. C. Roentgen

(Reprinted with permission from Dr. E. Lee Kitts, Sterling Diagnostic Imaging, Inc.)

calcium tungstate was found to have fluorescence approximately six times more intense than barium platinocyanide, the substance used by Roentgen during his historic discovery. Edison thought that fluoroscopic screens should be used for "real time" imaging, which he preferred to films made with dry plates.

In February 1896, Professor Michael Pupin of Columbia University made a radiograph using a fluorescent screen in combination with a film (Figure 1-5). An exposure lasting a few seconds was required, but this was shorter than direct exposures without intensifying screens because the film was exposed by light.

In 1897, Dr. Max Levy (Germany) was apparently the first to recommend and use a double-coated film between two intensifying screens. Until this time, emulsion was coated onto only one side of the first film bases.

The fluorescent intensifying screen, although introduced in the early years of radiology, did not gain much acceptance until several years later. Some reasons for its failure to win acceptance were: (1) the extreme amount of afterglow, (2) excessive

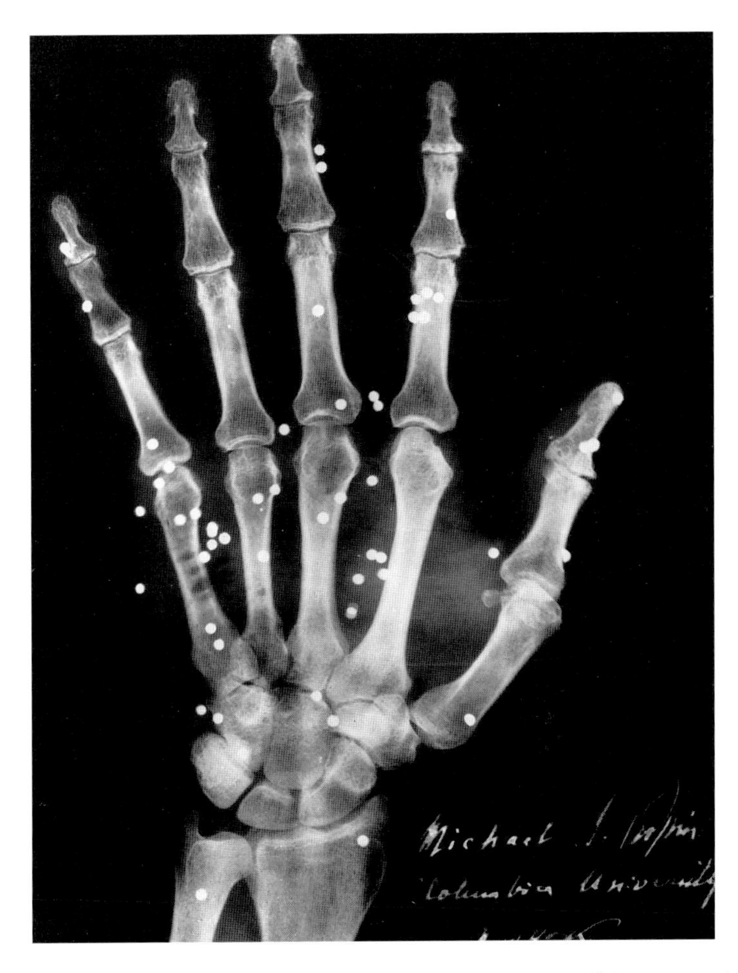

**Figure 1-5** The first radiograph made using a fluorescent intensifying screen in combination with a film, 1896. (Reprinted with permission from RadioGraphics 1989; 9:1208.)

"graininess" caused by the use of large-sized fluorescent crystals, and (3) nonuniformity of coating.

Carl V. S. Patterson was responsible for significant improvements in the manufacturing of fluorescent intensifying screens. In 1916, he produced screens of synthetic calcium tungstate that fluoresced brightly, had minimal afterglow, and overcame difficulties associated with impurities in the natural mineral.

The M. A. Seed Dry Plate Company of St. Louis, Missouri, and Wratten and Wainwright, Ltd. of Croyden, England, were early manufacturers of x-ray plates (1902-1912). In 1912, a glass plate coated with a thick emulsion that was rich in silver and contained a bismuth salt to make the emulsion more sensitive to the effects of x-rays was introduced (Figure 1-6).

Glass plates were popular in spite of several serious deficiencies. They were so fragile that shipment was extremely difficult. Glass plates were heavy—a 14 x 17-inch (35 x 43-centimeter) glass plate weighed about 2 pounds (907.18 grams); a 14 x 17-inch (35 x 43-centimeter) sheet of modern x-ray film weighs about 1.5 ounces

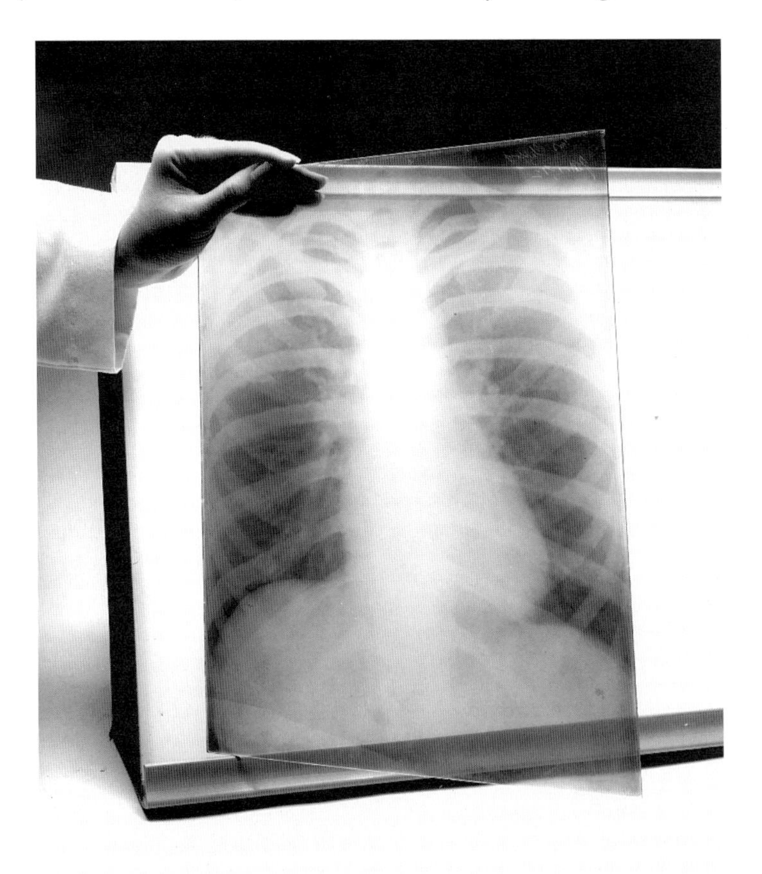

**Figure 1-6** This chest radiograph was recorded on a glass plate coated with emulsion in 1912. (Reprinted with permission from RadioGraphics 1989; 9:1209.)

(42.5 grams) (a factor of 20:1). Glass plates were expensive—a 14 x 17-inch (35 x 43-centimeter) plate cost about $1 in 1906. In that same year, a good suit of clothes cost $7, a pair of shoes cost $3, and steak was 15 cents a pound. In current U.S. dollars, a glass plate would cost approximately $100. Plates had other less obvious but serious defects. For example, because of the parallax introduced by the thickness of the glass, plates could be coated on only one side. Although better than those available before the turn of the century, these plates were still slow in speed and, when processed, lacked optimal contrast and sensitivity.

Many of the problems involved in using a glass plate as a support for the sensitive emulsion were recognized before Roentgen made his discovery, and roll films for photography had been marketed as early as 1889. The year 1914 marked the turning point. Because most of the glass needed for the manufacture of photographic plates was produced in Belgium, World War I halted the supply. Simultaneously, the demand for medical radiographs increased, making the need for suitable x-ray film urgent. In 1913, an x-ray film was introduced that consisted of a cellulose nitrate base coated on one side with an emulsion more sensitive to x-rays than any other previously available emulsion. Better radiographs could be produced with less exposure, and the new film permitted more extensive use of intensifying screens, with consequent reduction in exposure. Intensive research resulted in an improved product, and in 1916, a noncurling film base was introduced.

The year 1918 marked a significant advancement in x-ray film—the first film was introduced that was designed for radiographic purposes and had a high-speed emulsion coated on both sides of the support. It permitted the use of double intensifying screens, which further reduced exposure and exposure times. During the same year, the double-screen technique with "thin" front and "thick" back screens was developed.

Double-emulsion film had a profound effect on radiography; it made the use of the Potter-Bucky diaphragm (a device for the control of scattered radiation comprising a grid and a mechanism for moving it) practical for radiography of thicker body parts. The significant reduction in exposure provided by the combination of the new film and double intensifying screens made tolerable the increased exposure needed when the diaphragm was used. The improved diagnostic quality of the resulting radiographs was a significant factor in the growth of radiology in this period.

Strange as it may seem, it was not easy to convince radiologists and technologists that film offered any appreciable advantage over glass plates. There were years of prejudice to overcome, and film could not be processed by familiar techniques. Glass plates and film coated on one side could be processed in trays, but film coated with emulsion on both sides of the support could not be processed satisfactorily in this way (Figure 1-7). By 1920, however, film hangers and deep tanks for processing were readily available. Fortunately, excellent intensifying screens were also available, and the necessary cassettes and other film holders were soon placed on the market. Gradually, the reluctance to use film broke down, and the popularity of glass plates decreased.

In 1924, the first x-ray film with a cellulose acetate safety base was introduced, which removed the fire hazard of nitrate-based film. Some types of photographic film were available on cellulose acetate base before World War I, but efforts to use

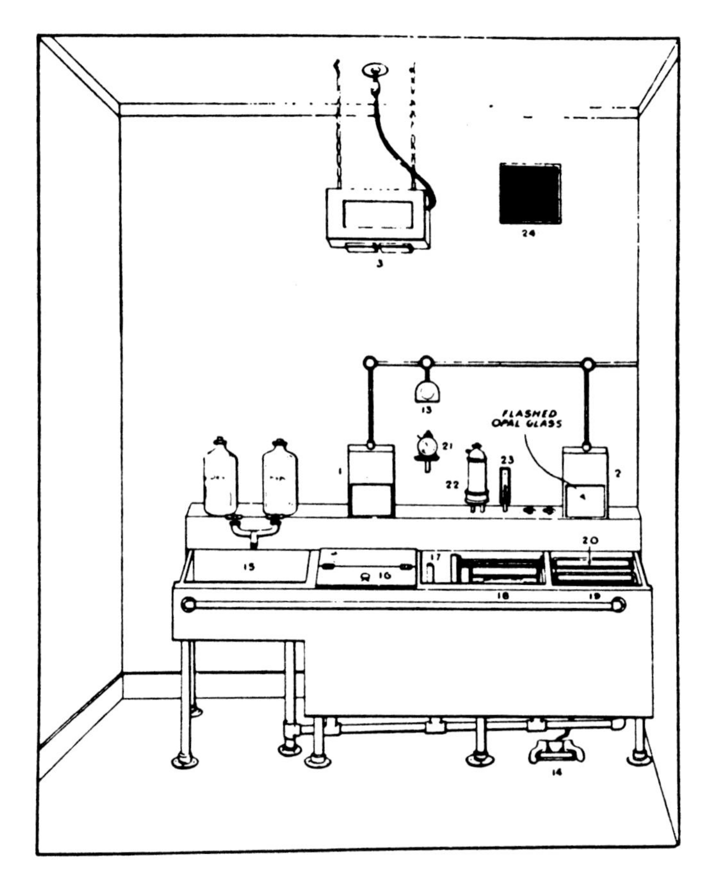


**Figure 1-7** Darkroom in 1918. Tray processing of glass plates coated with emulsion provided satisfactory results; double-coated film could not be processed in this way.

cellulose acetate as a base for x-ray film were unsuccessful until 1924. Even so, acceptance of the new safety film base was slow, and it was not until 1933 that the use of nitrate base was discontinued. This was prompted in part by the Cleveland Hospital fire in 1929 (Figure 1-8).

In 1933, an x-ray film with a blue-tinted base, which made medical radiographs more esthetically pleasing, was introduced. This change was adopted universally by all manufacturers of x-ray film.

A nonscreen x-ray film was introduced in 1936. Intended for x-ray exposure without fluorescent screens, this film had higher speed, contrast, and definition than screen-type films and was intended primarily for the examination of extremities.

In the 1930s and 1940s, most conventional medical radiographs were made using a pair of calcium tungstate intensifying screens with double-emulsion film (Figure 1-9). Figure 1-10 shows a double-emulsion film introduced in 1940 and widely used for the next 25 years.

In 1960, the first polyester-base film was demonstrated. It was thinner than the previously used base, reducing the parallax effect, and was fully commercialized in 1964.

In 1972, a high-resolution, single-screen, single-emulsion film combination was announced for mammography (Figure 1-11). With this combination, radiation dose

Complete Wire Reports of UNITED PRESS, the Greatest World-Wide News Service

# The Cleveland Press

STATE

CLEVELAND, FRIDAY, MAY 17, 1929 — PRICE THREE CENTS

# PLUMBER HELD IN BLAST PROB

## Blocked Fire Door Blamed for 126 Deat

### EXPERTS CLASSIFY DEATH-DEALING GAS

Declare Nitrogen Dioxide Formed by Burning X-Ray Films Caused Deaths After Blast

By DAVID DIETZ

BLOCKED FIRE DOOR BLAMED IN BLAST

### OFFICIAL FINDS PREVENTED SH

Steamfitter Questioned Ab ered Main; Doctors Fig Rising Death

Inquiry to determine if crimi responsible for the disaster tha at the Cleveland Clinic was beg Coroner A. J. Pearse.

The coroner, police and coun for more than an hour a steamf in the Clinic's X-ray plate st the first of two explosions.

**Figure 1-8** Cleveland newspaper headline after the 1929 Cleveland Hospital fire. (Reprinted with permission from RadioGraphics 1989; 9:1210.)

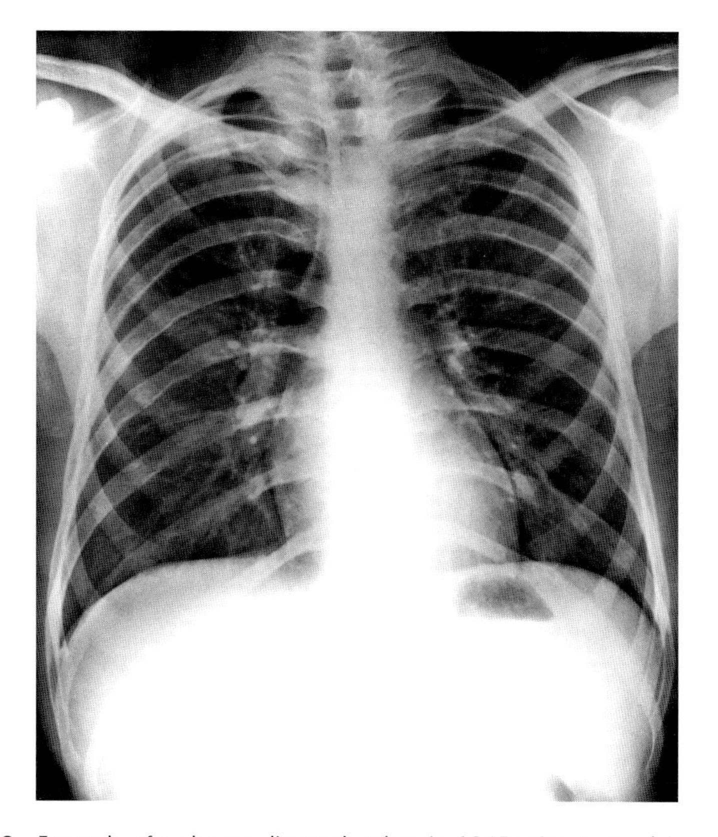

**Figure 1-9** Example of a chest radiograph taken in 1945 using two calcium tungstate intensifying screens with double-emulsion film. (Reprinted with permission from RadioGraphics 1989; 9:1211.)

**Figure 1-10** Double-emulsion Blue Brand film introduced by Eastman Kodak Company in 1940 and widely used for the next 25 years. (Reprinted with permission from Eastman Kodak Company.)

was significantly reduced compared with direct-exposure (nonscreen or industrial) film or xeroradiography, an imaging process also in use at this time, which employed electrostatically charged selenium plates.

Research with rare earth phosphors for color television tubes and image-intensifier tubes culminated in work by R. A. Buchanan, S. I. Finkelstein, and K. A. Wichersheim at Lockheed in the early 1970s, whose research suggested the use of rare earth phosphor screens for medical radiography. The rare earth phosphor screens, which were first marketed in 1973, significantly changed screen-film product offerings (Figure 1-12). The new screens used a high-absorption, green-emitting gadolinium oxysulfide phosphor coupled with orthochromatic films, primarily sensitive to green but also sensitive to blue light. Shortly thereafter, materials such as lanthanum oxybromide, yttrium oxysulfide, and barium fluorochloride, some having primary green and others having primary blue emission characteristics, were also used for fluorescent intensifying screens. Because of their absorption characteristics and increased efficiency in converting x-ray energy to light, rare earth phosphors made possible the production of faster screen-film combinations than those previously available. The ability to combine various screens and films selectively provides flexibility in techniques and improved diagnostic image quality.

Just as the use of rare earth phosphors of gadolinium and lanthanum oxysulfide provided a significant improvement in medical radiographic image quality, so did

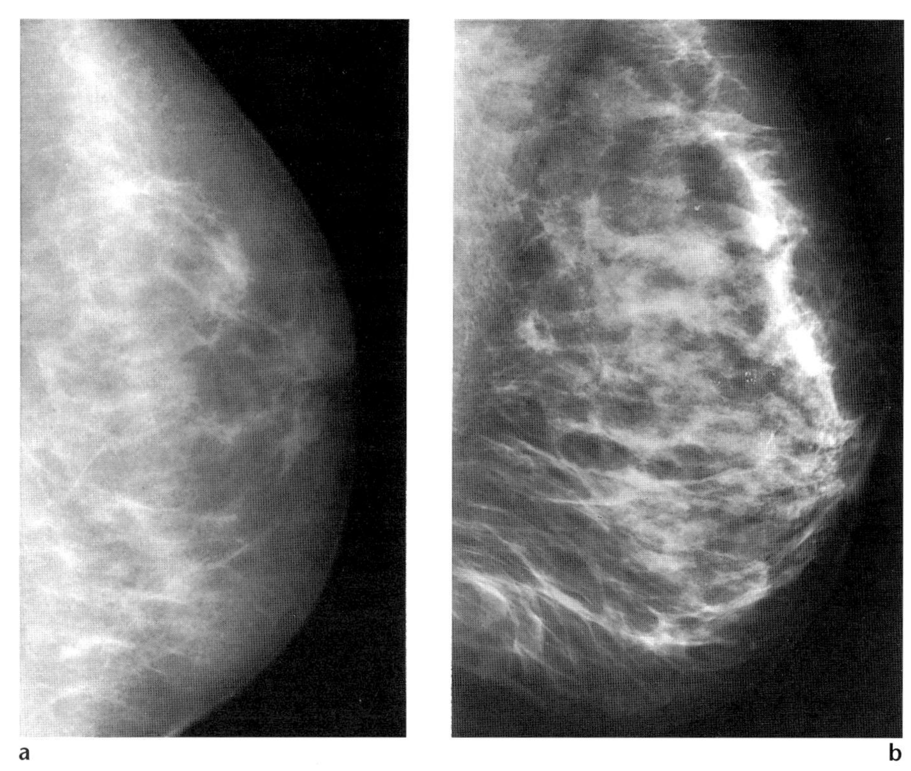


**Figure 1-11** (a ) Mammogram recorded on direct-exposure film using conventional, non-dedicated x-ray equipment—a system used until the late 1960s and early 1970s. (b) Mammogram recorded using dedicated x-ray equipment and a high-resolution single-screen, single-emulsion film combination designed for mammography. With this combination, radiation dose was significantly reduced compared with direct exposure. (Reprinted with permission from RadioGraphics 1989; 9:1215.)

the 1983 introduction of tabular-grain emulsions for green-sensitive films. Prior to tabular-grain emulsion technology, green-sensitive films featured conventional "pebble-shaped," or three-dimensional (3D), grains. Tabular grains were designed to offer a significantly greater surface area to volume ratio than 3D grains. Tabular-grain films provide images with greater sharpness and resolution without decreasing system speed.

In the early 1990s, newer technologies for double-emulsion film were developed to include zero-crossover systems that completely isolate one emulsion from the other. This dual-receptor technology produces a composite of two different screen-film images on a single sheet of film. A flexible, plastic film base is coated with a dye, which stops light crossover from exposing the other emulsion. This allows different clinically-targeted emulsions to be coated on the front and back surfaces of the same film base. For example, a slow emulsion may be coated on the front, a fast emulsion on the back, or any combination that may be of value for specific imaging requirements (Figure 1-13).

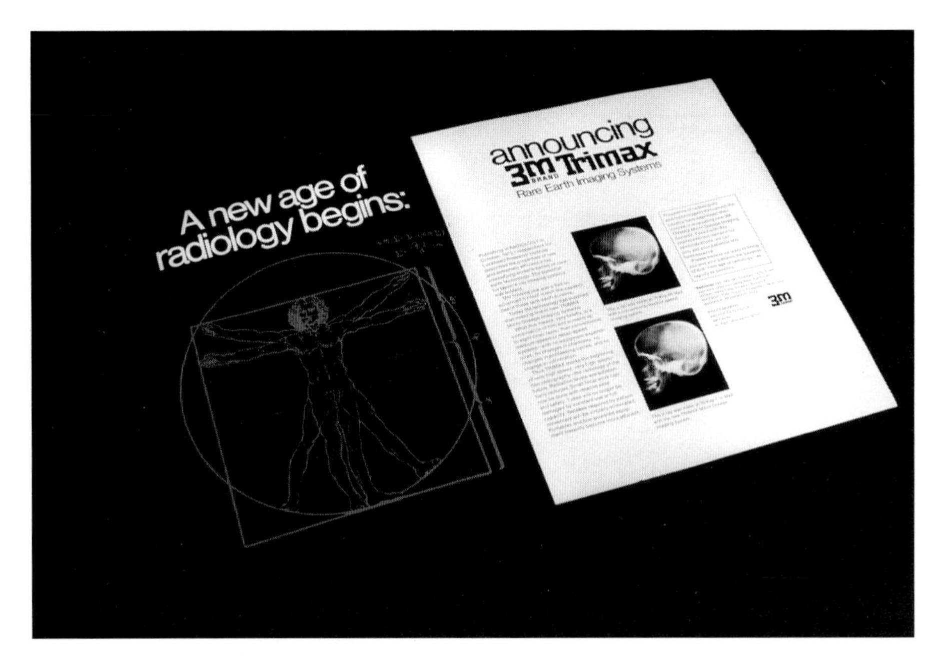


**Figure 1-12** Advertisement for the rare earth imaging system introduced by 3M in 1973. (Reprinted with permission from Edward Harder, Imation Corp.)

## Film Emulsion Characteristics

Films used for radiographic and other medical imaging procedures have either single- or double-emulsion coatings on a polyester film base (Figure 1-14).

### Single- and Double-Emulsion Films

Double-emulsion films have been in use since the late 1800s. Emulsion is usually coated on both sides of the base in a layer, each side identical in thickness to the other, from three to five micrometers thick. The use of double-emulsion film with a spectrally (color) matched pair of intensifying screens decreases the radiation exposure required to produce the appropriate film-blackening effect, thereby reducing radiation exposure to the patient. Dual-emulsion, dual-screen systems are said to be "faster" than single-emulsion, single-screen systems, but a compromise often comes with this increased speed—increased image blur. Examinations that require minimal image blur, such as mammography and imaging of the extremities, often use a single screen with single-emulsion film.

A comparison of features of a single-emulsion film used for mammography with a double-emulsion film illustrates several differences, including some advantages and some disadvantages.

Single-emulsion films usually exhibit higher low-scale contrast and a cooler image tone. They are currently limited to processing in either standard or extended processing cycles. Because they are coated on only one side of the film base, single-emulsion films are sensitive to minus-density artifacts caused by dirt and dust located

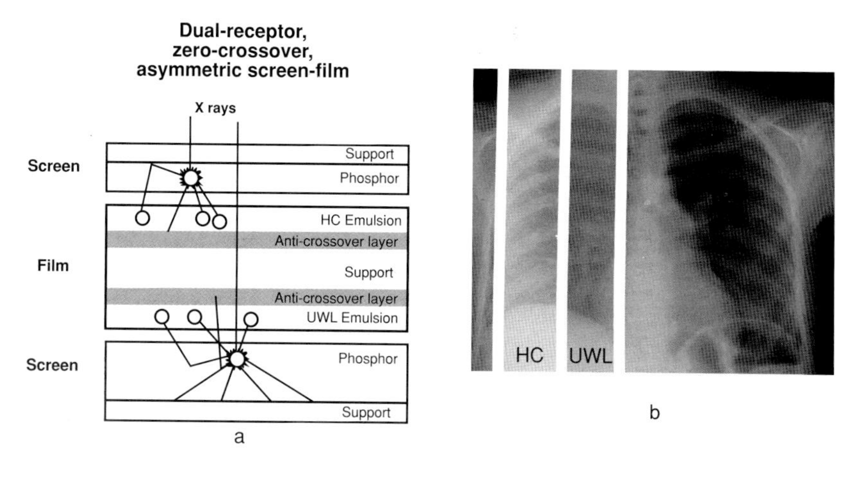


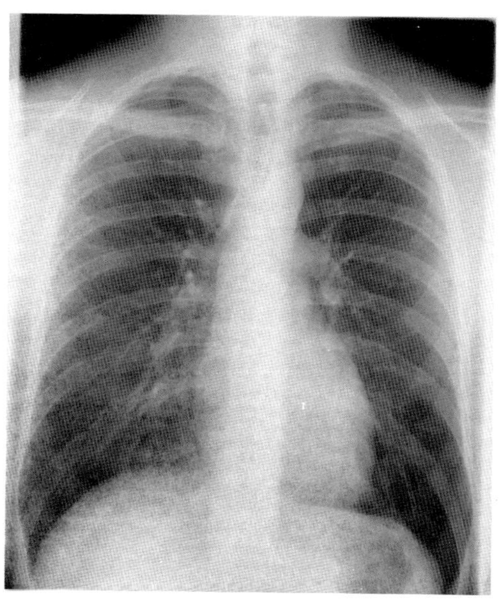


**Figure 1-13** Dual-receptor, zero-crossover technology introduced in the 1990s. (a) A dual-receptor system uses asymmetric screens with a thinner phosphor layer on the back screen, a thicker phosphor layer on the front screen, and a zero-crossover asymmetric film. The film consists of a high-contrast (HC) emulsion on one side of the support and an ultra-wide latitude (UWL) emulsion coated on the other side, with an anti-crossover layer in between. Conventional images on dual-emulsion film contain two essentially identical images back-to-back on the same support. With dual-crossover technology, an image on a single sheet of film can be a composite of the two separate images on each side of the support. (b) A chest radiograph made using dual-receptor, zero-crossover technology is shown, with strips of the front (HC) and back (UWL) emulsions removed. (c) Chest radiograph made using dual-receptor, zero-crossover technology without compromising the lung field. (Reprinted with permission from Eastman Kodak Company. EKC publication no. M1-18.)

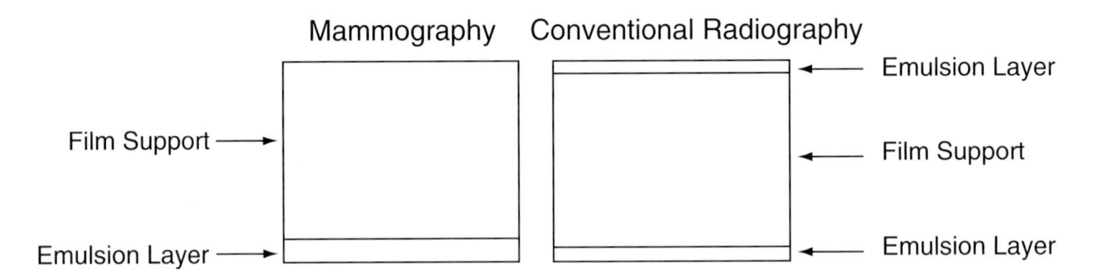


**Figure 1-14** Diagrams of the physical configurations of (a) single-emulsion film used for mammography and (b) double-emulsion film used for other radiologic procedures.

between the film emulsion and the intensifying screen. Cassette interiors, screens, and the darkroom itself must be kept extremely clean.

Double-emulsion films may be processed in multiple processing cycles, including those with very fast total processing times. They are not sensitive to minus-density artifacts from dirt and dust between the film emulsion and screen. However, they usually exhibit lower low-scale contrast and a higher minimum density.

## Base

The film base or support provides the appropriate degree of strength, stiffness, and flatness for handling and good dimensional stability. It is usually blue-tinted, with a thickness of about 180 micrometers. Made from a transparent plastic, the film base absorbs very little water (which is important in automatic processing), and it meets the safety requirements of the American National Standards Institute, Inc. (ANSI), including restrictions on flammability.

## Silver Halide Emulsions

Silver halide emulsion grains have been used as the image receptor on radiographic films for the past several decades. The term "silver halide" describes a compound of silver combined with a member of the halide family of elements: bromine, chlorine, or iodine. The emulsion is made up of innumerable tiny microcrystals of photographic grains of silver halide suspended in gelatin. Tabular grains are used predominantly in most medical x-ray films, and these microcrystals average about 2 micrometers in diameter and 0.13 micrometer in thickness.

## Gelatin

Gelatin is used in x-ray film to keep the silver compound (silver halide microcrystals or grains) well dispersed. Because it is relatively stable, it adds reasonable permanence to the emulsion before and after processing. The gelatin also permits rapid automatic processing because it is easily penetrated by developer and fixer solutions.

## Types of Grains

Until the early 1980s, silver halide emulsion grains were three-dimensional in shape and fairly nonuniform in size and distribution (Figure 1-15a). Research has led to

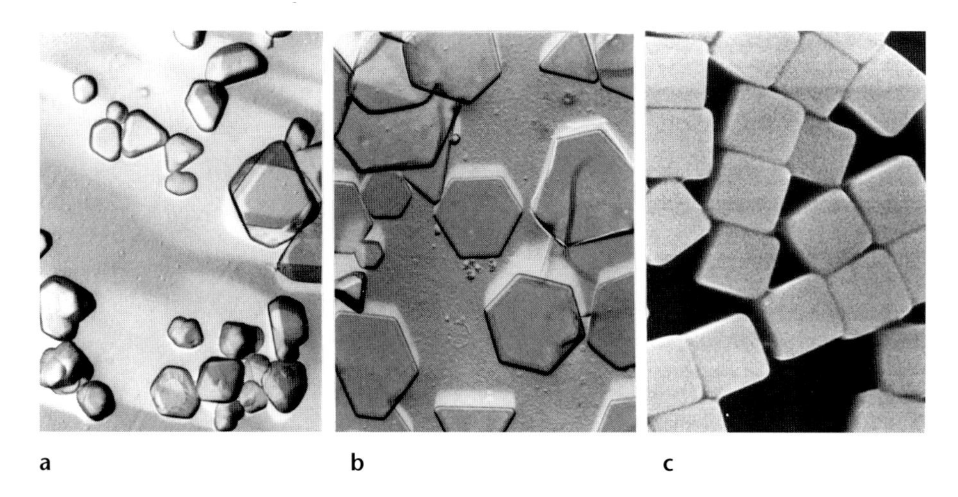


**Figure 1-15** Photomicrographs of (a) three-dimensional, (b) tabular, and (c) cubic-grain film emulsions. (Reprinted with permission from Eastman Kodak Company.)

improvements in emulsion precipitation that have resulted in tabular- and cubic-grain film emulsions (Figures 1-15b and 1-15c).

Tabular emulsions contain grains with increased surface area, which results in improved image sharpness through the reduction of screen-light crossover. Also, the high aspect ratio of these emulsions allows the films to be fully forehardened for very rapid processing and eliminates glutaraldehyde from film-processing chemistries. One limitation of tabular-grain emulsions is difficulty in achieving high enough contrast in the toe portion of the characteristic curve. This difficulty is due to the high projected area of the tabular-grain shape that shields other grains further down in the emulsion layer; these lower grains receive less exposure, which lowers the overall contrast.

On the other hand, the cubic emulsion is useful for film applications such as laser imaging films and mammography films. The uniform chemical and spectral sensitivity of these grains results in high contrast, especially in the toe portion of the curve, which is very useful in mammography. However, cubic emulsions also have limitations. Depending on the grain size of these cubes, forehardening can be difficult. As a result, very rapid processing and the elimination of glutaraldehyde from processing chemicals are presently not possible for emulsion grain sizes required for mammography films.

Another factor for consideration in the choice of an emulsion grain is the halide content. In the past, radiographic films were made with silver halide emulsions that contained bromide and iodide; more modern emulsions use silver bromide. Pure bromide emulsions allow faster processing cycles and less processing sensitivity than older silver bromoiodide emulsions.

### *Response to Processing Fluctuations*

Different film types will respond differently to variations in processing chemicals. Conventional or 3D grain films containing silver bromoiodide grains, in particular, will be more sensitive to processing changes from over-development or under-development.

Sensitometric changes such as higher fog and higher contrast can result when over-development occurs, and lower speed and lower film contrast will result when under-development occurs. Tabular- and cubic-grain films consisting of silver bromide grains, however, will show smaller changes in sensitometric characteristics than conventional films when either over-developed or under-developed. This is due to both the halide content of tabular-grain films and to the more uniform surface area of these crystals, which produces more constant development.

## Overcoat

Film emulsions have an overcoat of protective material to lessen the possibility of damage to the film's sensitive surface. Some protective overcoats also contain compounds to improve the transport of the film through serial film changers.

## Color Sensitivity

Silver halide films are used in conjunction with fluorescent intensifying screens. Intensifying screens are composed of phosphor particles that absorb x-ray quanta and reemit light, which exposes the film. They are either ultraviolet, blue-emitting, or green-emitting. The development of blue-emitting intensifying screens that consist of calcium tungstate phosphor particles led to the design of blue-sensitive films. Green-emitting screens containing gadolinium oxysulfide, terbium-activated phosphors are used with orthochromatic films. Film sensitivity must match the light emission of the intensifying screen (Figure 1-16).

## Film Sensitivity

### *Chemical and Spectral Sensitization*

Although silver halide microcrystals are inherently light sensitive, the addition of chemical sensitizers increases the microcrystals' sensitivity to light even further, providing high-speed films that allow low patient dosages to x-rays. Also, the use of spectral sensitizing dyes extends the sensitivity of silver halide to wavelengths longer than that of undyed silver halide.

### *Grain Size*

Sensitivity to light is also affected by emulsion grain size. Larger emulsion grains, due to larger exposed surface areas, increase the probability of absorbing light photons produced during exposure. As a result, films containing larger emulsion grains are faster or have more sensitivity than films containing smaller emulsion grains.

## Modifying Film Contrast

Films with different levels of contrast are designed to meet a wide range of imaging needs in medical radiography. Mammographic films, which are intended to image a relatively narrow range of anatomical structures with tissue densities ranging from fat to water, have very high film contrast. Films for chest radiography are low contrast

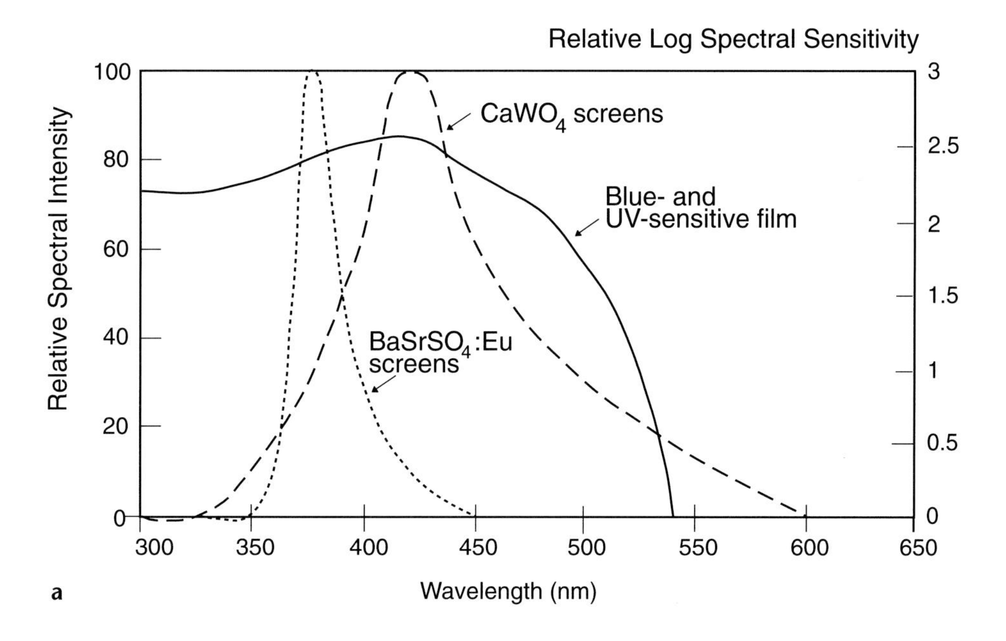


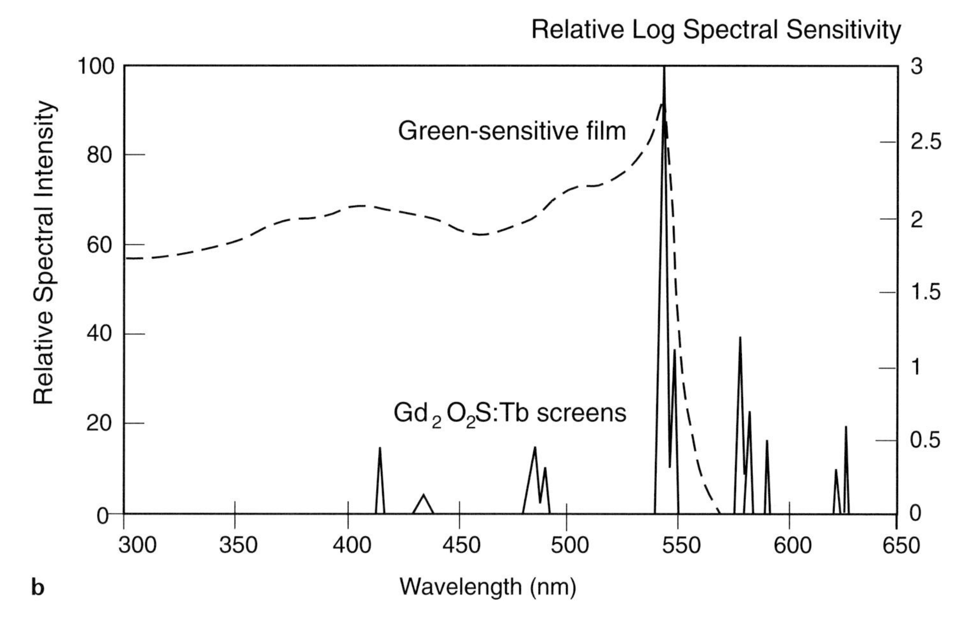


**Figure 1-16** Spectral sensitivity of typical x-ray film emulsions and spectral emission of corresponding intensifying screens. For optimal performance, the spectral sensitivity of the film must be matched to the emission characteristics of the intensifying screens used. (a) Relative spectral match of UV- and blue-sensitive film with UV- (barium strontium sulfate, europium activated [$BaSrSO_4$:Eu]) and blue- (calcium tungstate [$CaWO_4$]) emitting screens. (b) Relative spectral match of a green-sensitive film with green-sensitive (gadolinium oxysulfide, terbium activated [$Gd_2O_2S$:Tb]) film.

in order to image a wide range of structures in the thoracic cavity with tissue densities ranging from air to bone.

High-contrast films are generally designed by using emulsions with a narrow range of emulsion grain diameters or film sensitivity. Because all of the silver halide emulsion microcrystals have nearly the same sensitivity to light, a very narrow range of exposure is required to achieve the desired high contrast.

For illustration purposes, Figure 1-17 shows that a single silver halide crystal is sensitive to some level of exposure before a latent image is created and before development results in density. The amount of exposure required is a measure of the film speed, i.e., larger diameter grain emulsions require less exposure (and, therefore, are "faster") than smaller diameter grain emulsions.

Low-contrast films are designed by using a blend of emulsions of differing sensitivity or speed (Figure 1-18). Larger diameter grain emulsions that require less light exposure form the toe portion of the characteristic curve; medium diameter grain emulsions require a greater exposure and form the mid-scale portion; and small diameter grain emulsions require the greatest amount of exposure to create the shoulder of the curve. (Refer to Chapter four for information on characteristic curves.) Characteristic curves with low contrast are especially well suited for imaging a wide range of anatomy with differing x-ray attenuation. In a chest exam, relatively little radiation penetrates the very dense mediastinal and retrocardiac areas. Only the fastest emulsion grains are capable of being exposed with this low level of exposure. In the radiotransparent lungs, a much greater amount of radiation passes through the body and exposes the film. Slower speed emulsions are necessary to prevent the film from becoming too dark too soon. The resulting image has a wide range of grays—as opposed to the black-and-white image associated with a high-contrast film.

## Latent Image Formation and Amplification

When the microcrystals in an emulsion absorb energy from x-rays or light, a physical change takes place to form the latent image. This image is called "latent" because it cannot be detected by ordinary physical methods. However, when exposed film is processed in a developer solution, a chemical reaction called "reduction" takes place. Reduction of the exposed microcrystals of the silver compound causes tiny masses of black metallic silver to be formed, leaving the unexposed crystals essentially unaffected. The unexposed silver is removed from the film base as part of the fixing process.

The following is a discussion of the stages of latent image formation caused by light photons generated during x-ray exposure and subsequent emission of an x-ray intensifying screen. It is intended to be an overview and not a comprehensive explanation of latent image theory. The stages of latent image formation and amplification include exposure, nucleation, growth, and development.

Silver halide emulsions are manufactured with both chemical and optical sensitizers to form sensitivity specks that are efficient receptors for light photons produced during exposure (Figure 1-19a). On exposure, the energy produced from an absorbed light photon results in catalysis of a silver bromide molecule into a silver ion, a bromine atom (or photohole, as it is sometimes called), and an electron (Figure 1-19b).

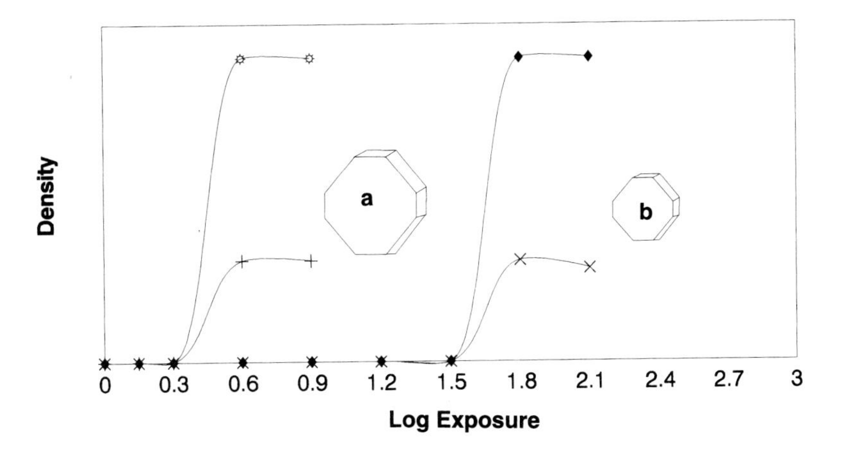


**Figure 1-17** Characteristic curves showing film speed as a function of emulsion grain diameter. A single silver halide microcrystal is sensitive to some level of exposure. (a) Larger diameter grain emulsions require less exposure to achieve the same optical density than (b) smaller diameter grain emulsions.

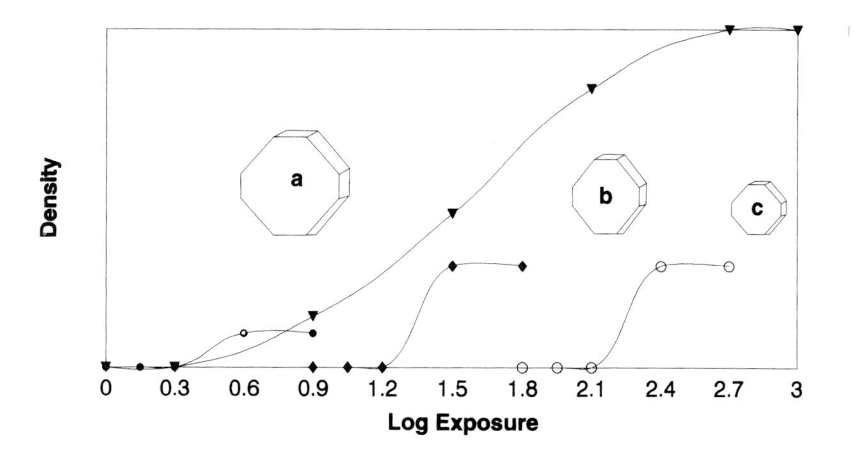


**Figure 1-18** Characteristic curves showing film contrast as a function of distribution of emulsion grain diameter. Low contrast films are designed by using a blend of emulsions of differing sensitivity or speed. (a) Larger diameter grain emulsions, which require less light exposure, form the toe portion of the characteristic curve; (b) medium diameter grain emulsions require a greater exposure and form the mid-scale portion; and (c) small diameter grain emulsions require the greatest amount of exposure to create the shoulder of the curve.

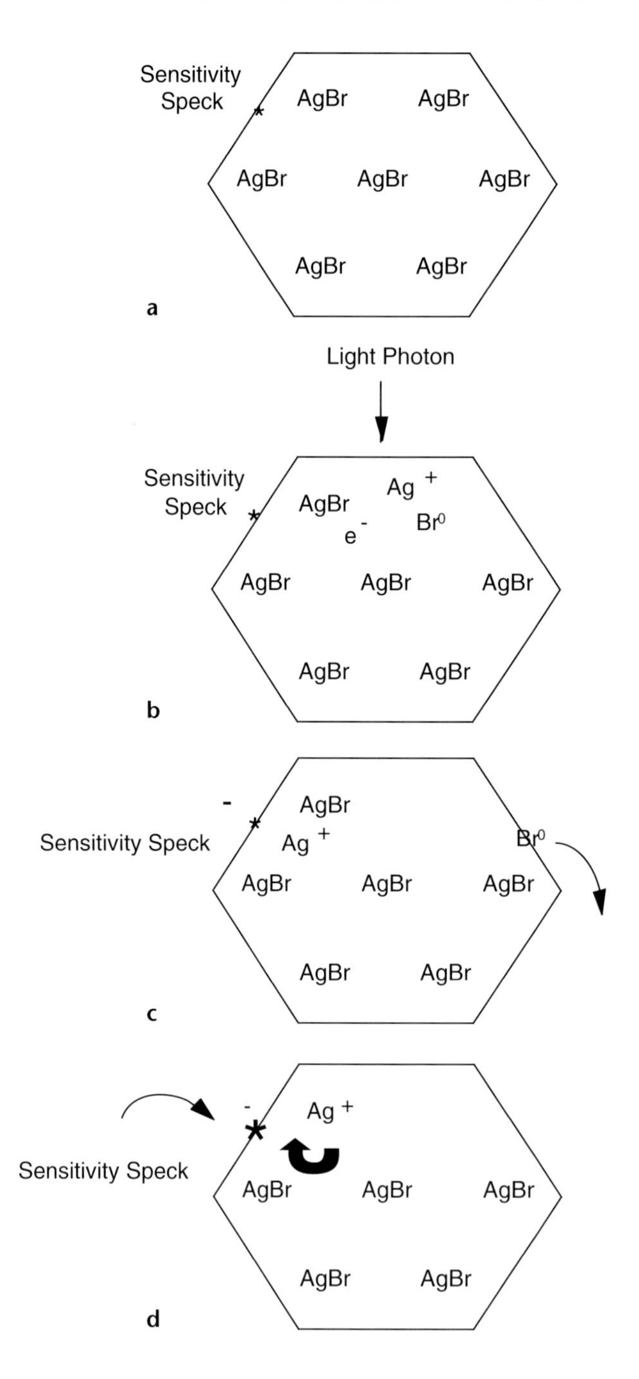


**Figure 1-19** Stages of latent image formation. (a) Sensitivity speck (*) on a silver halide microcrystal. (b) Exposure. Catalysis of a silver bromide molecule into a silver ion ($Ag^+$), a bromine atom or photohole ($Br^0$) and an electron ($e^-$). (c) Nucleation. Migration of the electron and absorption to the sensitivity speck produces a partial negative charge, which in turn attracts a positively charged ion. The bromine atom is removed. (d) Growth. If a bromine atom is not removed, it can combine with the nucleated sensitivity speck and destroy the latent image.

Migration of the electron and absorption to the sensitivity speck produce a partial negative charge, which in turn attracts a positively charged silver ion. This process is called nucleation (Figure 1-19c) and eventually will result in latent image formation. The literature reports that the minimum size of the latent image formation is at least four silver atoms. It is also important to deal with the bromine atom or photohole. If not removed, the photohole can recombine with the nucleated sensitivity speck and destroy the latent image (Figure 1-19d). Failure to remove the photohole can result in problems such as latent image fading and reciprocity law failure (Appendix A).

The formation of a latent image as a function of exposure does not in itself produce a useful image. It is the amplification of this latent image to a viewable size that forms the basis of medical radiography. This amplification process is called "development" and consists of several stages. These include: (1) diffusion of developer components, (2) adsorption of developer to latent image centers, (3) transfer of electron to latent image centers by electron transfer agents, (4) regeneration of electron transfer agent by hydroquinone, and (5) reduction of silver halide to metallic silver.

Figure 1-20a illustrates the diffusion of the developer components into the gelatin matrix, which consists of dispersed silver halide microcrystals containing latent image as a function of exposure. Figure 1-20b shows the next step that involves adsorption of the developing agents, which include both hydroquinone and an electron transfer agent such as phenidone. On adsorption, the developing agents transfer an electron preferentially to microcrystals containing latent image centers. Subsequent steps (Figure 1-20c) involve regeneration of the electron transfer agent by hydroquinone. This process is called "superadditivity"; it continues the process of development until the grains containing latent image centers are reduced to metallic silver, which forms the viewable image that radiologists have used for diagnosis for many years. The amplification factor from the latent image to the fully developed grain is $5 \times 10^9$. Few, if any, systems in nature provide this type of system gain or sensitivity. Figure 1-21 shows electron micrographs of (a) an undeveloped, (b) partially developed, and (c) fully developed grain.

## Storage

All film types must be properly stored with respect to environmental temperature and relative humidity. Users of photography film are aware of this and, for the most part, do a good job of adhering to the manufacturers' recommendations; it is unthinkable that the image quality of important photographs be compromised.

However, facilities using medical x-ray film frequently store film, both unprocessed and processed, improperly. Because improper storage may compromise the quality and stability of a radiographic image as well as affect processor quality control, it is important to follow the recommendations of the manufacturer. Proper storage at the supplier warehouse, facility storage area, and facility archive area is also an important consideration in preserving film life and reducing sensitometric changes.

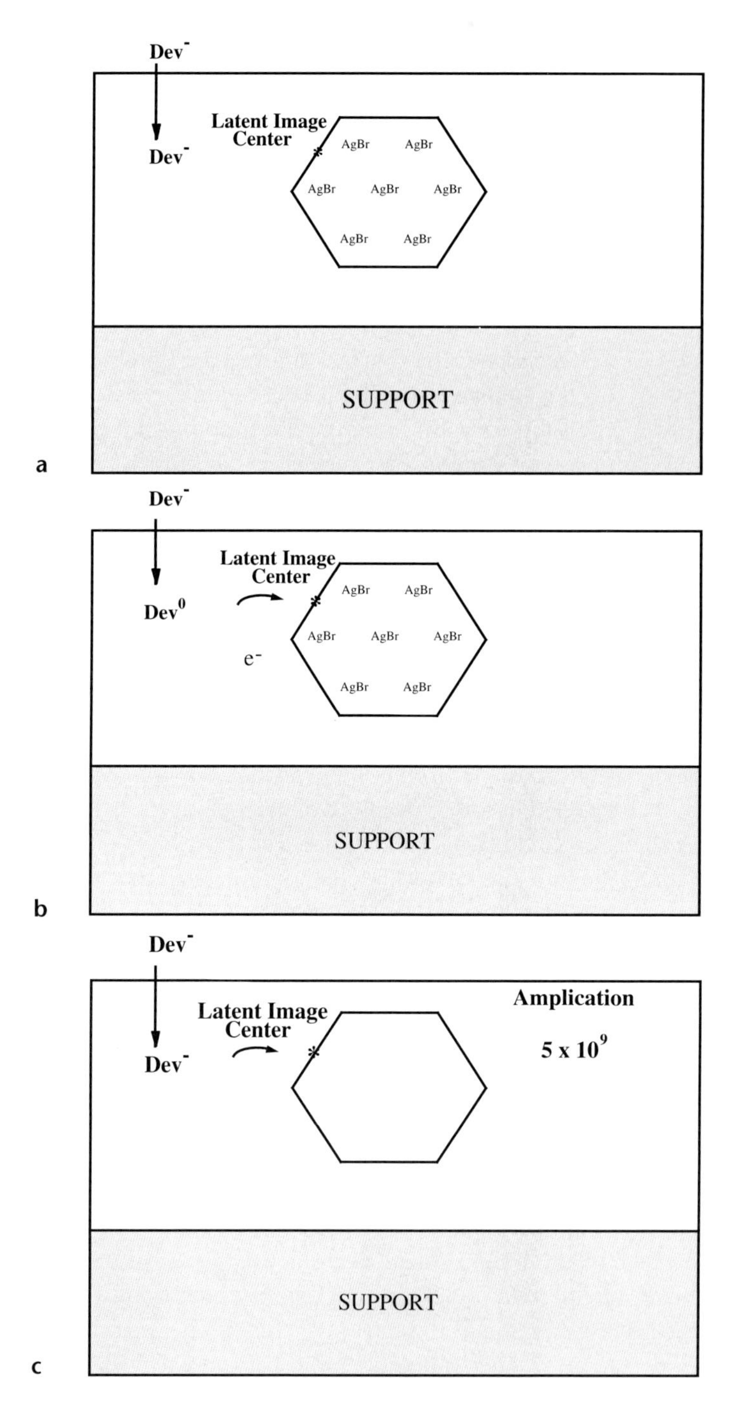


**Figure 1-20** Amplification (development) of the latent image. (a) Diffusion of the developer components into the gelatin matrix. (b) Adsorption of the developing agents to the latent image centers with electron transfer to the latent image centers by the electron transfer agent. (c) Regeneration of the electron transfer agent by hydroquinone, a process called "superadditivity," continues the development process until silver halide is reduced to metallic silver.

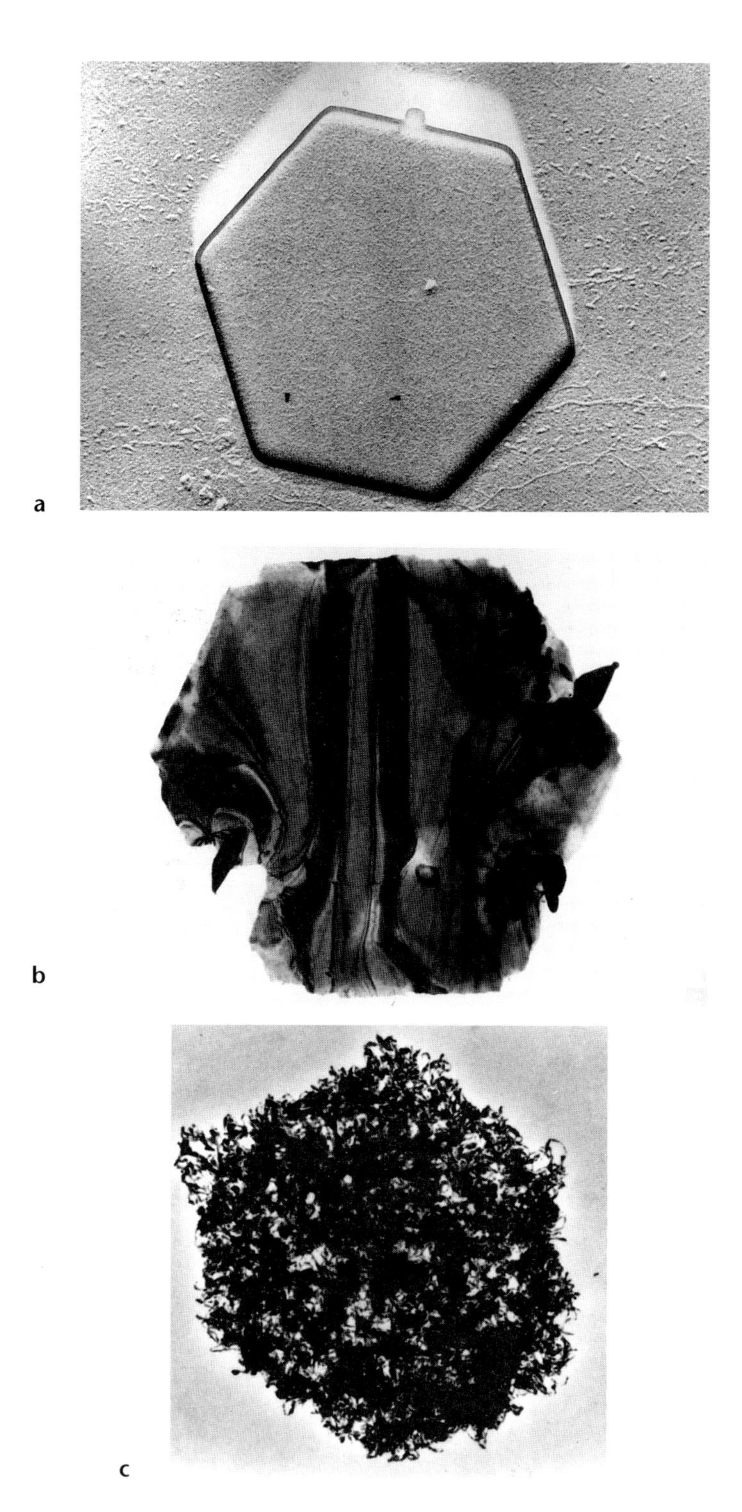


**Figure 1-21** Electron micrographs of (a) undeveloped, (b) partially developed, and (c) fully developed tabular grain.

## Storage of Unprocessed/Unexposed Film

Most medical x-ray films come in a sealed, moisture-proof inner wrap, which is packed in an outer cardboard box. Sealed packages of film are affected by heat; open packages are affected by both heat and humidity.

All packages of film should be stored away from heat sources in a cool, dry place at a temperature between 50 and 70°F (10 to 21°C). The National Council on Radiation Protection and Measurements (NCRP) Report No. 99 recommends that photographic material be stored at temperatures less than 75°F (24°C).

Keep open packages of film at a relative humidity between 30 and 50 percent. An inexpensive instrument called a sling psychrometer, commonly available through precision instrument supply houses, can be used to measure relative humidity. Hygrometers may also be used to measure relative humidity. Check the accuracy of the instrument being considered before purchasing.

Consult your film manufacturer(s) for temperature and relative humidity recommendations for the specific type(s) of film used in your facility. Also keep in mind that temperature and relative humidity values frequently fluctuate throughout medical facilities from season to season. Year-round control of temperature and relative humidity are necessary in order to achieve and maintain high-quality and stable radiographic images.

It is also important to avoid storing film near chemical fumes, which can fog film. Film may also be damaged by exposure to radiation from x-ray machines or radioactive materials. In addition, packages of sheet film should be stored on edge (like the books in a library). This allows for easy rotation of inventory. Always use older films first. The film's expiration date is generally printed on both the box's front and side panels so the date is visible when the boxes are stored on edge. Boxes of film should never be stacked horizontally because film on the bottom may show pressure artifacts from being weighted down by other boxes or cases of film.

## Storage of Processed Film

Proper storage and handling of processed film are imperative for stability of the radiographic image. It is especially important to treat single-emulsion mammography film with great care—the same care you would give to your most treasured family photographs.

Film should have been thoroughly washed during processing to remove residual fixer, which can cause staining and fading. Tests for fixer retention should be done at least every three months. Refer to Chapter five for additional information on the performance of the fixer retention test.

Since all films contain gelatin as one of the principal ingredients of their emulsion, it is important to maintain a constant temperature at about 70°F (21°C) and 40 to 60 percent relative humidity.

Subjecting radiographs to humidity below 30 percent and/or high temperatures can occasionally lead to emulsion cracking, an artifact that appears as a series of parallel lines in a D-max area of the film (Figure 1-22). This artifact can be avoided by storing film at constant temperatures and humidity, not overexposing film to "hot lighting," and using only the recommended bulb wattage for hot lights.

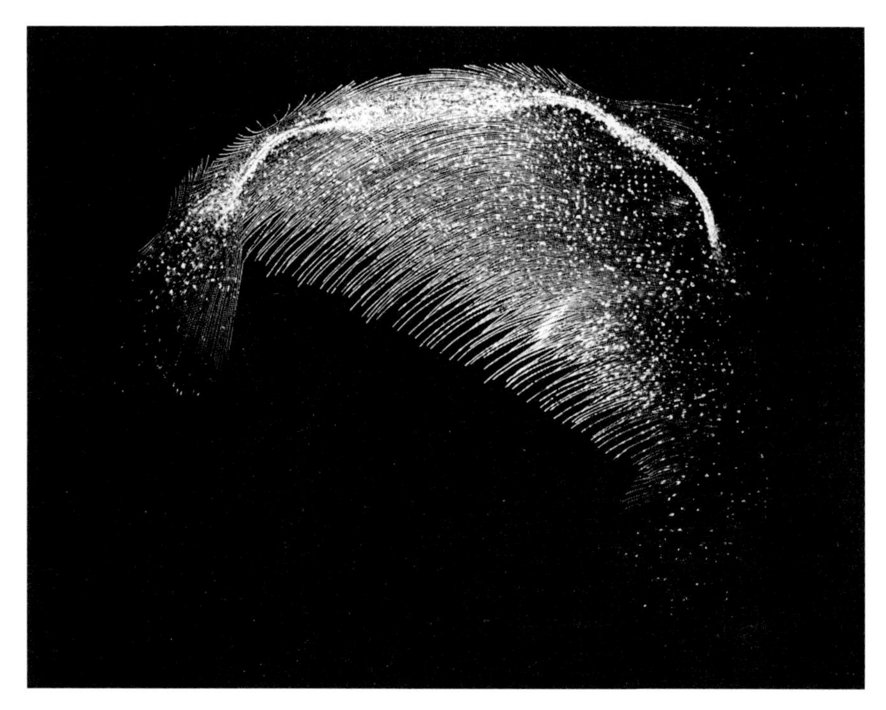

**Figure 1-22** Emulsion cracking artifact (25X). Subjecting radiographs to a combination of low humidity, high temperatures, and other stresses such as bending or kinking the film support can cause this handling artifact. Film, both unexposed and exposed, should be stored within the film manufacturer's temperature and relative humidity recommendations.

## References

1. Dickerson RE. Co-optimization of film and process chemistry for optimum results. In: Haus AG, ed. Film processing in medical imaging. Madison, WI: Medical Physics Publishing, 1993: 63–78.
2. Haus AG. Film processing systems and quality control. In: Gould RG, Boone JM, eds. A categorical course in physics: Technology update and quality improvement of diagnostic x-ray equipment. Oak Brook, IL: Radiological Society of North America, 1996: 49–66.
3. Haus AG. Historical developments in film processing in medical imaging. In: Haus AG, ed. Film processing in medical imaging. Madison, WI: Medical Physics Publishing, 1993: 1–16.
4. Haus AG, Cullinan JE. Screen-film-processing systems: A historical review. RadioGraphics 9: 1203–1224, 1989.
5. Introduction to medical radiographic imaging, Eastman Kodak Company publication no. M1–18, 1993.
6. James TH, Higgens GC. Fundamentals of photographic theory. 2nd Edition. Hastings-on-Hudson, NY: Morgan & Morgan, 1960.

7. James TH, ed. The theory of the photographic process. 4th Edition. New York: Macmillan, 1977.
8. Kitts EL. Physics and chemistry of film and processing. RadioGraphics 16: 1467–1479, 1996.

# 2

# Chemicals

- History
- Developer Components
- Developer Considerations
- Fixer Components
- Fixer Considerations
- Mixing
- Monitoring Chemicals
- Chemical Storage
- Fresh and Seasoned Chemicals
- Replenishment
- Use of Recommended Chemicals
- Medical Device Regulations
- Health, Safety, and Environmental Issues

Processing chemicals are required to convert an invisible latent image on film to a visible image and to remove unexposed silver halide crystals from the emulsion so that a permanent image is obtained. In addition to playing a key role in the formation and optical density of the image, the processing chemistry can also affect artifacts, drying, keeping characteristics, and other attributes. Ideally, the processing chemistry should help provide consistent, high-quality images with minimal patient exposure.

## History

In the 18th century, several investigators observed that some silver compounds blackened when exposed to light. At first there was no way to make a permanent record of what had been observed because areas that had originally been protected from light eventually darkened. Joseph Nicephore Niepce (France, 1826) is credited with the first permanent photograph with a camera. He exposed a light-sensitive material in a camera obscura and hand-traced the image reflected through a lens by a mirror onto a ground glass, thus fixing the image (Figure 2-1).

**Figure 2-1** The first camera, called a camera obscura, was constructed by making a small hole in a large box. An image formed by a lens and reflected by a mirror onto a ground glass was then traced by hand. (Reprinted with permission from RadioGraphics 1989; 9:1204.)

Louis J. M. Daguerre (France, 1839) is credited with establishing the basis for chemical development of a photosensitive material. Daguerre placed an exposed plate (silver salt fused with iodine to form a layer of silver iodide) in a cupboard. He later removed the plate and found that there was a well-defined positive image on the plate. Fumes from mercury previously spilled in the cupboard had developed the image completely. The unexposed silver iodide was removed by a solution of sodium chloride.

Many types of developers were employed in the days of glass plates (Figure 2-2). No two were alike. The quantities of chemicals employed were not standardized, and development was based on the user's experience.

After the plate had been developed, it was necessary to make a photographic print on sensitized paper. Direct interpretation of the plates was not possible. These early plates exhibited wide latitude with little contrast. The prints subsequently made from the plates were higher in contrast and detail.

Development in these early days was accomplished using a four-bottle photographic method. The material to be developed was first covered with a solution containing the developing agent. Preservative, accelerator, and bromide solutions were then used. As development proceeded, a little more of one solution or another was

**CONCENTRATED IMPROVED DEVELOPER**

**(IN ONE SOLUTION).**

This developer is made by an improved formula, especially adapted for both glass and paper negatives.

PRICE, per bottle, sufficient to develop 100 5x8 paper negatives, - 50 cts.

**Figure 2-2** Page of the 1886 price list of the Eastman Dry Plate and Film Company showing the listing for a concentrated developer solution

added in order to "bring out the desired anatomical detail" until development was complete.

The more successful developers contained pyrogallic acid, metol, or hydroquinone as reducing agents. Various quantities of alkali, sodium sulfite, and potassium bromide were used to accelerate or retard developer activity.

Many of the finished negatives also underwent further treatment to increase the maximum density of the image. Early radiographic images were inherently low in density and contrast and slow in speed.

In 1929, standardized tank development—using a constant time of development for a given temperature based on the rate of exhaustion of the developer—became available. This "exhaustion system" ensured uniformity of results and made it possible to check exposure time.

In 1947, the "replenisher system" of development was described. This technique was designed to maintain the activity of the developer by the addition of developer replenisher solution at a constant rate. It made possible the use of a constant time of development for a given temperature for the life of the solution.

In 1965, 90-second rapid processing was introduced. This advancement combined a new type of roller transport processor with new film emulsions and, for the first time, specialized chemicals for processing medical radiographs.

In 1990, specialized chemicals that allowed for even faster processing cycles (approximately 60- and 45-second total processing times) were developed. As in 1965, advancements in film and processing systems were concurrently introduced.

Throughout the 1990s, chemicals have generally been modified for environmental and health-related exposure considerations. These issues are of increasing importance, and efforts to optimize the total processing system to minimize the impact of processing chemicals on the environment continue.

In the future, new film technologies will offer opportunities for silver recovery and further reductions in chemical replenishment rates and water use. Understanding processing chemicals and how they are optimally used will continue to be a key element in achieving high-quality medical radiographs for some time. (Figure 2-3 shows that the relationship of processing chemicals to radiographic image quality has been an important issue for a long time.)

## Developer Components

Development, or detection and amplification of the latent image, requires a developer solution, which is the first step in the photographic process (Figure 2-4). Subsequent steps include fixing, washing, and drying. Figure 2-5 illustrates development and fixation.

A typical developer solution contains several chemicals. The key components in typical developers used for medical imaging are:

- Developing agents (hydroquinone and a pyrazolidone-type compound).
- Buffering agents (usually carbonates and hydroxides).
- Preservatives (usually sulfite).
- Restrainers (bromide and/or organic antifogging agents).
- Hardeners (usually glutaraldehyde).
- Solvents (usually water).
- Sequestering agents.

**Figure 2-3** Advertisement from 1934 issue of Radiography and Clinical Photography discussing the importance of processing chemicals for radiographic quality

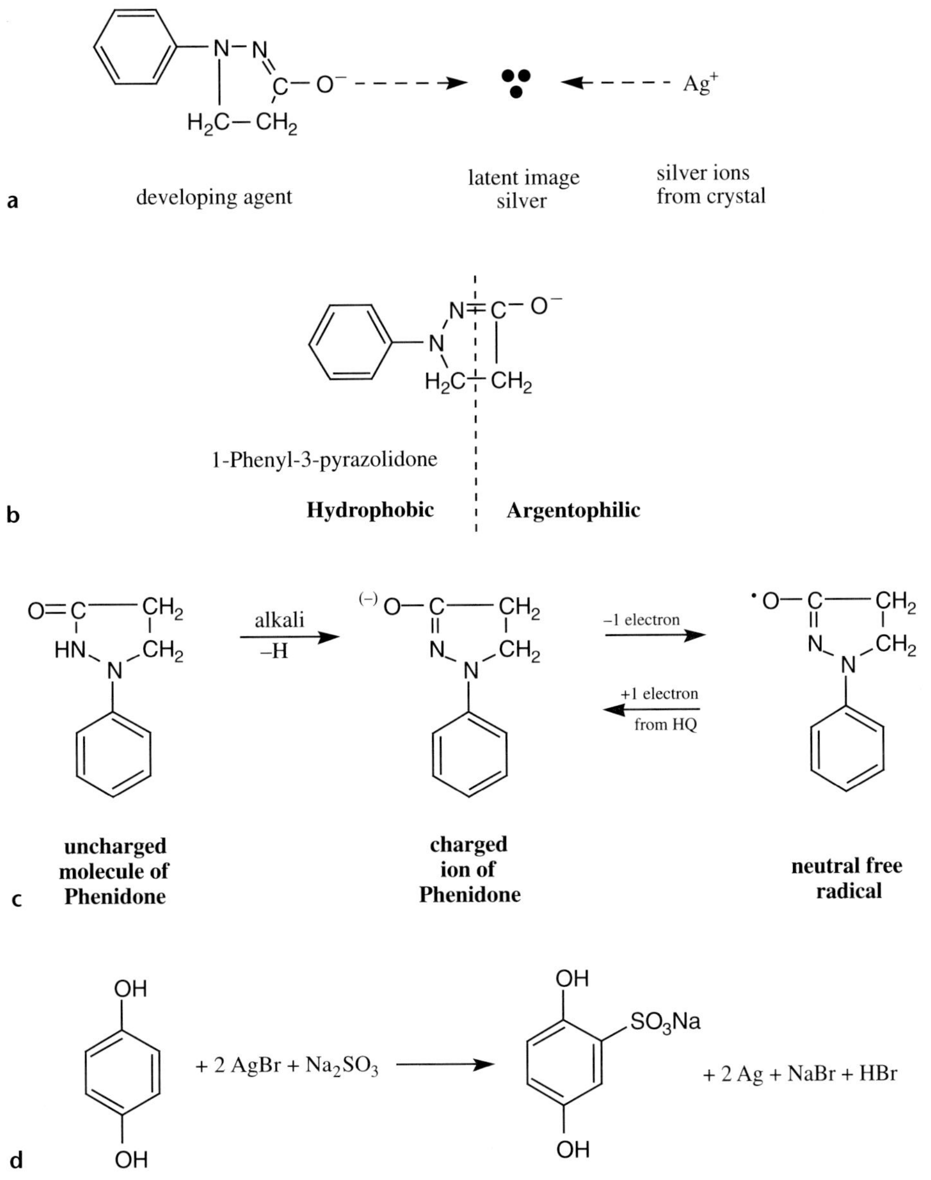

a

developing agent — latent image silver ← silver ions from crystal ($Ag^+$)

b

1-Phenyl-3-pyrazolidone

**Hydrophobic** | **Argentophilic**

c

**uncharged molecule of Phenidone** —alkali, –H→ **charged ion of Phenidone** —–1 electron→ / ←+1 electron from HQ— **neutral free radical**

d

hydroquinone + 2 AgBr + $Na_2SO_3$ → (hydroquinone with $SO_3Na$) + 2 Ag + NaBr + HBr

**Figure 2-4** (a) The critical complex for photographic development, using Phenidone as an example. (b) The strength of adsorption in forming this complex is determined by the "surfactant" nature of the developing agent. (c) In alkaline solution, Phenidone is believed to ionize and then transfer one electron. Hydroquinone "regenerates" the Phenidone molecule. (d) Net equation for the reaction of hydroquinone and silver halide in the presence of sulfite. (Reprinted with permission from Eastman Kodak Company. EKC publication no. N-327.)

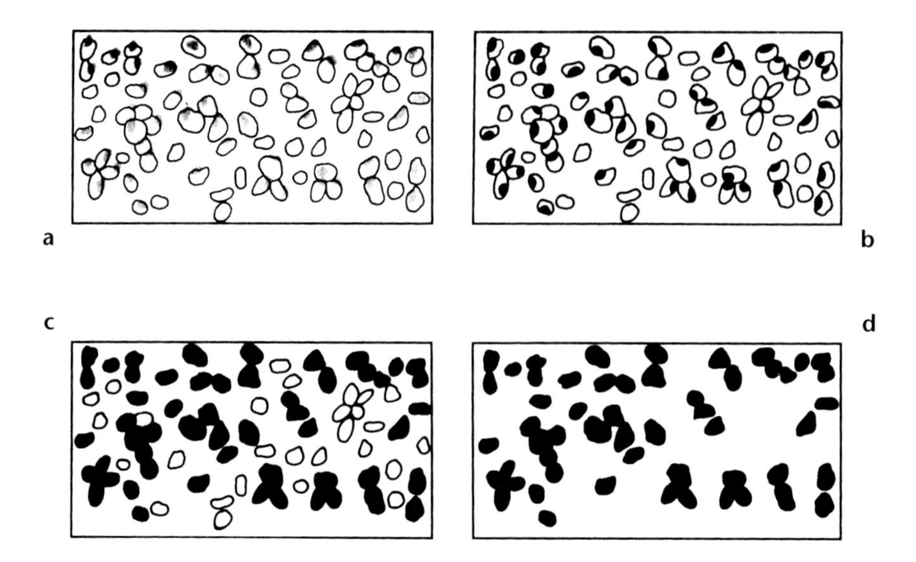


**Figure 2-5** Diagrams of processing action in x-ray film emulsion. (a) Schematic distribution of silver halide grains. The gray areas indicate latent image produced by exposure. (b) Partial development begins to produce metallic silver (black) in exposed grains. (c) Development completed. (d) Unexposed silver grains have been removed by fixing. (Reprinted with permission from Eastman Kodak Company. EKC publication no. N-327.)

## Developing Agents

The developing agents—usually hydroquinone and a pyrazolidone-type compound—convert exposed grains of silver halide into metallic silver. The developing agents used should not have a significant effect on the unexposed grains in the emulsion.

The presence of codevelopers in the same developing solution can produce a cooperative action in which the rate of silver development is greater than the rate produced by each agent individually. This enhancement of development is also known as "superadditivity of development." Many pairs of developing agents exhibit this phenomenon to varying degrees. Typically, one of the agents is a nitrogenous compound and the other is a phenolic compound.

Rates of development vary depending on the film type, as well as on the class and the concentration of the particular developing agents used in the developer. In most black-and-white developers, hydroquinone is used as a developing agent and is present in the highest concentration. However, many other compounds are known to be effective as well. Common codeveloping agents used with hydroquinone include Elon (N-methyl-p-aminophenol) and Phenidone (1-phenyl-3-pyrazolidone), along with the substituted pyrazolidone compounds.

## Buffering Agents

Since development is accelerated at alkaline pH levels, buffering agents, such as carbonates and hydroxides, are added to the developer to elevate and maintain the pH. Maintaining a balance is important. The pH must not drop significantly during seasoning, or an activity loss will be observed. If the pH gets too high, the chemical stability of developer components may become a concern because of unwanted side reactions, such as oxidation and polymerization.

## Preservatives

A preservative (or antioxidant) retards the oxidation of the alkaline developer solution, which helps maintain the development rate. In black-and-white developer solutions, sulfite salts are commonly used as preservatives. By decoloring any oxidized developer that forms over time, sulfite helps prevent staining of the emulsion layer of the film. Sulfite also decreases the solubility of oxygen in developer solutions, thereby reducing the likelihood of oxidation.

## Restrainers

Restrainers consist of bromide and/or organic antifogging agents. Bromide ions minimize fog growth by protecting unexposed grains from the action of the developer. Although halide ions diffuse into the developer from the film during processing, bromide is often added to the developer intentionally to minimize changes in development rate during processing. The bromide ions can be added as part of a starter solution or can be directly incorporated into the developer concentrates. Starters are often employed to minimize chemical and sensitometric differences between fresh chemistry (or replenisher solutions) and seasoned chemistry (or tank solutions).

In addition to bromide, organic antifoggants are often added to black-and-white developer solutions. These compounds have a powerful effect on development and film sensitometry at very low levels. In addition to reducing fog, antifoggants may exhibit different effects on development, depending on the compound used and the concentration of the antifoggant. For example, slight variations in antifoggant level may drastically affect speed or contrast for a particular film.

## Hardeners

A hardening agent, such as glutaraldehyde, is frequently used in developers intended for automated processing of medical imaging films. During processing, this agent in the high-temperature alkaline developer solution hardens the film to prevent excessive swelling of the gelatin and damage to the film as it passes through the rollers. It is necessary to maintain a hardener level that allows for acceptable transport, without inhibiting emulsion swell to the point that development rates are adversely affected.

Although glutaraldehyde has a relatively low threshold limit value (TLV = short-term exposure, 10 minutes or less) of 0.2 parts per million (ppm) at the alkaline pH of most developer solutions, it does not seem to be very volatile. This is most likely due to forming a complex of the aldehyde with the excess sulfite present in developer solutions to form the bisulfite adduct. Any free glutaraldehyde is likely to react quickly with free amino groups in the gelatin protein chains. This cross-linking reaction provides the mechanism for hardening the emulsion.

Developer concentrates should be handled with care if they contain free glutaraldehyde at high concentrations. Even in concentrates where glutaraldehyde is complexed as glutaraldehyde bis-bisulfite, free glutaraldehyde is present and can be generated at significant levels at neutral pH.

### Solvents

Water is the primary solvent in the developer solution. Water dissolves and ionizes the developer components, as well as the chemicals that are transported from the emulsion during processing. In the case of developer concentrates, organic solvents are often necessary to assist the dissolving of the developer components, especially at lower storage temperatures. Glycols are commonly used as solvents for components such as hydroquinone.

### Sequestering Agents

Sequestering agents are used to counteract the effect of soluble salts or metal impurities that may be present in the developer. These salts and metal ions can come from the water, or they may be impurities in the developer components. They can also be introduced into the developer by diffusion of chemicals from the film emulsion. Common ions of concern include calcium (which can precipitate in developer solutions) and metals, such as iron and copper. These heavy metals can accelerate the oxidation of hydroquinone and thus have an impact on developer stability. In radiographic developers, it is desirable to have good stability and to minimize the degree of hydroquinone oxidation. Typically, sequestering agents function by forming stable complexes with the metallic ions, leaving low concentrations of free metallic ions.

## Developer Considerations

### Developer/Film Interactions

Because of the many differences between film types and film technologies, all x-ray film types do not exhibit the same developer component sensitivities. For example, subtle differences in the concentrations of antifoggants and developing agents can have a dramatic effect on sensitometry for some films, but a negligible effect on other films. The data in Table 2-1, which are derived from developer models designed to study the effect of developer components on sensitometry, demonstrate the effect bromide ion concentration would have on the contrast (average gradient) of different representative film types.

Film manufacturers develop new films for optimal performance in a specific chemistry or set of chemistries. The chemistry is often an existing developer used to process other films, but it can also be a new developer designed for a specific system.

An optimal chemistry would be one that is manufactured according to product specifications or to tolerances set by the manufacturer with an understanding of the sensitivity and needs of the most critical films. These tolerances would include the

**Table 2-1** Developer Component/Film Sensitivity Data: The Effect of Bromide Concentration on Contrast (Average Gradient) (Simulated Data from Modeling Studies)

| *Bromide Concentration* | *Contrast* | | |
|---|---|---|---|
| | *Film Type A* | *Film Type B* | *Film Type C* |
| 1.5 | 2.75 | 2.89 | 3.16 |
| 2.5 | 2.81 | 2.88 | 3.15 |
| 3.5 | 2.85 | 2.88 | 3.13 |
| 4.5 | 2.91 | 2.88 | 3.09 |
| 5.5 | 2.96 | 2.87 | 3.04 |

(Reprinted with permission from Eastman Kodak Company. EKC publication no. N-327.)

manufacture of the solutions and the filling tolerances used for product packaging. In addition, the manufacturer must set specifications for the suppliers of the raw materials. All of this can be done by a film and chemistry manufacturer that understands how these chemical variables impact film quality and the total variability of the system.

There are advantages to using a developer that is capable of providing high-quality radiographs for a full portfolio of films, or of supporting a range of diagnostic film types. These specialized developers, designed for optimized processing of a subset of films or of a specific film type, must be used in a dedicated processing environment or in a processor for these specific films. Specialized developers have the disadvantage of perhaps adversely affecting the quality of films they were not designed to process.

## Developers and Replenishment

A number of films can be adequately processed with developers that have lower than normal levels of key components, such as developing agents. These developers take advantage of the film manufacturer's robust film design. In reality, reduced component levels often result in a less active developer or in a developer that is more sensitive to processing variability.

A developer with reduced buffer capacity results in a solution that exhibits increased process sensitivity to seasoning effects as more films are processed. In addition, if sulfite levels are reduced, the solution has less protection from oxidation. This could limit the length of developer life in the processor or replenisher holding tank. Another possible result of reduced sulfite levels can be increased silver sludging in the processor.

These problems can be overcome with increased replenishment rates, which would neutralize seasoning effects. When replenishment rates are considered, developers with reduced levels of key components may not have the advantages that are demonstrated by a more robust developer.

## Fixer Components

The significant active components in fixers used for medical imaging are:

- Fixing agent (thiosulfate).
- Preservative agent (sulfite).
- Hardener (usually aluminum salts).
- Buffer (usually acetate).
- Sequestering agents (usually carboxylic acids).

The following equation describes the fixing process:

$$2AgX + Na_2S_2O_3 \rightarrow Ag_2S_2O_3 + 2NaX$$

where X designates halide ion content (chloride, bromide, or bromoiodide).

### Fixing Agent

The most common fixing (or clearing) agent is thiosulfate (hypo). Thiosulfate ions can be present as ammonium or as sodium salts; however, ammonium thiosulfate is the more active of the two compounds.

The fixing agent complexes with the undeveloped silver ion. This soluble complex is then free to diffuse from the gelatin. If film is improperly cleared, the remaining unexposed crystals darken upon exposure to light and can obscure information on the radiograph. The literature indicates that fixers with thiosulfate concentration below 150 grams per liter exhibit fixation rates largely dependent on the thiosulfate concentration.

The American National Standard IT9.1-1989 sets limits for retained thiosulfate for radiographic films (Table 2-2). The thiosulfate concentrations are reported as the "maximum permissible concentration of thiosulfate ion."

### Preservative Agent

A preservative, such as sulfite, is necessary to protect thiosulfate from oxidation and to maintain a consistent concentration of thiosulfate. Sulfite is particularly important for concentrate storage and electrolytic silver recovery applications.

### Hardener

The most common hardeners are aluminum salts. The aluminum or metal ions can complex with gelatin to harden emulsions during the fixing step. Control of swell in the fixing step is necessary to reduce the swelling in the final wash. Swell reduction minimizes mechanical damage to the emulsion during washing and drying. The hardener also reduces water absorption during the washing stage and makes drying easier.

**Table 2-2** Retained Thiosulfate Limits for Radiographic Films (for each side of film)

| | *Thiosulfate Ion Concentration* | |
|---|---|---|
| *Film Classification* | *grams per square meter* | *micrograms per cubic centimeter* |
| Medium Term | 0.100 | 10 |
| Long Term | 0.050 | 5 |
| Archival | 0.020 | 2 |

(Reprinted with permission from Eastman Kodak Company. EKC publication no. N-327.)

### Buffer

A buffer, usually acetate, is present in the fixer to keep the pH constant. The fixer solution requires a sufficient buffer capacity to be resistant to the adverse effects of developer seasoning transported into the fixer. The pH is also affected by the amount of excess acetic acid present in the solution. The pH of the fixer is very important because fixing is a diffusion-controlled process, and the solution pH has a large influence on diffusion, which in turn affects fixing, dye stain, and hypo retention.

### Sequestering Agents

In fixer solutions, sequestering agents are usually carboxylic acids (or acid salts). Boric acids or borate salts can also be used. At lower pH levels, these compounds are excellent complexing agents for aluminum ions and, to some extent, for other metals. Their function is to prevent localized formation of aluminum hydroxide as the developer carries over into the fixer solution. If an inappropriate sequestering agent is used at a high level, it may combine with aluminum and have an adverse effect on drying.

## Fixer Considerations

Unlike development, the consequences of incomplete fixing become manifest only after the image has been stored for some time. The silver complexes with thiosulfate ions in the emulsion, and this complex needs time to diffuse into the fixer solution. If the silver content is too high, these complexes are not readily removed by subsequent washing.

The degree of exhaustion affects the rate of fixation. A fixer becomes exhausted in three main ways:

1. The silver content increases. Silver is complexed with the thiosulfate. High solution concentrations slow down the rate of diffusion from the gelatin.
2. The concentration of halide ions builds up, which slows the fixing rate. Iodide has the greatest effect.
3. Developer is carried into the fixer from the emulsion, and the fixer is carried out. The net effect is the dilution of the thiosulfate level.

In addition, most forms of silver recovery (for example, electrolytic silver recovery) involve some destruction of thiosulfate and can lead to more rapid fixer exhaustion.

Replenishment is one way to compensate for the effects of fixer exhaustion. Replenishment decreases silver and halide levels and minimizes the dilution of thiosulfate from developer carryover and fixer carryout. However, the more robust a fixer is, the less susceptible it is to fixer exhaustion. In fact, when fixer replenishment and silver recovery are considered, it may be more cost-effective to use a more robust fixer. Fixers with lower concentrations of key ingredients often require higher replenishment rates to compensate for these seasoning effects.

There are definite advantages to using a fixer manufactured to specifications set with an awareness of the importance of these factors in relation to the films that it needs to process. Given the variety of radiographic emulsions currently on the market and the ever-increasing demands of processing high-quality radiographs as efficiently as possible, it makes sense to use a film manufacturer's fixer (or a comparable product) to process these films.

## Clearing

While most films can be adequately fixed and cleared with fixers that have reduced component concentrations, a more robust fixer may be desirable to provide an acceptable fixer environment for current and future film technologies. In addition, a robust fixer is particularly important in the processing of mammography films, which are more difficult to clear and have specific requirements for hypo retention.

## Silver Recovery/Reuse

There are many types of silver recovery units or fixer manager units on the market. Most of these units effectively remove silver, but if they are not set up and used properly, they can destroy some of the fixer components, such as thiosulfate and sulfite, and weaken the fixer. A fixer with adequate thiosulfate and sulfite levels is necessary as replenishment rate reduction and fixer reuse become more prevalent.

Silver recovery methods are discussed in greater detail later in this chapter.

## Odor

The odor threshold is the concentration at which the smell of a chemical can be detected in the ambient air.

Sulfur dioxide, which can be irritating to the nose and throat, can be detected by smell at concentrations as low as 0.5 ppm. Low-odor fixers are specifically formulated at a pH and sulfite concentration that will not adversely impact processing (drying, dye stain, fixing, surface quality, etc.), while significantly reducing the volatile sulfur dioxide—0.15 to 0.30 ppm under worst case processing conditions. Because these fixers are formulated for environmental considerations, they must be used as specified to obtain desired results.

Unpleasant odors can be masked with perfumes or fragrances. However, some risk is associated with this practice. The sense of smell is a sensitive and relatively accurate indicator of the presence of many potential fixer fumes, in most cases at levels well below the TLV (threshold limit value) for short-term exposure. If people in the processing area were no longer able to smell fixer-related odors, they would lose the capability to recognize a possible exposure issue.

### Drying

An optimized level of hardener and pH is needed to provide adequate drying for critical films in a wide range of processing conditions (different processors, film volumes, etc.).

### Buffer Capacity

A fixer with sufficient buffer capacity will become more advantageous as replenishment rate reduction and fixer reuse become more prevalent (to counteract the effects of developer carryover into the fixer).

## Mixing

Many different manufactured formats of processing solutions are available: liquid concentrates, powdered concentrates, mixtures of both liquid and powdered concentrates, and ready-to-use solutions. Some are packaged as single-part concentrates. Others are multiple-part concentrates. Most manufacturers supply mixing directions with their products. These directions should be followed closely because factors such as the mix order, temperature, and stir times can be essential to obtaining a good solution.

The one ingredient in a processing solution that is usually supplied by the facility is the water. The water should be relatively pure. Table 2-3 may be used as a guideline to determine the water purity needed to mix processing solutions. Processing solutions usually contain a sequestering agent that complexes heavy metals or

**Table 2-3** Maximum Contaminant Levels in a Water Supply Used for Mixing Solutions

| *Impurity* | *Maximum or Range (ppm)* |
|---|---|
| Color and Suspended Matter | None by appearance |
| Dissolved Solids | 250 |
| Silica | 20 |
| pH | 7.0 to 8.5 |
| Hardness as $CaCO_3$ | 40 (preferable) to 150 |
| Copper, Iron, Manganese (each) | 0.1 |
| Chlorine, as Free Hypochlorous Acid | 2 |
| Bicarbonate | 150 |
| Sulfate | 200 |
| Sulfide | 0.1 |
| Chloride | 25 |

(Reprinted with permission from Eastman Kodak Company. EKC publication no. N-327.)

impurities from a hard water supply. As previously discussed, heavy metal ions present in the water can cause developer oxidation. Some ions, such as calcium, form insoluble hydroxide compounds at the alkaline pH levels that are present in developer solution.

Although most developers on the market have an excess of sequestering agents, the level of these agents may not be high enough to clean up a supply of water with contaminant levels higher than those shown in Table 2-3 and to react with emulsion by-products to inhibit salt formation in the solution during processing.

Multiple-part concentrates are packaged as such for stability reasons. The majority of general-use, machine-processing developers are supplied as three-part liquid concentrates for a number of reasons. Developer solutions are alkaline. However, the Phenidone that is typically used in these developers is unstable in alkaline conditions over long periods of time and must be separated from the alkaline portion. Glutaraldehyde is unstable in alkaline conditions and will also react with Phenidone when packaged together in an acidic environment. Therefore, these two components need to be separated, accounting for the three-part concentrate format.

Some of the newer niche market developers contain derivatives of Phenidone that are more stable in alkaline conditions and allow for two-part concentrate products. If the glutaraldehyde can be omitted from the process, as is the case when using forehardened films, a one-part concentrate is possible.

## General Steps of Mixing

Whether using a one-part liquid concentrate or a multiple-part, hybrid liquid/slush powder mix, the mixing process is important for obtaining a good solution.

### *Metering Water for Mixing*

With current technology, the water added should be within 3 percent of the stated amount. This can be achieved with meters or by weighing the water carefully and accurately calibrating mixing tanks. Either technique is acceptable as long as the desired accuracy is obtained. Differences greater than 3 percent could lead to sensitometric and physical properties that are out of control in the process and difficult to correct. These situations are more critical in a process that uses minimum replenishment rates and/or solution reuse.

Adhere to the temperature stated in the instructions. If circumstances do not allow this, follow a trial-and-error process, noting the dissolution of the concentrates. If a temperature colder than the recommended temperature is used, a longer stir time may be needed. Always consult the manufacturer if this situation arises.

### *Mixing Concentrate Parts*

The agitation system used should be strong enough to provide adequate mixing, but not so strong that excess oxidation can occur.

Following the proper mix order is important. If the mix order is not followed, some components may not dissolve, or they may undergo a chemical reaction that is harmful to the solution. The stir time should be observed; if it is not, make sure that the components are completely dissolved before going to the next step.

### *Weighing Components*

When weighing components or splitting concentrates to make a fractional batch, use an accurate scale. The scale must be kept in good working order. Never split a blend of powders any further than the unit package because the powders tend to shift during storage, making an accurate split impossible.

### *Filtering and Packaging Solutions*

The solutions should go through a clarifying filter to remove any particulate matter introduced during the mixing process. Filtering also removes any insolubles introduced into the solution from the concentrates, especially from the powders. To avoid contamination, containers should be dedicated to one particular solution. The containers should be clean, filled to the top, and completely sealed (airtight). To inhibit any aerial oxidation, package solutions and seal the containers as soon as possible.

## Automixers

Follow the manufacturer's mixing directions when using an automixer. Two basic types of commercial automixers are the electronic probe and the specific gravity mixer. Both types have been proven to be reliable. When using the gravity mixer, make sure that the mixer is set to the proper specific gravity for the solution being mixed.

Also, be aware that some dealers and manufacturers of automixer chemistry purchase bulk products from other manufacturers and repackage them into an automixer format. Be sure that the repackaged material contains the correct quantities of concentrates intended by the original manufacturer and that the material has been handled so that there has not been premature oxidation of the solutions. To verify that the chemistry is acceptable, test the pH, specific gravity, and process control sensitometry of the mixed solution, as described in the next section.

# Monitoring Chemicals

Typically, if the manufacturer's mix directions are followed, analysis of the solutions is not necessary. Because the composition of the processing solutions can affect the sensitometric and physical properties of the film, it is sometimes of interest to verify that the solution has been mixed correctly. Several relatively common approaches to this are:

- pH measurement.
- Specific gravity measurement.
- Laboratory analysis of components.
- Processor quality control (QC).

## pH Measurement

A pH measurement determines the concentration of hydrogen ions in solution. Measurement of pH can be used as a measure of chemical activity because development generally decreases in activity as pH decreases. This measurement, however, is not very accurate and is useful only for finding trends or large changes in developer concentration. pH measurements, or detection of hydrogen ion concentration, are difficult to achieve in high salt solutions, such as developers, unless carefully calibrated electrodes are used.

An incorrect pH value can indicate that a solution was improperly mixed. However, a correct pH measurement by itself does not necessarily mean that a solution was properly mixed. Further testing is required.

A common problem with the pH measurements is the lack of proper maintenance of the total system. The electrodes need to be calibrated for each use, and the temperature must be accurately maintained. Along with the proper equipment, a fully trained technician is essential to obtaining useful results.

## Specific Gravity Measurement

Specific gravity—a unitless value—is the ratio of the mass of a body to the mass of an equal volume of water at a specified temperature. A specific gravity measurement for a processing solution is the relative weight of a solution as compared with the density of water at 4°C.

Density is the measurement of mass per unit volume, usually expressed in grams per milliliter.

Specific gravity measurements are useful to demonstrate that the proper dilution has been made. It will only give a good indication of over- or under-dilution. However, if used alone, this test is not specific enough to determine whether critical components, such as developer antifoggants, are missing or removed from the solution. Hydrometers can be used to measure specific gravity and can be purchased at a reasonable price. Be sure to measure specific gravity at the temperature at which the hydrometer has been calibrated, or the measurements will be incorrect.

## Laboratory Analysis of Components

Many facilities would like to verify that their developer and fixer are properly mixed. As previously discussed, the best way to measure the levels of components in the developer and fixer is through chemical analyses. Analysis of solutions by an analytical laboratory is the most accurate and predictive approach. Proper analytical equipment and technical skill are required to determine the component levels. This approach, however, is costly and time-consuming.

If it is determined that chemical analysis should be performed (because a problem with the chemicals is suspected), four samples are necessary to obtain a complete picture of the chemical environment. Samples should be taken from the developer and fixer tanks in the processor ("working solutions") and from the developer and fixer replenisher in the holding tanks or automixer. Each sample bottle must be completely filled to exclude all air, which would affect results by accelerating chemical oxidation. Samples must be shipped according to regulations.

Taking samples as discussed above will provide information on initial chemical quality and on the state of the working solutions; both are necessary for determining if the chemicals are contributing to a particular problem.

### Processor Quality Control (QC)

Probably the easiest, most widely used, and most cost-effective approach to monitoring the chemicals and verifying that solutions were properly mixed is through processor QC. By monitoring changes in the sensitometry of a process control strip in correlation to chemicals being added to the replenisher holding tanks or freshly mixed in an automixer, changes due to chemistry can be detected. If the sensitometric parameters are out of specification for the process when other sources of variability are under control, the quality of the processing solutions should be considered.

Processor QC is the most effective method for checking whether the developer was mixed properly. A well-run quality control program can determine whether the processor is within control limits. If so, then obtaining the same sensitometric responses indicates that the process remains in control.

There are three indicators for testing whether a fixer mix is acceptable:

1. Determine whether undeveloped silver is removed from the film by processing an unexposed sheet of film and visually inspecting it. If the sheet has opalescent (milky) streaks, the fixer has a low level of hypo or no hypo and is unacceptable.
2. Measure the retained hypo in the film using a hypo estimator and chemical solution specifically designed for this purpose. A fixer retention test (hypo test) provides a relatively simple method of estimating the amount of thiosulfate retained in a processed radiograph. Assuming that washing is not a problem, very high retained hypo values suggest that the fixer may not be mixed properly.
3. Check film drying. The need to increase the processor's dryer temperature to obtain dry film may indicate an improperly mixed fixer.

Table 2-4 lists the four methods of testing chemical quality.

## Chemical Storage

Concentrates of chemicals are relatively stable when stored properly. At 70°F (21°C), unmixed concentrates are stable for as long as two years.

The recommended temperature range for proper storage of mixed chemicals is between 40 and 85°F (5 and 30°C). Ideally, the temperature should not exceed 70°F (21°C) for a prolonged period of time.

Note that environmental temperature for chemicals should be controlled 24 hours per day, seven days per week.

Studies of mixed solutions stored at room temperature in replenisher holding tanks with a floating lid have shown that the sensitometric properties stay relatively stable

**Table 2-4** Testing of Chemical Quality

pH measurement
- Indicates the acidity or alkalinity of a solution at a specified temperature.
- pH values outside the acceptable range indicate the chemicals were not mixed properly.
- pH values within the acceptable range do not guarantee properly mixed chemicals.

Specific gravity measurement
- Indicates the ratio of the mass of a body to the mass of an equal volume of water at a specified temperature.
- Gives a good indication of over- or underdilution only.
- Values within the acceptable range do not guarantee the presence of all required components.

Laboratory analysis of components
- A service available from some companies.
- Verifies chemical components and proper mixing.
- May be costly and time-consuming.

Processor QC
- May be used to monitor chemical quality besides monitoring the processor.
- Can quickly alert you to check chemical quality if changes occur on the chart in correlation to chemicals being added, etc.
- Easiest and most cost-effective way to monitor the chemicals.

for up to six weeks under perfect conditions. If the solutions are properly mixed and stored, there should not be a significant difference in performance between a fresh automixed solution and a six-week-old premixed solution.

As the temperature increases, or if no floating lid is used, the stability or lifetime of the solutions diminishes. The general recommendation for mixed chemicals is that they should be used within two weeks.

## Fresh and Seasoned Chemicals

Newly mixed chemicals are frequently referred to as "fresh." Fresh chemicals should be used in the replenisher holding tanks or automixer for the initial startup of a processor, especially when first establishing a processor quality control program and after a complete chemical change during the performance of a preventive maintenance procedure. (Note that seasoned developer may be reused under certain circumstances, as discussed in Chapter seven.)

For best results, it is important to follow the manufacturer's recommendations by (1) using the proper chemicals, (2) mixing to the correct concentrations, and (3) adding the appropriate amount of starter solution.

"Seasoning" refers to the changes that take place in the developer solution as films are processed after fresh chemicals have been added to the processor.

The developer is considered fully seasoned when the developer tank in the processor has "turned over" three times. In other words, an amount of fresh developer replenisher equal to three times the volume of the developer tank has been pumped into the processor.

The time required to go from a fresh to fully seasoned state in a particular processor will vary depending upon the number of films processed and the amount of fresh developer replenisher (developer replenishment rate) pumped into the developer tank as each film passes through the entrance detector crossover of the processor.

The time can be calculated by knowing the volume of the developer tank in milliliters, the developer replenishment rate (also in milliliters), and the number of films processed in a given period of time.

The occurrence of seasoning is the basis for averaging the sensitometric values from five processor quality control films, one per day. Averaging is necessary to reflect a meaningful processor quality control baseline against which the values (speed and contrast) from all daily sensitometric films will be compared to determine if the processor is in control.

As film development occurs, developing agents are consumed, and bromide ions and other by-products are released into solution. As more of the developing agents are consumed, the sensitometry of processed films and developer activity changes. Slight changes in speed and contrast occur; developer activity decreases. The sensitometric effect of seasoning is very dependent upon the type of film being processed, the chemicals in use, and whether correct replenishment rates are set and adjusted for changes in film volumes.

At some point in a seasoned process, film speed and contrast will stabilize, provided everything is balanced. Balance is best achieved by following the manufacturer's recommendations, as previously discussed.

## Starter

The addition of starter solution—a clear, odorless liquid—begins the seasoning process, mainly by adding bromide ions to the developer.

As film and chemical formulations vary significantly from manufacturer to manufacturer, the starter solution sold by the chemical manufacturer should be used.

Additionally, the use of a graduated cylinder with milliliter markings is recommended to ensure that the correct amount of starter solution specified for the developer tank volume is added.

After a complete chemical change, the correct amount of starter should be added to a half-filled developer tank. The balance of the developer may then be added. (Note that the fixer tank should first be filled and the fixer rack replaced before refilling the developer tank; additionally, a splash guard should be used.)

If flooded replenishment is used (recommended for low film volumes and discussed later in this chapter), starter should be added to the developer replenisher holding tank. The typical volume of starter is 3 fluid ounces (89 milliliters) per gallon (3.8 liters) of developer replenisher. Additional starter should not be added to the processor developer tank.

# Replenishment

As exposed film is processed, the chemicals in the processor (developer and fixer) gradually weaken, and their effect on film (activity) slows down. To compensate for this progressive loss of activity, chemicals must be replenished. This occurs automatically in automatic roller transport processors, as film passes the entrance detector crossover assembly.

If replenishment did not occur or if the replenishment rate is too low, the gradual reduction in chemical activity will cause a loss of quality in the processed radiographs; the grays become softer and blacks become lighter. In addition, it will be necessary to increase radiation exposure to patients to achieve the proper film optical density.

Replenishment is important for maintenance of stable developer and fixer activity. Proper replenishment provides stable sensitometric results (film contrast, film speed, and base plus fog), reduces processing artifacts, and provides archival keeping.

## Replenishment Rate Guidelines

Film and chemical manufacturers provide extensive replenishment rate guidelines.

Published replenishment rate guidelines (Table 2-5) are sometimes categorized based on daily film volumes (high, medium, and low) in an eight-hour day, film types (general-purpose films, mammography, etc.), number of films fed at a time, and whether the processor is dedicated to a particular film type (such as mammography) or used for all film types (intermixed).

A low film volume in an eight-hour day requires higher replenishment per sheet of film. Processors with very low film volumes (such as operating rooms) are very difficult to stabilize and control for consistent image quality. A method called flooded replenishment is recommended under these conditions. Flooded replenishment is discussed later in this chapter.

A high film volume per eight-hour day requires lower replenishment per sheet.

Note that published replenishment rate guidelines should be used as initial starting points only. It may be necessary to set higher or lower rates based on a number of factors, such as the type of chemicals used, whether or not the target film volume is spread throughout the day, whether films are processed everyday, whether one or two 18 x 24 centimeter mammography films are typically fed at the same time, etc.

Once replenishment rates have been set and the processor quality control program established, the goal is maintenance of a controlled processing environment while minimizing chemical use (lower costs) and the impact on the environment. It may be possible to reduce the rates systematically as long as the processor remains in control, a regulatory requirement in many cases.

Additionally, it is important that facility personnel responsible for overseeing the processor be in close communication with the representatives of the processor service company concerning changes in film volume. As film volume decreases, it may be necessary to increase the replenishment rates; as film volume increases, it may be necessary to decrease the replenishment rates. Such communication is frequently overlooked and may be the source of an upward or downward trend and an out-of-control processor.

**Table 2-5** Typical Published Replenishment Rate Guidelines Based on Film Type and Volume (a) for General-Purpose Films and (b) for Mammography Films

(a) General-Purpose Films, Standard-Cycle Processing

| *Processed film* | *Volume (Use)* | *Sheet Film Volume in 8 hours* | *Replenishment Rates mL per 35 x 43 cm, for 35 or 43 cm of film travel* | |
|---|---|---|---|---|
| | | | *Developer* | *Fixer* |
| Intermixed film sizes | High | 115 or more | 50 | 70 |
| (18 x 24 cm to 35 x 43 cm) | Medium | 40 to 115 | 54 | 85 |
| | Low* | 40 or less | 80 | 100 |

(b) Mammography Films, Standard- or Extended-Cycle Processing

| *Processed film* | *Volume (Use)* | *Sheet Film Volume in 8 hours* | *Replenishment Rates mL per 18 x 24 cm, for 24 cm of film travel* | |
|---|---|---|---|---|
| | | | *Developer* | *Fixer* |
| Intermixed film sizes | High | 150 or more | 20 | 30 |
| (18 x 24 cm | Medium | 60 to 150 | 27 | 35 |
| to 24 x 30 cm) | Low* | 60 or less | 35 | 40 |
| Single film feeding | | | | |

Note: Data are for KODAK X-OMAT M35A-M Processor and KODAK RP X-OMAT Processors, models M6 and M7.
*Fewer than 30 sheets of film may require flooded replenishment.
(Reprinted with permission from Eastman Kodak Company.)

It is also important to note that different films may require different replenishment rates. Figure 2-6 shows two different mammography films. Both films are stable in speed regardless of whether a developer replenishment rate of 20 or 50 milliliters per 18 x 24 centimeter film is used. Notice, however, that the contrast of the film in (b) is significantly reduced at 20 milliliters; it requires 50 milliliters per sheet to achieve the proper contrast.

## Defining Replenishment Rates

"Replenishment" may be defined as the amount of fresh developer and fresh fixer that is added to the processor at a set rate, called the "replenishment rate." Replenishment rates are commonly recorded by users as the number of milliliters of developer and fixer. Because it is very easy to miscommunicate regarding replenishment rates and replenishment is often an important consideration when troubleshooting an out-of-control processing environment, merely noting two numbers does not pro-

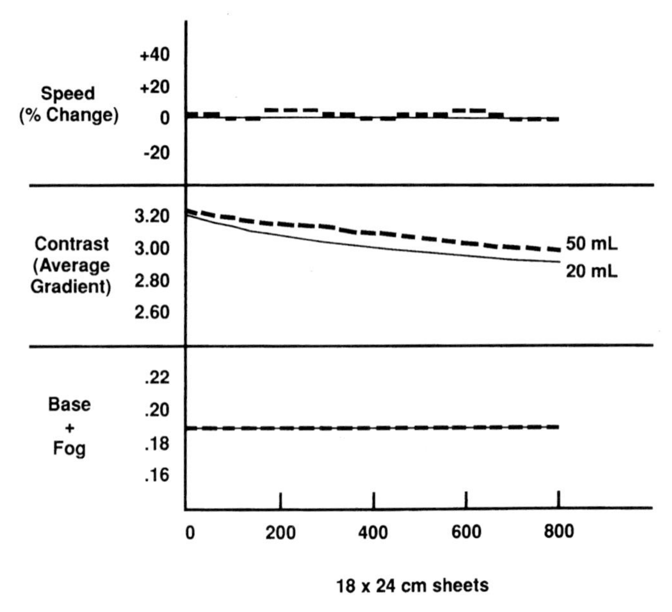


a

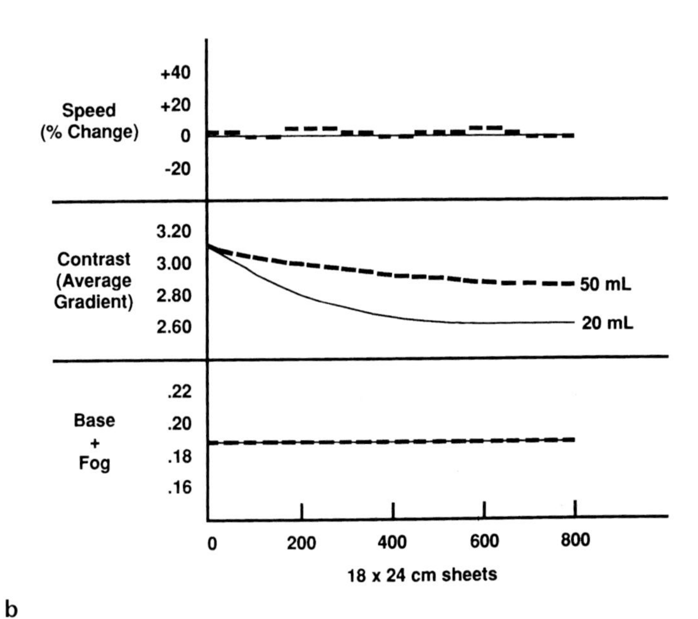


b

**Figure 2-6** Effects of two developer replenishment rates (20 and 50 milliliters per sheet) on two different mammographic films. Different film-emulsion formulations may have different sensitometric responses to changes in replenishment rates. The film depicted in (a) is less sensitive than the film depicted in (b) to differences in replenishment rates. Note that the film in (b) requires substantially higher replenishment to maintain proper contrast. Also note the seasoning effects (little change in speed, gradual downward shift in contrast) that typically occur in going from a fresh start (complete change of chemicals) to a seasoned processing environment achieved after processing several hundred films.

vide enough information about replenishment rates; rates should be fully defined in terms of the number and size of films and length of film travel.

For intermixed general-purpose films, a 35 x 43 centimeter sheet of film is usually the benchmark. The length of travel is either 35 centimeters for a processor with a 17-inch (43-centimeter) throat, or 43 centimeters for a processor with a 14-inch (35-centimeter) throat. Using the high-volume replenishment rate information available in Table 2-5a for general-purpose films, the developer replenishment rate is 50 milliliters and the fixer replenishment rate is 70 milliliters per 35 x 43 centimeter film. The length of film travel could be either 35 centimeters (14 inches) or 43 centimeters (17 inches) depending upon the model processor used.

For mammography films, typically one or two 18 x 24 centimeter films are processed. The length of film travel is 24 centimeters.

Using the high-volume replenishment rate information available in Table 2-5b for a mammography film, the developer replenishment rate is 20 milliliters and the fixer replenishment rate is 30 milliliters per one 18 x 24 centimeter film fed singly, for 24 centimeters of film travel. In other words, 20 milliliters of developer and 30 milliliters of fixer should have been pumped into the processor after an 18 x 24 centimeter film with the short dimension as the leading edge was fed into the processor.

Miscommunication most frequently occurs in mammography processors if the replenishment rates are set based on 35 or 43 centimeters of film travel, rather than 24 centimeters, or on an incorrect number of films typically fed.

In the above mammography example, if two 18 x 24 centimeter films are usually processed at one time, an initial developer replenishment rate of 40 milliliters and a fixer replenishment rate of 60 milliliters would be required. Note that feeding two 18 x 24 centimeter films at a time is the norm. However, single film feeding is recommended for some processors with narrower film feed trays. Processor modifications may be available to widen the film feed tray so an adequate space separates two 18 x 24 centimeter mammography films fed at the same time.

In addition, for mammography, film feeding protocols should be posted in the darkroom and all personnel trained to process film the same way. Processing mammography film consistently is important to help standardize replenishment rates.

## Flooded Replenishment

In low volume per eight-hour situations, sufficient replenishment does not occur as films are being processed and undesired chemical components accumulate (e.g., bromide, a by-product of the development process). As a result, the processor quality control chart may indicate that the processor is out of control.

To rectify the situation, it is necessary to add fresh chemicals to the processor at a rate that will ensure constant chemical activity. A method first introduced in 1979, called "flooded (or flood) replenishment," can be used to help stabilize the processing environment for low-volume use.

Flooded replenishment is accomplished by not only pumping fresh developer and fixer replenisher into the processor when film is fed (moving through the entrance detector crossover assembly) but also on the basis of the amount of time the processor is operational. A timer installed on the processor activates the replenishment pump periodically (e.g., 20 seconds every 5 minutes at a rate of 65 milliliters per actuation). The additional amounts of developer and fixer replenisher will help maintain the developer and fixer activity at a stable level.

Note that when flooded replenishment is used, the starter solution is added to the developer replenisher holding tank at the rate of 3 fluid ounces per gallon (89 milliliters per gallon or 25 milliliters per liter). Additional starter should not be added to the developer tank in the processor. Also note that the use of flooded replenishment provides a stable fresh process; seasoning does not take place.

Flooded replenishment should be set up on a processor by qualified service personnel.

## Replenishment Reduction

### *Dedicated Systems*

Manufacturer-recommended replenishment rates are set in such a way that a variety of film types can be processed through processors with good sensitometric results. Currently, many processors are dedicated to processing a specific set of films. Under these conditions, replenishment rates can be reduced, especially if the manufacturer's recommended chemicals are used. In some cases, control measurements may indicate that replenishment rates should be increased.

If a processor has a good quality control program, the rates can be lowered slowly and the sensitometry can be monitored. The fixer rates can also be reduced if proper testing determines that complete fixation is taking place.

### *Fixer and Wash Reuse*

When using an electrolytic silver recovery cell plumbed inline with the processor, replenishment rates can be reduced by essentially reusing the fixer. The fixer can also be desilvered in a batch mode, and a percentage of the fixer can be reused in the next fresh mix. Wash water reuse kits, which can reduce water consumption, are also available. Any of these options must be verified, monitored, and possibly modified to fit the particular application in which it is to be used.

## Oxidation

Chemicals exposed to air (oxygen) become oxidized over time. Mixed chemicals should be used within 10 to 14 days.

The use of oxidized chemicals is detrimental to image quality and to the stability of the processor QC chart.

Chemical oxidation may be accelerated by the following factors:

- Storage and use of chemicals at temperatures higher than recommended.
- Using replenisher holding tanks larger than needed to contain the volume of chemicals typically used in a week.
- Not using a floating lid on top of the developer solution in the developer replenisher tank, which is a frequent cause of chemical oxidation.

The proper and consistent use of a floating lid, in particular, is recommended as the best way to control chemical oxidation (Figure 2-7).

## Use of Recommended Chemicals

All film manufacturers have recommended chemicals (or equivalent) for their films. Many users consider chemicals from various manufacturers to be interchangeable. However, surveys have documented that film speed, film contrast, and base plus fog respond differently to the various types of chemicals used (Figure 2-8). These effects also depend on the type of film being processed.

Chemical manufacturers distribute chemicals as concentrates. Solution service providers add water locally to complete the mixture. In some cases, chemicals are not mixed to the appropriate concentration in accordance with the manufacturer's recommendations.

Processing chemical variability can occur in the medical imaging marketplace because of a number of factors. Although most manufacturers use similar processing chemicals to achieve development and fixing, the concentration of these chemicals can vary, either initially or after being mixed by solution services dealers. This concentration variation can result in changes in film response of differing magnitudes for certain film types. In addition, variability can also result from improper replenishment. Either over-development or under-development can occur, depending on the degree of replenishment or initial chemical concentration.

**Figure 2-7** A floating lid should always be used on top of the developer replenisher to reduce evaporation and chemical oxidation. (Reprinted with permission from Eastman Kodak Company.)

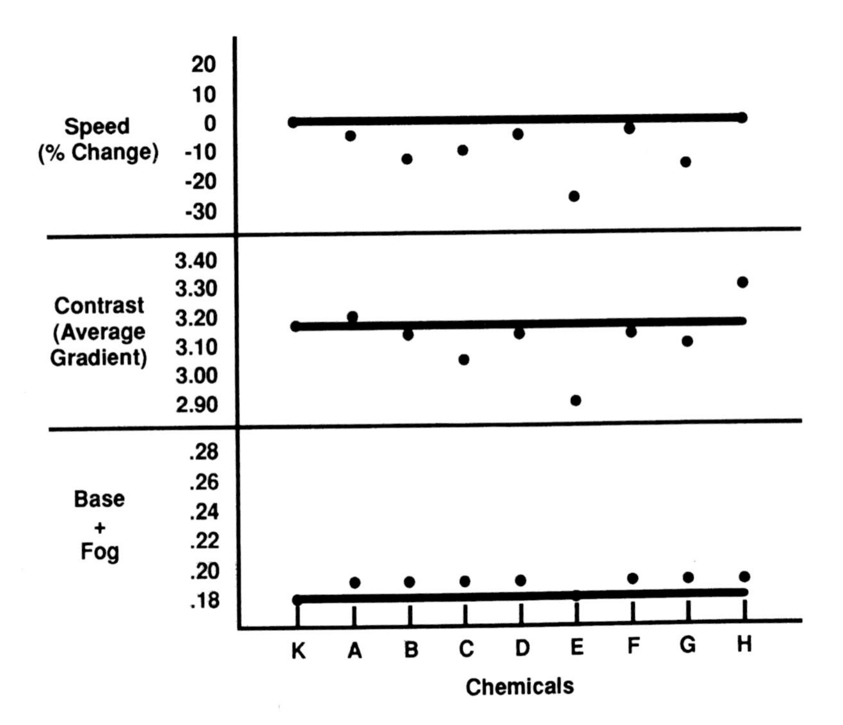


**Figure 2-8** Chart produced from film-processing survey data. The chart shows film-processing variations due to use of different chemicals for a single-emulsion mammography film. K indicates processing data (and expected values) obtained by using the film manufacturer's processor and chemicals. A horizontal line is drawn at the letter K data point. A to H are data for different brands of chemicals. Data were obtained by using film strips sensitometrically preexposed to light that simulates the light spectrum from a mammographic screen. Film-speed differences, film contrast (average gradient), and base plus fog values were determined from the sensitometric data.

## Medical Device Regulations

Medical x-ray processing chemicals are an integral part of the x-ray film processing system. The importance of the consistent quality of processing chemicals is underscored by the fact that both the United States and Europe have incorporated medical x-ray processing chemicals into medical device regulations.

The United States regulates medical x-ray processing chemicals in its Code of Federal Regulations (CFR) publication. The Food and Drug Administration (FDA) outlines its requirements in CFR21, parts 800-1299, and classifies medical x-ray processing chemicals as a medical device accessory. This regulation requires that medical x-ray manufacturers follow "good manufacturing practices" (GMP) for both medical devices and medical device accessories such as medical x-ray processing chemicals.

In Europe, the European Commission (EC) listed its classification for x-ray photochemicals in its EC Medical Device Directive 93/42, entitled "Classification of Diagnostic Imaging Recording Devices." The directive categorizes x-ray photochemicals as a Class 1 Photoprocessing Device.

## Health, Safety, and Environmental Issues

Medical imaging facilities, both large and small, are required to comply with a variety of government regulations and standards that address everything from patient rights to Medicare reimbursements. In addition, medical imaging administrators are challenged to manage a maze of health, safety, and environmental (HSE) laws that affect their facilities and occupants. In the United States, these laws are created at the national, state, and local level.

At the international level, the European Commission has a variety of environmental directives, and Australia has maintained Photographic Universal Regulations for the Environment (PURE) for several years. HSE standards vary depending on where the medical imaging facility is located. Country-specific standards may include strict laws about waste discharges to the environment or even demand that all process effluent be hauled away. Other countries may choose not to address these issues at all. While it is important to realize that government-mandated HSE activities and programs could impact most healthcare facilities, this section will focus on HSE issues as they relate to the medical imaging facility.

In the United States, an awareness of the necessity of preserving the quality of the environment has increased since the 1960s. The growth in technology placed unprecedented demands on the environment, both directly and indirectly. The problems created by these demands have increased an awareness of the need to protect air, land, and water resources. To protect these resources, legislation was enacted at many levels of government, requiring businesses and industry to manage these assets properly. The result was strict environmental laws regulating air emissions, waste water discharges, and solid waste, both hazardous and nonhazardous.

For the medical imaging facility, the focus of environmental compliance issues is the x-ray processor and the used processing solutions, or effluent, that it generates. Regulations require that these used solutions be managed properly due to the silver content, but medical imaging administrators should recognize that it is not rocket science to accomplish this task. Used solution management techniques, such as silver recovery and the use of environmental service companies to manage the effluent, are readily available to assist in achieving regulatory compliance. For additional assistance, some chemical and film manufacturers also have HSE experts who have the resources to provide the facility with information regarding how to comply with regulations. These experts will typically have the technical background and the product knowledge necessary to provide the medical imaging facility with sound regulatory guidance.

To address the health and safety issues encountered in the U.S. workplace, in 1970 the Occupational Safety and Health Act was passed by Congress. This law was designed to protect the nation's human resources by requiring employers to establish a safe and healthy work environment. Employers have a general duty responsibility to provide a workplace free from recognized hazards that are causing, or likely to cause, death or serious physical harm to employees. To assist employers in meeting this responsibility, numerous regulations have been promulgated by the Occupational Safety and Health Administration (OSHA), many of which have applicability to the medical imaging facility. Examples include Employee Emergency Plans, Fire Prevention Plans, Personal Protective Equipment, Means of Egress, Bloodborne Pathogens, and Hazard Communication. In this chapter, we will focus on the OSHA Hazard Communication and Personal Protective Equipment standards. Note that regulations can be modified at any time. Annual workplace citation summary reports assembled by OSHA indicate that the requirements in these regulations are frequently violated at medical imaging facilities.

From an enforcement standpoint, OSHA has the authority to inspect the workplace. An inspection may be the result of an employee complaint or a scheduled onsite visit. OSHA Compliance Officers are trained to observe those conditions that may be harmful to employees. If such violations are uncovered, the employer will receive a citation and a notice of proposed penalties. The severity of the violation determines the penalty amount. For a serious violation, a mandatory fine of $7,000 for each violation will be proposed. Larger penalties can be assessed, up to $70,000, for each willful violation. These fines can compound quickly. Noncompliance is costly and, more importantly, places employees at risk. Maintaining a safe and healthy workplace is beneficial to all.

From an international perspective, health and safety requirements in the developed countries are similar to those found in the United States. The developing countries are just beginning to address workplace safety issues. They will need to focus initially on the most hazardous situations and activities.

## Hazard Communication

Making employees aware of chemical hazards is an important part of providing a safe workplace (Figure 2-9). While a medical imaging facility is typically considered a "low-hazard" workplace, there are certain materials and operations that can present hazards to employees. An understanding of those potential hazards and measures to protect employees who handle those chemicals should be a key element in the health and safety program at your facility.

OSHA has developed a framework of federal regulations that govern workplace safety. These regulations are based on the following principles:

- Every employee has a need and right to be made aware of the hazards in his or her workplace.
- Every employee has a need and right to be protected from hazards in his or her workplace.

The OSHA Hazard Communication Standard was established to ensure that employers and employees know about workplace chemical hazards and how to protect themselves from those hazards. This standard is different from other OSHA standards in that it broadly covers all aspects of hazardous chemical manufacturing, movement, and use rather than focusing on a single operation or chemical. The standard includes information on hazard evaluation, Material Safety Data Sheets (MSDSs), chemical lists, container labels, and employee training to ensure that chemical hazards are properly communicated to users of those materials.

**Figure 2-9** Making employees aware of chemical hazards is an important part of providing a safe workplace. (Reprinted with permission from Eastman Kodak Company.)

**Table 2-6** Hazardous Chemical Descriptions

| *Hazard Type* | *Examples* |
|---|---|
| Physical Hazard | Compressed gases (cylinder of nitrogen), oxidizers (chlorine bleach), combustible liquids (kerosene), and flammable materials (cleaning solvents) |
| Health Hazard | Chemicals that may cause acute or chronic health effects, such as irritants (ammonia), corrosives (acetic acid), carcinogens (formaldehyde), and sensitizers (color developing agents) |

(Reprinted with permission from Eastman Kodak Company. EKC publication no. J-311.)

This section reviews how the OSHA Hazard Communication Standard applies to medical imaging facilities. A recommended compliance process and a self-assessment checklist are also included to assist in reviewing and improving your current program.

## *Hazardous Chemicals*

To determine if the OSHA Hazard Communication Standard applies to your facility, you must first determine if there are hazardous chemicals in your workplace. OSHA estimates that there are more than 575,000 chemical products that can be identified as hazardous. A hazardous chemical is defined by OSHA as being either a physical hazard or a health hazard (Table 2-6).

Determining whether a chemical is hazardous or not is the responsibility of the manufacturer or importer of the chemical (Table 2-7). You do not need to make a separate assessment of chemicals in your workplace—you can rely on the evaluation of the manufacturer or importer. When a manufacturer has determined a product is or contains a hazardous chemical, it is required to provide you with an MSDS. Information on the physical and/or health hazards of a chemical is in the MSDS.

### *Exemptions*

There are some materials that are specifically exempt from some or all of the requirements of the Hazard Communication Standard. Two examples are consumer products and articles.

Consumer products are exempt from the standard when used in a frequency, quantity, and duration equal to that reasonably experienced by consumers outside the workplace. For example, if you are using a household cleaner for occasional cleaning, this chemical would not need to be included in your program. However, if you are using this product to clean processors, which is not its intended use (as labeled), this chemical must be included in your program.

If you have an employee whose job requires daily use of this product for janitorial services, the exposure would be greater than experienced by a typical consumer of the product and this material would need to be included in your program. Even if the consumer products in your facility are not covered by the standard, it is a good idea to obtain the MSDSs for these products and make them available to employees.

Articles are also exempt from the standard. An article is a manufactured item that meets the following criteria:

- Is not a liquid or a particle.
- Is formed to a specific shape.
- The end use depends on the product's shape or design.
- Does not release more than traces of hazardous chemicals during use.
- Does not pose a physical hazard or health risk to employees.

**Table 2-7** Chemical Hazardous Communication Information Flow

| *Manufacturers/Importers* | *Distributors (Stockhouse/Retail Sales)* | *Employers (Medical Imaging Facility)* |
|---|---|---|
| • Determine hazards | • Transmit MSDS | • Obtain and maintain MSDS |
| • Develop and provide MSDS | • Maintain container labels | • Maintain and provide labels and chemical list |
| • Label shipping containers | | • Communicate information to employees |

(Reprinted with permission from Eastman Kodak Company. EKC publication no. J-311.)

### *Employees Covered by the OSHA Hazard Communication Standard*

The purpose of the Hazard Communication Standard is to communicate known potential hazards to employees who may be exposed to hazardous chemicals (Table 2-8). The definition of "employee" includes anyone who is compensated for performing work at your facility. This definition includes family members who are paid for employment but are not owners of the business. There are no exemptions for small workplaces. Any business with at least one employee is covered by this standard if hazardous chemicals are present in the workplace.

### *Employees of Other Companies*

If you have employees of other companies, such as maintenance workers or vendors, working at your facility, you are responsible for ensuring that they are informed of any chemical hazards they might come into contact with. Hazard communication training of those employees is the responsibility of their employers.

You must identify chemical hazards to which employees of other companies may be exposed. They must also have access to MSDSs and information on your facility's hazard labeling system. You must also inform them of any protective measures, such as personal protective equipment (PPE) or ventilation, that are required during normal operating conditions or foreseeable emergencies.

You can provide the worker's employer with a copy of your written program as a means of complying with these requirements.

### *State and Local Requirements*

In addition to the OSHA requirements for Hazard Communication, there may be additional state and local requirements that apply to your facility.

Some states may have Worker Right-to-Know laws that have additional requirements. These typically include additional information on chemical labels or requirements for annual employee training. Local laws administered by fire departments or county agencies may also have additional requirements, such as facility hazard signs.

**Table 2-8** Employees Covered by the OSHA Communication Standard

| *Personnel* | *Examples* |
|---|---|
| Employees who work with hazardous chemicals | Photographic processor operator operator, chemical mixing |
| Employees who work around and may have contact with hazardous chemicals | Sales person who works in or near processing area |
| Employees of other companies who work in your facility who may have contact with hazardous chemicals | Contract maintenance or janitorial staff, vendors |

Each state has an OSHA consultation service and OSHA area offices that can help you with the specific requirements for your area.

### Health and Safety Program

Your Hazard Communication Program is a key element in an effective Health and Safety Program. There are several things you can do to ensure that your program is effective:

1. *Assign responsibilities.* Defining and assigning responsibilities for key activities, such as maintaining labels, obtaining MSDSs, or employee training, will help ensure that these tasks are being performed.
2. *Maintain the program.* Make sure that you keep your program up-to-date on any changes that take place in your facility. Also, make sure that you keep up-to-date on any new requirements.
3. *Ensure that everyone in your facility works safely.* This recommendation not only includes your employees and yourself, it also includes contract workers who work at your facility. You should have procedures to ensure that everyone at your facility follows your Health and Safety Program.

### Compliance Requirements

A medical imaging facility is required to have a Hazard Communication Program that includes the following elements:

- Written Hazard Communication Program.
- List of hazardous chemicals that are known to be present in the facility.
- Labels that identify hazardous chemicals and hazard warnings.
- Maintenance of Material Safety Data Sheets (MSDSs).
- Employee information and training.

### Written Program

Each facility that uses hazardous chemicals must develop, implement, and maintain a Written Hazard Communication Program. The written program does not need to be lengthy or complicated. Sample written programs are available from OSHA. The written program must be kept in the workplace and available to employees. If you operate multiple facilities, each location must have a copy of a written program specific to that location.

The program must describe:

- The system of labels or other forms of warning that you use at your facility. This could be warning labels available from the manufacturer or another system, such as the Hazardous Materials Identification System (HMIS).
- The location of and system for maintaining MSDSs.
- How employee training is accomplished.
- How employees are informed of the hazards of nonroutine tasks. Nonroutine tasks are things that an employee does not normally do; therefore, the employee may not have received training on the potential hazards of that task. This can include things such as occasional processor tank cleaning.

- How employees are informed of the hazards from unlabeled pipes. This protects employees in the event of a leak from the piping. Some facilities choose to label all piping as a means of complying with this requirement.
- Provisions for employees of other companies who work at your facility. This includes providing them with the following:
  - Access to MSDSs.
  - Information on employee protection requirements during normal operating conditions and foreseeable emergencies.
  - Information on your facility's labeling system.

### *List of Hazardous Chemicals*

As part of the written program, each facility must compile a list of all hazardous chemicals in the workplace. The list should use the chemical identity, such as the product or trade name that is used on the MSDS. The list does not have to include individual components of mixtures or indicate the hazards of each chemical. You can compile a list for the entire facility or lists for individual work areas in the facility. You may also want to include another reference, such as product catalog number or manufacturer, to make it easier for employees to find information on the MSDS or product label.

You may want to include all chemicals in the workplace on your chemical list and designate those that are hazardous chemicals.

The list must be kept in the workplace and be readily available to employees. The list can be posted, or you may want to use it as an index to the MSDSs.

### *Labels and Other Forms of Warning*

Manufacturers, importers, and distributors are required to label any hazardous chemicals properly before they are shipped to your facility. These labels include the product identity, appropriate hazard warnings, and the name and address of the manufacturer or other responsible party.

Employers must make sure that the arriving container labels are intact and that any container a hazardous chemical is transferred into is labeled with the chemical identity and the appropriate hazard warnings. Containers are broadly defined as anything that can hold a hazardous chemical. In a medical imaging facility, this includes storage and replenisher tanks, processors, and silver-recovery systems.

Portable containers are not required to be labeled if the hazardous chemical is transferred from a labeled container and is used within the workplace by the same employee who makes the transfer. However, if other employees are involved in the transfer or the material will be used over several work shifts, a warning label is required. It is a good idea to label all containers in the workplace with their contents and hazard information to make sure that employees are aware of any potential hazards.

Labels must be in English, legible, and prominently displayed. Torn or discolored labels should be replaced to ensure that employees can easily read the information. You can use individual container labels, or in some specific situations you can use signs or placards in lieu of affixing individual labels to containers; be sure the same information that is on the label is conveyed on the sign or placard. Labels in other languages may be used in addition to the English labels if they will assist in better communicating the hazard information.

Your labeling system must include the chemical identity and appropriate hazard warnings, and you will need to select a hazard warning system. Hazard warnings that convey the hazards associated with the container contents can be words, pictures, or symbols. Labels using plain English hazard warnings are available from manufacturers for many processing chemicals. You can also create your own labels for products by referring to the section of the MSDS that contains label statements for the products.

Some facilities may choose to use a hazard warning system such as the HMIS. This system uses a numeric scale to depict the severity of associated hazards. If you use a system of this type, be sure to train employees adequately on how to read labels properly. Placing a poster with a definition of the system in the work area is a good way to help employees understand and remember the numeric hazard scales.

HMIS labels are available through safety equipment suppliers.

## *Material Safety Data Sheets (MSDSs)*

Chemical manufacturers and importers are responsible for developing an MSDS for each hazardous chemical that they produce or import. Each MSDS must be in English but, like the hazard warnings, can be produced in other languages as supplemental information to help users better understand the hazards associated with the chemicals in the workplace. Each MSDS must contain the following:

- Identity of the hazardous chemicals in the product.
- Physical and chemical characteristics.
- Physical and health effects, and physical and health information.
- Exposure limits.
- Primary routes of entry.
- Precautionary and control measures.
- Emergency and first aid information.
- Identification of preparer.
- Date of preparation.

Manufacturers and importers must provide the MSDS at the time of initial shipment to distributors or users. Distributors also have requirements to provide MSDSs to chemical users.

You must make sure that you have an MSDS for each hazardous chemical you receive at your facility. If your supplier does not provide you with an MSDS, contact the supplier and request the MSDS.

If you have questions about the information contained in an MSDS, contact the manufacturer or importer of the chemical.

The MSDS for each hazardous chemical you are currently using must be readily available to employees during each work shift. You can keep MSDSs in a single location for the entire facility or place them in the individual work areas where the chemicals are used (Figure 2-10). MSDSs are typically maintained in paper form, but you can keep them on microfilm or in an electronic format as long as employees have ready access to them.

You may have to keep MSDSs for products that are no longer used as records of employee exposure.

## *Employee Information and Training*

All employees who may be exposed to hazardous chemicals while performing their job must be trained (Figure 2-11). This includes not only those individuals who work with the chemicals but also others who may enter the areas where the chemicals are used, such as delivery and maintenance personnel. If your film-processing area is not segregated from other work areas in your facility, you should consider training everyone who works in the facility.

Training must be conducted at the following times:

- Prior to initial assignment for new employees. Training must be done before the new employee works with any hazardous chemicals. You may want to incorporate this training into new employee orientation.
- Prior to introducing new hazards into a work area. This may include any new processes or chemicals that had not been included in previous employee training. If a new process uses similar chemicals to those already covered in existing employee training, new training would not be required.
- Prior to reassigning employees to a new work area with hazards for which the employee has not received training. Training must be done before the employee works in the new area. You may want to incorporate hazard communication into your process of retraining reassigned employees.

Annual Hazard Communication training is not required by OSHA, but it may be required by your state's Worker Right-to-Know laws. You may want to conduct

**Figure 2-10** Material Safety Data Sheets (MSDSs). The MSDS for each hazardous chemical that you are currently using must be readily available to employees during each workshift. (Reprinted with permission from Eastman Kodak Company.)

**Figure 2-11** All employees who may be exposed to hazardous chemicals while performing their job must be trained. (Reprinted with permission from Eastman Kodak Company.)

retraining routinely to be sure that all employees understand the hazards of the chemicals they are working with.

Training should be conducted for each employee to a level that is appropriate for the hazards that he or she is likely to encounter while performing his or her job. An employee who mixes chemicals or works around processors may require more extensive training on chemical hazards than an employee who has infrequent contact with chemicals or works in an area adjacent to where chemicals are used.

Training must include the following:

- The requirements of the OSHA Hazard Communication Standard.
- Operations in the employee's work area where hazardous chemicals are present.
- Location and availability of the written program, list of hazardous chemicals, and MSDSs.
- Methods and observations used to detect the presence or release of a hazardous chemical in the work area, such as visual appearance or odor. Odor is usually a good method for detecting the presence of processing chemicals.
- The physical and health hazards of chemicals in the work area. The physical and health hazards of each chemical are described on the product label and in the MSDS.
- Measures to protect employees from workplace hazards such as work practices, emergency procedures, and personal protective equipment. Information on the safe handling of chemicals is included on the MSDS.

- Details of the hazard communication program including labeling system, MSDSs, and how employees can obtain and use appropriate hazard information. Your facility's written program should include this information.

There is no requirement under the OSHA standard to maintain training records. However, it is recommended that you keep training records so you can verify that training has been conducted. If you keep records, include the employee's name, the training date, and verification of the effectiveness of training. The verification of effectiveness can be accomplished by using a simple test that the employee takes following the training.

Table 2-9 summarizes the hazard communication record-keeping requirements for medical imaging facilities. Table 2-10 summarizes the hazard communication training requirements for medical imaging facilities.

**Table 2-9** Summary of Hazard Communication Record-Keeping Requirements for Medical Imaging Facilities

Maintain current copies of:

- Written Hazard Communication Program.
- List of Hazardous Chemicals.
- Material Safety Data Sheets (MSDSs) for hazardous chemicals used.

Maintain training records (if documented).

(Reprinted with permission from Eastman Kodak Company. EKC publication no. J-311.)

**Table 2-10** Summary of Hazard Communication Training Requirements for Medical Imaging Facilities

Conduct training:

- Prior to initial assignment for new employees.
- Prior to introducing a new hazard into the workplace.
- Prior to reassigning employee to new area with new hazards.
- Annual retraining is recommended, but not required.

Training covers:

- OSHA Hazard Communication Standard.
- Operation where hazardous chemicals are present.
- Location and availability of Written Program, Chemical List, MSDSs.
- Physical and health hazards of chemicals in the work area.
- Methods and observations used to detect the presence or release of a hazardous chemical in the work area.
- Measures to protect employees from workplace hazards.
- Detail of your facility Program including labeling system, MSDSs, and how employees can obtain and use hazard information.

Training records that verify effectiveness of training are recommended, but not required.

(Reprinted with permission from Eastman Kodak Company. EKC publication no. J-311.)

## Self-Assessment Checklist for Hazard Communication

The following checklist is provided to help managers of medical imaging facilities assess their compliance and identify areas for improvement with their program.

| | Yes | No |
|---|---|---|
| • The medical imaging facility has a written hazard communication program that meets all requirements. | ☐ | ☐ |
| • The facility has a process for informing employees of other companies working at the facility of hazard communication requirements. | ☐ | ☐ |
| • A list of all hazardous chemicals at the facility has been compiled. | ☐ | ☐ |
| • An MSDS for each hazardous chemical is available at the facility. | ☐ | ☐ |
| • Warning labels are on all containers of hazardous chemicals at the facility. | ☐ | ☐ |
| • All current employees received hazard communication training. | ☐ | ☐ |
| • Hazard communication training is given to all employees prior to initial assignment, when hazards in the workplace change, or when reassigned to an area with new hazards. | ☐ | ☐ |
| • State or local hazard communication requirements, if any, have been implemented at the facility. | ☐ | ☐ |

## Personal Protective Equipment (PPE)

The OSHA PPE standard was designed to ensure that employers evaluate hazards in the workplace so that they can select the proper safety equipment to protect employees from those hazards.

Employers have the responsibility to implement and maintain an effective PPE program. They also need to provide protective equipment whenever it is necessary to protect employees from process hazards, workplace environment, or chemical hazards that are capable of causing injury to any part of the body through absorption, inhalation, or physical contact.

Employers have the responsibility to do the following:

- Assess the workplace to determine what hazards are present.
- Select and provide appropriate PPE that fits properly and protects against the hazards identified for each affected employee.
- Communicate to employees which PPE was selected and why it is required.
- Ensure that employees use the PPE provided.
- Maintain PPE in a sanitary and reliable condition.
- Replace PPE when it becomes damaged or defective.
- Train employees.

It is a good idea to post signs in the facility to communicate where and when PPE is required. Signs serve as an effective communication tool for those employees who work in the area, as well as for employees or other people (such as visitors, vendors, and contractors) who do not routinely work in the area.

### *Eye and Face Protection*

Protective equipment for the eyes and face (Figure 2-12) is required when employees are exposed to eye or face hazards from flying particles, liquid chemicals, chemical gases and vapors, and potentially injurious light radiation (such as lasers and welding operations). Employees who work with or around processing chemicals (both concentrates and working strength) need to protect their eyes and face from possible chemical splashes that could result in burns. Also, maintenance and janitorial employees may need to protect their eyes and faces from chemicals.

Typical eye and face protection includes:

- Safety goggles (with side shields) that protect against splashes in chemical mix and silver-recovery areas.
- Safety glasses (with side shields) in processing areas that do not involve processor cleaning or chemical mixing.
- Eye and face protection with side shields that protect against flying objects, light, or chemical exposure for maintenance operations.

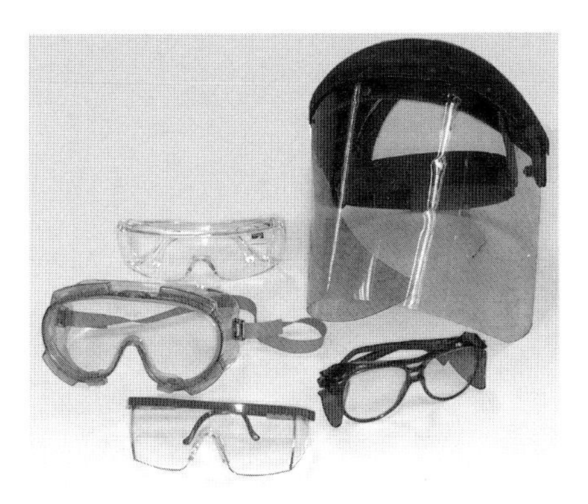

**Figure 2-12** Typical eye and face protection for employees exposed to chemicals (Reprinted with permission from Eastman Kodak Company.)

When purchasing PPE from your safety equipment vendor, be sure that all eye and face protection meets or exceeds the requirements of ANSI Standard Z87. 1-1989, "American National Standard Practice for Occupational and Educational Eye and Face Protection." The eye and face protection must be distinctly marked to identify the manufacturer of the PPE.

Employees required to wear eye and face protection who also need prescription lenses have the option of (1) incorporating the prescription into the design of the PPE or (2) wearing eye protection that can be worn over the prescription lenses that does not interfere with the protective equipment.

Eye and face protection should be maintained in sanitary conditions. Use a damp soft cloth to clean glasses or goggles. When cleaning, avoid the use of abrasive materials to prevent scratches. Store in a clean container.

Inspect eye and face protection routinely. Worn, broken, and scratched surfaces weaken the material and obstruct vision. Replace PPE when it displays any of these defects. Normally, you can discard eye and face protection with the regular trash. However, if it is chemically contaminated and cannot be cleaned, handle it using the disposal procedure that applies to that specific chemical.

### *Hand Protection*

PPE (Figure 2-13) is required when employees' hands are exposed to potential skin absorption of harmful substances; severe abrasions, cuts, punctures, or lacerations; chemical burns, thermal burns, or harmful temperatures. Employees who work with processing chemicals (both concentrate and working strength) need to protect their hands from any chemical contact that could lead to dermatitis and possible chemical burns.

Select gloves based on an evaluation of how well the glove will hold up relative to the job function being performed. Consider the following factors when selecting the appropriate PPE for chemical exposure:

- *Thickness:* The thicker the hand protection, the longer it will take for the chemical to break through or permeate the inside of the PPE.
- *Chemical resistance:* Chemicals can react with or permeate materials at different rates. Select materials that are impervious to the chemicals that employees use.
- *Quality of construction:* It is important to inspect gloves prior to use. Check for pinholes or physical defects that could have an impact on user safety.

**Figure 2-13** Gloves should be used to protect hands from chemicals. (Reprinted with permission from Eastman Kodak Company.)

- *Degradation:* It is important to check gloves for changes in their physical properties as a result of contact with a chemical. This can include discoloration, loss of physical strength, or deterioration.

Neoprene is the glove material typically used with processing solutions (both concentrate and working strength). However, neoprene may not be the appropriate glove material for use with all chemicals in your facility. If you use solvent-based chemicals or formaldehyde-based solutions, check with your glove manufacturer to determine the appropriate glove material.

Keep gloves clean, and store them out of direct sunlight to prevent degradation. Store chemically resistant gloves flat, not folded, so they do not develop kinks or cracks that could allow chemicals to penetrate more easily.

Inspect gloves before each use for cracks, holes, and leaks. Replace gloves whenever they show physical defects or degradation or become permanently stained.

Normally, you can throw out uncontaminated gloves with the regular trash. If gloves are chemically contaminated and cannot be cleaned, handle them using the disposal procedure that applies to that specific chemical.

### Protective Clothing

While not specifically included in the OSHA PPE standard, OSHA provides guidance on body protection. The body should be protected from:

- Fire and heat exposures.
- Cuts and abrasions.
- Chemicals.
- Dusts and other splashes.

Typically, chemically resistant aprons (materials composed of neoprene) (Figure 2-14) are recommended for employees working with processing chemicals who could potentially come into contact with chemical splashes.

**Figure 2-14** Chemically resistant apron and lab coat are examples of protective clothing. (Reprinted with permission from Eastman Kodak Company.)

## Self-Assessment Checklist

Use this checklist as an aid for assessing your compliance and identifying areas for improvement within your medical imaging facility's PPE program.

| | Yes | No |
|---|---|---|
| • The facility has a written hazard assessment and certification. | ☐ | ☐ |
| • PPE is available as determined by the facility hazard assessment. | ☐ | ☐ |
| • The facility has up-to-date documentation verifying that all employees required to use PPE are trained. | ☐ | ☐ |
| • PPE training records include employee's name, date of training, subject of training, and verification of training effectiveness. | ☐ | ☐ |
| • PPE in use properly fits employees and is properly maintained. | ☐ | ☐ |

## Air Quality

A study was done by a film manufacturer[1] to measure the practical volatility of several chemicals used in processing medical imaging films—acetic acid, ammonia, glutaraldehyde, and sulfur dioxide.

Acetic acid can be a component of developer and fixer concentrates. It may be present as a solvent in developers and as a buffer in fixers.

Free ammonia may be present in fixer (in the ammonium thiosulfate as a result of the manufacturing process).

Glutaraldehyde is necessary as a hardener in developer solutions.

Very small amounts of sulfur dioxide can be emitted from a working solution of a fixer at low pH. It is released from sulfite and thiosulfate under acidic conditions.

To determine the emission levels of the above chemicals, several processors, the manufacturer's chemicals, and the mixing operation were tested under specific conditions, such as concentration of chemicals, temperatures, etc. The processing room was approximately 9 X 12 X 10 feet (2.7 X 3.6 X 3.0 meters). To maximize the accumulation of vapors, room ventilation was turned off and the processor was exhausted directly into the room.

The study found that the volatility of these chemicals is significantly reduced at pH levels of typical developer and fixer solutions.

Tables 2-11 and 2-12 show the results of the air sampling studies and the comparative OSHA and American Conference of Governmental Industrial Hygienists (ACGIH) occupational exposure limits, as well as the odor threshold value.

[1]Eastman Kodak Company, Rochester, NY.

**Table 2-11** Results of X-Ray Processor Air Monitoring Studies

| *Processor* | *KODAK Chemicals (Developer/Fixer)* | *Acetic Acid (ppm)* | *Ammonia (ppm)* | *Glutaraldehyde (ppm)* | *Sulfur Dioxide (ppm)* |
|---|---|---|---|---|---|
| KODAK X-OMAT Multiloader 300 | RP X-OMAT/ RP X-OMAT | 0.4–0.5 | 0.07–0.1 | <0.005 | 0.38–1.1 |
| KODAK X-OMAT RA Processor | RP X-OMAT/ RP X-OMAT | 0.2–0.6 | 0.09 | <0.005 | 0.8 |
| KODAK X-OMAT RA Processor | RP X-OMAT/ RP X-OMAT | 0.2–1.3 | 0.14 | <0.005 | 0.04 |
| KODAK X-OMAT RA Processor | RA/30/RA/30 | <0.1 | 0.09–0.11 | <0.005 | 0.04 |
| Mixing Operation | RP X-OMAT/ RP X-OMAT | 0.9–2 | 0.14 | <0.02 | 0.2 |

(Reprinted with permission from Eastman Kodak Company. EKC publication no. N-327.)

**Table 2-12** Comparative OSHA and ACGIH Occupational Exposure Limits

| *Agency* | | *Acetic Acid (ppm)* | *Ammonia (ppm)* | *Glutaraldehyde (ppm)* | *Sulfur Dioxide (ppm)* |
|---|---|---|---|---|---|
| OSHA | 8-hour TWA PEL | 10 | not available | 0.2(C) | 2 |
| OSHA | STEL | not available | 35 | 0.2(C) | 5 |
| ACGIH | 8-hour TWA PEL | 10 | 25 | 0.2(C) | 2 |
| ACGIH | STEL | 15 | 35 | 0.2(C) | 5 |
| | Odor Threshold | 0.01 | 0.04 | 0.04 | 0.3 |

Key:
OSHA Occupational Safety and Health Administration
ACGIH American Conference of Governmental Industrial Hygienists
TWA Time-Weighted Average
PEL Permissible Exposure Limit
STEL Short-Term Exposure Limit

(Reprinted with permission from Eastman Kodak Company. EKC publication no. N-327.)

The maximum amount of acetic acid detected was 1.3 and 2 ppm for processor and mixing operations, respectively. Both values are well below the OSHA (10 ppm) and ACGIH (10 ppm) levels. The lowest level detected, <0.1 ppm, is above the odor threshold value of 0.01 ppm.

The highest level of ammonia detected was 0.14 ppm for both the processor and mixing operations. This value is below the OSHA and ACGIH levels of 35 and

25 ppm (respectively) but is significantly higher than the odor threshold (0.04 ppm). The highest level of glutaraldehyde was <0.005 for the processor and <0.02 for the mixing operation. Both values are well below the OSHA and ACGIH limits of 0.02 ppm and the odor threshold value of 0.04 ppm.

Sulfur dioxide levels of 1.1 ppm were measured for the processors using RP X-OMAT fixer and replenisher and 0.30 ppm for the LO fixer, which was designed to reduce sulfur dioxide emissions. Values are well below minimum OSHA and ACGIH levels of 2 ppm for sulfur dioxide and equal to the odor threshold (0.3 ppm).

Even under these stringent test conditions, the measured levels of the volatile chemicals in processing solutions are significantly below the OSHA and ACGIH limits. Under some conditions, the odor threshold values are exceeded for acetic acid, ammonia, and sulfur dioxide. Glutaraldehyde levels are well below the odor threshold for all test conditions.

According to these studies, there is a low probability that employees would be exposed to hazardous levels of air contaminants during routine mixing operations or from using chemicals for the processing of medical radiographs.

Note that the processors in these studies were intentionally misused to simulate poor ventilation conditions, so the results indicate that gases and vapors given off from processors and processing chemicals should be below occupational exposure limits when processors are installed to recommended specifications.

Employees who have certain medical conditions, such as asthma or other respiratory diseases, may require special consideration. Individuals who experience adverse health symptoms when handling processing chemicals should seek medical advice.

In poorly ventilated areas, some of the chemicals given off by the processors in this study may be detectable by odor because the chemicals have low odor thresholds. Although the odor of a chemical is not indicative of a potential health hazard, strong odors or odors that are increasing in intensity may be an indication of inadequate ventilation. The situation should be reviewed, and a qualified health and safety professional should be consulted.

In summary, the following processor chemistry safe handling strategies should be employed to minimize worker exposure to processing chemicals:

1. Provide employee training on the potential hazards in the workplace as a means to militate against such hazards.
2. Encourage employees to read and understand hazard communication vehicles, e.g., MSDSs and product labels.
3. Make certain that all equipment is installed and maintained to manufacturers' specifications.
4. Employ engineering controls, e.g., ventilation systems.
5. Substitute potentially less hazardous or less odorous products, as available.
6. Use PPE when necessary.
7. Consider health and environmental screening as a verification of effective safe handling strategies.

## The Regulation of Silver and Other Wastes in Medical Imaging Facilities

The regulations that apply to silver, silver-bearing materials, and other wastes generated at medical imaging facilities are contained in many of the environmental laws that are administered by the United States Environmental Protection Agency (USEPA). These laws are designed to protect air, water, and land resources. Most state and local agencies also regulate waste materials. In some cases, state or local regulations may be more stringent or may include additional requirements not found in federal regulations.

Waste materials generated in a medical imaging facility are typically regulated by the USEPA under the following federal laws:

- Resource Conservation and Recovery Act.
- Clean Water Act.

### *Silver and the Resource Conservation and Recovery Act*

The Resource Conservation and Recovery Act (RCRA) protects human health and the environment by making sure that hazardous waste management practices are conducted in a sound manner. RCRA is a cradle-to-grave system for tracking and regulating hazardous waste from the time it is generated to its ultimate disposal.

A material must be defined as a solid waste before it can be defined as a hazardous waste. As defined by RCRA, a solid waste can be a liquid, semisolid, sludge, or contained gas. Solid wastes are generally those materials that have been used and discarded. However, recycled materials can also be solid wastes, depending upon the nature of the materials and the manner in which they are recycled.

Once a material is defined as a solid waste, it can be classified further as either a listed or characteristic hazardous waste. Silver and silver-bearing wastes generated by a typical medical imaging facility (Table 2-13) are not listed. Solid wastes that possess certain characteristics, such as containing greater than or equal to 5 ppm of leachable silver, are considered hazardous wastes due to their toxicity characteristic (TC). Leachability is determined for solid materials using the Toxicity Characteristic Leaching Procedure (TCLP), which subjects the material to a dilute acid solution. Liquid wastes that contain greater than or equal to 5 ppm of silver are also classified as characteristic hazardous wastes (EPA Hazardous Waste Number D011).

Medical imaging facilities that generate hazardous waste are divided into different categories depending on the amount of hazardous waste that is generated. These categories are determined by counting the amount of hazardous waste that is generated at a facility in one month. Some silver-bearing hazardous wastes are not counted because they are managed in a wastewater treatment unit or because they are not stored or accumulated.

Facilities that generate hazardous waste are subject to a number of requirements for the management of their wastes depending on their generator category. Silver-bearing hazardous wastes that are reclaimed to recover economically significant amounts of silver are regulated under specific provisions of RCRA. These provisions

**Table 2-13** Identifying Silver-Bearing Hazardous Wastes

| *Material* | *Hazardous Waste Determination* |
|---|---|
| Radiographic Films | Processed and unprocessed medical x-ray films will generally not leach silver at greater than or equal to 5 ppm; contact the film manufacturer for specific TCLP test result information. |
| Processing Solutions | Unused photographic processing solutions do not contain silver and would therefore not be regulated as a silver-bearing hazardous waste when discarded or recycled.<br><br>Used photographic processing solutions that typically contain greater than 5 ppm of silver include fixers, bleach-fixes, activators, and low flow washes, as well as bleaches and stabilizers from washless processes. Developers generally do not contain greater than 5 ppm of silver; however, in processes that have very high volume or employ regeneration of developers, concentrations of silver can reach levels greater than 5 ppm. Wash waters can also develop concentrations of silver at levels greater than 5 ppm. This silver is usually the result of carryover from the preceding tank. |
| Regenerated or Recycled Processing Solutions | Materials can be excluded from being a hazardous waste by being managed in a certain way, such as being reclaimed and returned to a process in a closed system. This exclusion requires that the recovery process and associated storage be in fixed tanks and that the materials be transferred to the recovery process using fixed piping. Materials that are collected and transferred in portable containers do *not* qualify for this exemption. Typical inline electrolytic silver-recovery, ion-exchange wash water treatment, and bleach-fix regeneration systems are examples of where the materials in the recovery loop are not considered hazardous wastes unless they are removed from the recovery process and contain $\geq$5 ppm silver. |
| Silver-Bearing Materials from Recovery | High-purity flake silver from electrolytic silver-recovery units is considered to be a product by the USEPA and is not regulated as a hazardous waste under RCRA when sent for refining.<br><br>Other technologies used to recover silver, such as metallic replacement or precipitation, are wastewater treatment units and result in the generation of silver-bearing materials, which are sludges. These materials are sludges because they are residues from pollution-control processes. Characteristic sludges being reclaimed are exempt from the definition of solid waste and therefore cannot be a hazardous waste. Silver-bearing materials such as metallic replacement cartridge (MRC) sludge that are sent to a silver refiner to complete the silver-recovery process are exempt from being classified as hazardous waste. |

(Reprinted with permission from Eastman Kodak Company. EKC publication no. J-214.)

include a limited number of generator requirements and an exemption from needing an RCRA permit for the treatment of those hazardous wastes.

Many states have been approved by USEPA to operate their own RCRA program. Most states regulate the management of silver-bearing hazardous wastes in the same manner as USEPA; however, some states may have more stringent requirements. For example, some states may have different generator categories or may include different management requirements for reclaiming silver from a hazardous waste.

### *Compliance Requirements*

To determine the compliance requirements for your medical imaging facility, you must identify those materials at your facility that could be classified as hazardous waste. Once you have identified those materials, you must determine which of them should be counted to determine your generator category. Certain silver-bearing materials may be exempt or excluded based on how they are generated or how they are managed. Once you have determined your generator category, you can identify those compliance requirements that apply to silver-bearing hazardous wastes at your facility.

It is important to note that you determined your generator category by counting the total amount of hazardous waste generated at your facility. This total quantity also includes other hazardous wastes, such as solvents from maintenance operations or corrosive cleaning liquids.

### *The Clean Water Act*

The goal of the Clean Water Act (CWA) is to minimize the discharge of pollutants to surface waters. It prohibits the discharge of pollutants directly into surface waters except in compliance with a National Pollution Discharge Elimination System (NPDES) permit. Wastewater treatment facilities or Publicly Owned Treatment Works (POTW) must have NPDES permits that establish discharge limits for the POTW's effluent quality. A medical imaging facility that discharges directly to surface water must also have an NPDES permit.

### *Silver Issues*

Silver is classified as a "priority pollutant" under the CWA, and the discharge of silver is regulated by many NPDES permits. The discharge limitation of silver in an NPDES permit is often based on state water quality standards for silver. Water quality standards can vary greatly depending on the stringency of the state's standards and the use of the water. These differences, coupled with design and wastewater loading variations at each POTW, result in the need for site-specific silver limitations in each NPDES permit. These limits for silver are typically well below concentrations achievable using the best available silver-recovery technologies, making it virtually impossible for a medical imaging facility to discharge processing effluents directly to surface waters.

Most facilities discharge their wastewater to a municipal sewer collection system that conveys the material to the POTW for further treatment. The POTW must limit the chemical constituents in the incoming wastewater to ensure that the treatment plant will be able to maintain compliance with their NPDES permit. These limitations are referred to as pretreatment standards or sewer use codes. Each POTW has its own pretreatment standards (or sewer use codes) that are unique to its regulatory situation. Silver is usually regulated in discharges from facilities at concentrations that require silver recovery prior to discharge.

### *Compliance Requirements*

To determine your sewer discharge compliance requirements, you must identify all materials that your facility discharges to a POTW. You may need a discharge (pretreatment) permit or approval from the POTW. Specific compliance requirements, such as analytical testing or reporting, may also be included in the permit or authorization.

### *Sewer Use Permits*

If your facility does not have a wastewater discharge permit or you are unsure if a permit is required, you should contact the wastewater treatment authority that manages the POTW. Some states also have their own pretreatment permitting program in conjunction with the wastewater treatment authority. You should also request a copy of the local sewer use code that applies to your facility.

Some wastewater treatment authorities will require you to complete a discharge survey or application before they will issue a permit. If your facility is not required to obtain a wastewater discharge permit, request a letter that specifically states that no permit is required. Ask the authority to notify you when a change is proposed to the wastewater discharge requirements for your facility.

### *Silver Discharge Limits*

In addition to understanding your POTW's permitting requirements, you should also determine what pretreatment limits in the local sewer code apply to your facility. Silver is commonly regulated with restrictive discharge limits. If the POTW regulates silver in the pretreatment program, you may be required to demonstrate compliance by periodically collecting samples of the processing effluents for analysis. Sample collection points for processing effluent will often be specified in your wastewater pretreatment discharge code or permit. If no sampling point is specified, sample at a point where all the facility's wastewater is collected and discharged. This point will result in a sample that is representative of the total discharge from the entire facility.

### *Other Effluent Parameters to Consider*

In addition to silver, there are several other materials and effluent characteristics that POTW authorities will typically regulate in the waste stream. To verify which parameters are of concern to your local municipality and to understand the sewer discharge limitations for your facility, review your local sewer use code. If you have difficulty understanding the requirements in the code, you can contact the local pretreatment coordinator for a detailed explanation of your responsibilities. Other sources of assistance include local or national healthcare associations and radiographic film and processing chemical suppliers.

The following terminology is widely used in sewer use code documents:

- Biochemical Oxygen Demand (BOD)—a measurement of the amount of dissolved oxygen that an effluent consumes. It is usually measured over a five-day period and is referred to as BOD5.
- Chemical Oxygen Demand (COD)—an analytical method for measuring the oxygen demand of an effluent. It is faster than the BOD5 test and responsive to a broader range of components. A chemical can have a high COD and still be nonbiodegradable.
- Heavy Metals—metals with a specific gravity greater than 5.0. This includes some metals found in processing effluents such as iron, silver, and chromium.

The sewer limitations imposed on medical imaging facilities will vary between communities. Some POTWs will regulate more substances than others depending on the pollutant removal capabilities of the treatment plant and the water quality standards it must meet.

Tables 2-14 and 2-15 outline the most frequently regulated materials and effluent characteristics and typical sewer limits encountered in the United States. These values are provided as ranges. Your local sewer code must be consulted to obtain the correct discharge parameter limits for your facility.

**Table 2-14** Typical Municipal Sewer Code Limits

(All units, except pH and temperature, are specified in ppm [milligrams/liter].)

| *Parameter Regulated* | *Typical Code Limits* | *Normal Range of Code Limits* |
|---|---|---|
| Maximum temperature | 150°F (65.5°C) | 120–160°F (48.9–71.1°C) |
| Minimum pH | 5.5 | 4.5–6.5 |
| Maximum pH | 9.5 | 8.0–12.0 |
| Suspended solids | 350 | 200–1000 |
| Oil and grease | 100 | |
| Flammables | None | |
| BOD5 | 300 | 200–1200 |
| COD | 500 | 400–2500 |
| Chlorine demand | 30 | 10–40 |
| Cyanide | 2.0 | 0–10.0 |
| Phenols | 0.1 | 0.002–20.0 |

(Reprinted with permission from Eastman Kodak Company. EKC publication no. N-327.)

**Table 2-15** Typical Municipal Sewer Code Limits for Heavy Metals

| *Heavy Metals* | *Typical Code Limits (ppm)* | *Normal Range of Code Limits (ppm)* |
|---|---|---|
| Cadmium | 0.5 | 0.002–20.0 |
| Chromium (total) | 5.0 | 0.02–25.0 |
| Chromium (hexavalent) | 0.5 | 0.1–10.0 |
| Copper | 3.0 | 0.1–20.0 |
| Iron | 20 | 3.0–100.0 |
| Lead | 0.1 | 0.05–5.0 |
| Mercury | 0.0005 | 0–5.0 |
| Nickel | 5.0 | 0.1–12.0 |
| Silver | 2.0 | 0.025–20.0 |
| Zinc | 5.0 | 0.25–25.0 |

(Reprinted with permission from Eastman Kodak Company. EKC publication no. N-327.)

Under typical medical imaging processing conditions, maintaining effluent temperature and pH values below maximum limits should not be difficult. In normal use, no flammable or explosive material, detergents, or oils and grease should be present in the effluent. There should also be no phenolic material or cyanide in the effluent.

The biochemical and chemical oxygen demand of the effluent will vary, depending on the type of chemistry used, as well as the types and volumes of films, replenishment rates, and silver recovery equipment.

The levels and types of heavy metals present in the effluent could be influenced by seasoning effects, replenishment rates, and water quality.

## *Wastewater Sampling*

The type of sample required is also important. A grab sample is a single one-time sample that can be a good indicator of the character of the wastewater. For processes that are not continuous, such as batch silver recovery, this sampling technique may not be representative of the actual total discharge. A composite sample is the accumulation of a series of grab samples collected at regular intervals over a long period of time, usually 24 hours. The interval of sampling can be proportional to either time or flow. A flow-proportional composite sample will result in a sample that most accurately characterizes the discharge.

## *Analytical Laboratories*

The laboratory you select should have a good understanding of, and experience in, analyzing all the parameters on your sewer use permit, especially silver. This is due to the complexity of analyzing silver. Some states require certification of laboratories that perform analysis for regulatory compliance. Make sure you are using a certified laboratory if your state requires certification.

Strict adherence to EPA-approved testing methods and laboratory certification yield accurate analytical results that you can submit for regulatory compliance demonstration purposes in most states.

### *Silver Analysis*

Silver can be measured in several ways. The most common is through an analytical method known as total recoverable silver. This method measures all silver that is present in a sample. Some POTWs may require dissolved silver to be measured or may include a translator to estimate the portion of dissolved silver from a total measurement. It is important that you determine which analysis is required and request the appropriate analysis from your analytical laboratory.

Keep records of all monitoring and sampling results for three years unless local requirements specify a longer period of time.

### *Discharges of Hazardous Wastes*

If the wastewater you are discharging to your local POTW has a concentration greater than or equal to 5 ppm of silver at the point where it leaves your facility and enters the sewer system and you are discharging more than 15 kilograms/month of hazardous waste, you must submit a one-time written notification to your local POTW, the USEPA, and your state hazardous waste authority. This notification must include the following:

- Hazardous waste name (i.e., silver).
- EPA hazardous waste number (i.e., D011).
- Type of discharge (e.g., continuous, batch).
- Certification that a waste minimization program exists, to the extent that it is economically practical, to reduce the volume and toxicity of hazardous wastes generated.

If the discharge is greater than 100 kilograms/month, the notification must also include the following:

- Identification of the hazardous constituents in the waste.
- An estimate of the mass and concentration of the hazardous constituents discharged.
- An estimate of the mass of the constituents in the waste stream expected to be discharged during the course of the following year.

## Using the Code of Management Practice to Manage Silver

A number of municipalities across the United States are adopting—or are considering adopting—a code for managing silver in the wastes that enter their sewage treatment plants.

The purpose of the code is to develop a consensus among the regulated and regulatory communities for controlling silver discharges in a cost-effective and environmentally sound manner.

The code includes recommendations on technology, equipment, and procedures for controlling silver discharges from processing facilities.

Pollution prevention is another important part of the code. Recommendations include several activities for preventing pollution before it can occur—i.e., the use of processes, practices, materials, and energy in ways that avoid or minimize pollutants and waste.

### *The Benefits*

By implementing the Code of Management Practice, you help your business by managing your processes more efficiently, collecting a nonrenewable resource (silver), and obtaining a cash value for the recovered silver.

The POTW benefits because it saves the large capital expense required to remove silver from huge volumes of water. Also, your municipality will not have to resort to implementing unnecessarily strict discharge limits for silver and expensive enforcement programs.

As a processor of films used for medical imaging, you may already be working with your POTW and have an effective program of silver management in place. You may need to change or add some management procedures. Or you may need to begin a program of silver management.

The Code of Management Practice provides an easy-to-implement means of promoting environmental performance by enhancing silver recovery. It gives industry a framework for developing cost-effective and sound environmental management systems.

If your POTW chooses to adopt the Code of Management Practice, it can apply the code in a number of ways:

- It can adopt the code exactly as it is and require your business to follow code recommendations for silver recovery and management.
- It can adopt the code but modify specific requirements in it.
- If your POTW already has specific effluent limitations on silver concentrations, it can test the code to see if implementing it will achieve compliance with the existing limits.

## The Technology of Silver Recovery

As previously discussed, unexposed silver is removed from the film during fixing. It is carried out in the solution and the wash overflows, usually in the form of a silver thiosulfate complex. Unlike free silver ion ($Ag^+$), which is toxic to microorganisms, silver thiosulfate complex is relatively nontoxic and is not detrimental to a secondary waste treatment plant. Although most silver is not harmful and is removed during secondary treatment, it is a good practice to recover it before discharging the effluent. In addition to conserving a natural resource, recovering silver provides a source of revenue.

A number of techniques are available to remove silver from silver-rich x-ray processing solutions. Of these, three are used in virtually all practical methods of silver recovery. They are the following:

- Electrolysis.
- Metallic replacement.
- Precipitation.

Additionally, ion-exchange technology can be used to treat wash waters to remove silver. This technology is typically used when you must meet stringent discharge requirements, and capital and operating costs are a secondary concern.

Other technologies, such as reverse osmosis, distillation, and evaporation, can produce a silver sludge; however, they only alter the silver concentration and do not actually remove silver from solution.

Table 2-16 presents a comparison of silver recovery and wash water treatment technologies.

### *Electrolysis*

In the process of electrolysis, or electrolytic silver recovery, a direct current is passed through a silver-rich solution between a positive electrode (the anode) and a negative electrode (the cathode) that are submerged in silver-bearing fixer (Figure 2-15).

During this electrolytic process, an electron is transferred from the cathode to the positively charged silver, converting it to its metallic state, which adheres to the cathode. In a simultaneous reaction at the anode, an electron is taken from some species in solution. In most silver-rich solutions, this electron usually comes from sulfite.

An overview of the reactions is:

Cathode:

$Ag(S_2O_3)_2^{-3} + e^- \rightarrow Ag^0 + 2(S_2O_3)^{-2}$

or

Silver Thiosulfate Complex + Electron → Metallic Silver + Thiosulfate

Anode:

$SO_3^{-2} + H_2O \rightarrow SO_4^{-2} + 2H^+ + 2e^-$

or

Sulfite + Water → Sulfate + Hydrogen Ions + Electrons

Electrolysis produces a nearly pure metallic silver, contaminated only slightly by some surface reactions that also take place. The plated silver should be greater than 90 percent pure.

One important advantage of electrolytic recovery is that, when done properly, this method permits fixer reuse. Recovery efficiency and percentage of fixer reuse will vary according to the equipment manufacturer's recommendations and processing conditions.

**Table 2-16** Comparative Silver Recovery and Wash Water Treatment Technologies

| | *Terminal Electrolysis* | *Inline Recirculation Electrolysis* | *Metallic Replacement* | *Sulfide Precipitation* | *Ion Exchange (only for washwaters)* |
|---|---|---|---|---|---|
| Typical Initial Silver Concentration | 2,000 to 12,000 mg/L | 500 to 7,000 mg/L | Variable | Variable | <30 mg/L |
| Typical Final Silver Concentration | 50–250 mg/L | Adjustable: usually 250–1,000 mg/L | 0.5–15 mg/L | 0.1–1.0 mg/L | 0.1–1.0 mg/L |
| Treatable Solutions | Most silver-rich solutions | Fixers in certain processes | Most silver-rich solutions | Most silver-rich solutions | Washwaters and very dilute hypo-containing solutions |
| Capital Cost | $2,000 to $30,000 depending on size and sophistication of equipment | $1,500 to $8,000 depending on size and sophistication of equipment | $50 to $3,000 needs a tank and pumping system for best results | $2,000 to $10,000 "off-the shelf" equipment not available | $10,000 to $100,000 |
| Operating Cost | Low | Low | High | Medium | High |
| Advantages | Can produce >90% pure silver | Minimizes silver carry-out to following wash; reduces replenishment rates | Can be relatively inexpensive | Consistently low silver concentration | Provided low silver concentration for washwaters |
| Disadvantages | Relatively high final silver concentration; usually requires "tailing" (secondary recovery) | May require electronics adjustment; requires terminal treatment system | Difficult to know when to replace; discharges iron, limited by some sewer codes | Requires professional care to avoid potentially hazardous off-gases; difficult to filter | Expensive; requires expert maintenance |
| Application | All photographic processing facilities except very small labs | All photographic processing facilities except very small labs | All photographic processing facilities | Solution service companies | Large photographic processing facilities |

(Reprinted with permission from Eastman Kodak Company. EKC publication no. J-212.)

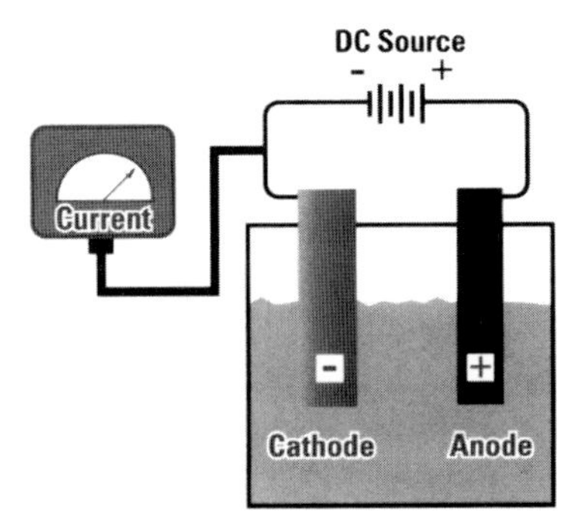


**Figure 2-15** A typical electrolytic silver recovery cell. (Reprinted with permission from Eastman Kodak Company.)

## *Terminal Electrolysis*

The electrolytic recovery process is efficient and cost-effective, using reusable equipment. The efficiency of the system is dependent, among other things, on the availability of silver-rich solution at the cathode surface. In current commercial recovery equipment, this is accomplished in one of two ways:

- The cathode is moved within the solution. One application is the rotary cathode cell. The negative current is applied to a rotating drum in the solution and the silver plates onto the drum.
- The liquid is rapidly pumped over the stationary cathode. This design often tends to be somewhat less efficient than rotating cathode cells; however, these units usually require less maintenance.

Electrolytic silver recovery has its disadvantages. Attempts to accelerate the recovery process or to desilver to silver concentrations below 200 milligrams/liter—by extending the residence time in the cell or increasing the current density (amperage/cathode surface area) on the cathode—will produce an inferior, black, crumbly silver-sulfide-contaminated plate, which reduces the cell efficiency dramatically.

You can use electrolysis only as a primary treatment. Postelectrolysis silver concentrations are generally in the several-hundred-milligram-per-liter (ppm) range. If you must achieve a low regulatory limit, use some other type of secondary silver recovery such as metallic replacement.

## *Inline Electrolysis*

Inline electrolytic fixer recirculation is used in some medical imaging facilities (Figure 2-16). With this technology, fixer is recirculated between the processing tank and a specially designed electrolytic silver-recovery unit. Some of these units have electronics that automatically control the silver concentration in the recirculating solution, usually in the range of 250 to 1,000 milligrams/liter. Since the silver concentration in the fixer is significantly decreased, silver in the following wash

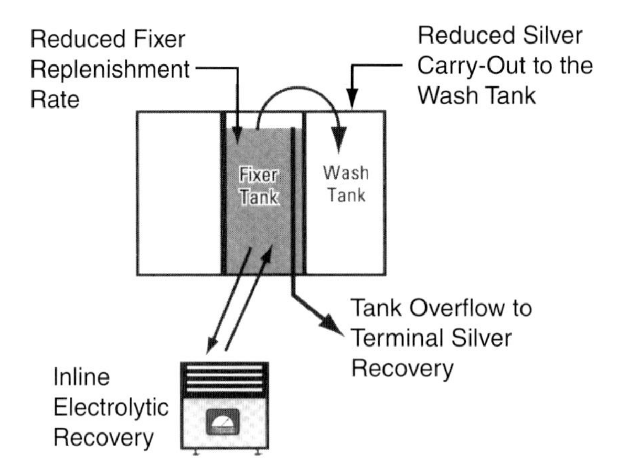


**Figure 2-16** Inline electrolytic fixer recirculation. Inline electrolytic fixer recirculation is used in some medical imaging facilities. (Reprinted with permission from Eastman Kodak Company.)

tank is also reduced due to less silver "carry-out" with the film to the wash during processing.

It is important to use a good quality fixer when employing inline electrolysis because sulfite depletion does occur. In certain processes, up to a 50 percent reduction in fixer replenishment rate is possible because properly adjusted equipment does not significantly degrade most fixer components.

You must optimize recirculating systems by monitoring the silver concentration in the flow returning to the processor tank. You can easily do this by using a colorimetric silver test kit or silver estimating test paper.

When using recirculating electrolytic equipment, you will need to perform additional silver recovery on the fixer-tank overflow to recover silver to a level sufficient to meet local regulatory limits.

## *Metallic Replacement*

Silver is recovered when a silver-bearing solution is passed through a cartridge containing steel wool. The more active metal (in this case, iron) reacts with the silver complex and goes into solution. The less active metal (silver) is deposited on the steel wool. Recovery efficiency can be high—as much as 90 percent of the available silver recovered in a single pass—if the equipment is routinely monitored for correct operation.

The basis for metallic replacement is the reduction by metallic iron (usually present as steel wool) of the silver-thiosulfate complex to elemental silver. The commercial equipment you can use for the recovery is often referred to as a metallic recovery cartridge[2] (MRC) or a chemical recovery cartridge (CRC).

[2]Cartridges used in the metallic replacement process for recovering silver have been described as chemical recovery cartridges (CRCs), metallic recovery cartridges (MRCs), and silver recovery cartridges (SRCs). The photographic industry has avoided the term "SRC" to prevent theft of the cartridges during shipment. The term "CRC" is closely associated with the original product developed by Eastman Kodak Company, which was protected by a U.S. patent. Therefore, we will use "MRC" as a generic term to refer to metallic replacement.

The most common source of iron is fine steel wool, chosen for its surface area. The steel wool is wound on a core or chopped up and packed into a cartridge. Some cartridge manufacturers use other forms, such as iron particles glued to fiberglass or iron-impregnated resin beads, or wound iron screening material. For best results, the silver-rich solutions are slowly metered into the cartridge and through the iron medium. The silver is left behind in the cartridge, while dissolved iron is carried out by solution.

Like the electrolytic process, metallic replacement is a reduction-oxidation process. An overview of the reaction is:

$$2Ag(S_2O_3)_2^{-3} + Fe^0 \rightarrow 2Ag^0 + Fe^{+2} + 4S_2O_3^{-2}$$

or

Silver Thiosulfate Complex + Metallic Iron → Metallic Silver + Ionic Iron + Thiosulfate

The final silver concentration is affected by flow rate, iron surface area, contact time, pH, original silver concentration, thiosulfate concentration, and the volume passing through the cartridge. If the MRC is operating properly, the silver concentration may be reduced to less than 5 milligrams/liter.

As the cartridge is used, particularly with steel wool, channels or bypasses may start to form. These grow with time, and eventually the steel wool may begin to collapse internally, and silver breakthrough can occur well before the iron is consumed. Iron consumption in an MRC is related to the volume of solution passing through the cartridge, as well as the solution's silver content and pH.

To prevent undesirable silver losses and discharges from an exhausted cartridge, a system usually consists of two cartridges in series. You should test the output from the first cartridge on a regular basis for silver breakthrough using a colorimetric silver test kit or silver estimating test paper. When the first MRC is exhausted, remove it from the sequence, move the second cartridge into the lead position, and place a fresh cartridge in the second position (Figure 2-17).

You can use metallic replacement as either a primary or secondary (tailing) treatment for solutions treated primarily by electrolysis. You cannot reuse solutions passed through metallic replacement cartridges for further film processing because the dissolved iron and other reaction by-products will contaminate solutions in the processor tank.

Like electrolysis, metallic replacement has its drawbacks. It may not consistently reduce silver concentrations to very low compliance levels. Without good flow-rate control and proper system maintenance, random variations in effluent silver concentrations may occur. Silver sludge from cartridges is relatively expensive to refine, and the recovered silver frequently barely pays for the materials and equipment used to collect it. How you decide to use MRCs depends on your desired silver-recovery efficiency and the discharge codes you must meet. Some codes limit iron discharge, which may restrict your use of MRCs.

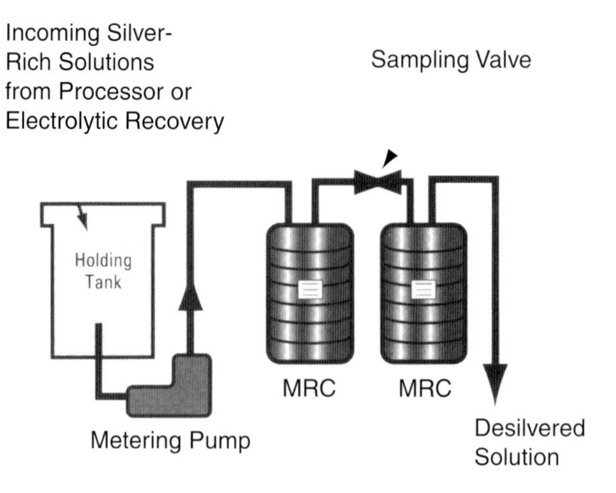


**Figure 2-17** Two metallic replacement cartridges (MRC) with holding tank and metering pump. When the first MRC is exhausted, remove it from the sequence, move the second cartridge into the lead position, and place a fresh cartridge in the second position. (Reprinted with permission from Eastman Kodak Company.)

## *Precipitation*

Precipitation can remove silver from silver-rich solutions (fixer and wash), reducing it to very low levels. When properly applied, it can reduce levels to the low ppm range. Precipitation has not been as widely used as a silver-recovery technique in medical imaging facilities. Common precipitating agents classically have been alkali metal salts of sulfide (sodium sulfide, potassium sulfide, etc.), which will form silver sulfide in solution; the silver sulfide is removed by filtration.

The lack of acceptance of the silver sulfide precipitation-filtration process can be attributed primarily to two factors:

- You must measure the solution silver concentration accurately prior to sulfide addition to prevent overdosing and the discharge of toxic hydrogen sulfide gas.
- The silver sulfide precipitate is difficult to filter; it tends to plug filter media.

Sulfide desilvering is most effective in centralized facilities when used by trained personnel.

Other precipitating procedures generally involve converting silver in solution to the metallic state by adding strong reducing compounds such as borohydrides. These techniques are best used by solution service companies or centralized treatment facilities staffed with technical professionals. There are serious safety considerations when handling chemicals like borohydrides.

Technology advances are being made for onsite silver recovery by precipitation. Commercial units are available for other photographic industry segments (e.g., photo processing labs). These units, however, do not provide cost-effective silver recovery for the medical imaging sector at this time.

## *Ion Exchange*

Ion exchange is used mainly to recover silver from dilute solutions, such as wash water (Figure 2-18). Ion exchange refers to the substitution of an ion in solution for one that is bound on a large polymer molecule. The most familiar use of ion exchange is as a water softening technique. A water softening system typically consists of a resin column and a tank containing a salt solution used for resin regeneration. The polymer, or resin, supplied as small beads and packed in columns, is chosen for its affinity for certain dissolved chemicals to be removed from solution.

The resin is treated, or "activated," by filling all its exchange sites with an exchange ion such as chloride or sodium during the regeneration cycle. As the solution to be treated is passed through the packed resin column, the resin releases its exchange ion for an ion of higher preference. In the water softener, that ion might be a water-hardening ion such as calcium. In processing wash waters, it can be the silver thiosulfate complex.

Ion exchange works best on dilute solutions like wash water. Typically, you would use it when you must treat wash waters to meet very low silver regulatory limits. A well-controlled ion-exchange system can remove silver to about 0.1 to 0.5 ppm.

To treat more concentrated solutions, you can combine wash water and electrolytically desilvered solutions and treat them by ion exchange. The concentrated solutions, after primary silver recovery, are metered into the wash water tank to maintain a conductivity in the wash of about 2000 to 2500 microseimens per centimeter. If the thiosulfate level is allowed to get too high, it will compete for exchange sites on the resin, resulting in low capacity and silver passing through the column.

There are two ion-exchange scenarios that differ only in the regeneration step, where the silver is removed from the active sites on the resin.

- *Elution Regeneration*—When the ion-exchange column is exhausted (i.e., unable to absorb additional silver), a concentrated solution of thiosulfate (hypo) is passed through the column and collected separately. This hypo solution elutes the silver off the resin beads and carries it from the column.

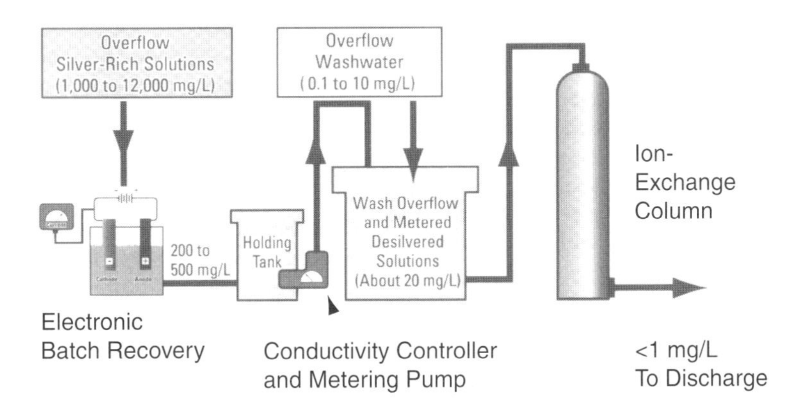


**Figure 2-18** A typical wash water treatment system using ion exchange. Ion exchange is used mainly to recover silver from dilute solutions, such as wash water. (Reprinted with permission from Eastman Kodak Company.)

The thiosulfate solution is then desilvered electrolytically. The excess thiosulfate is rinsed from the column and ready to remove more silver. You can save the thiosulfate regeneration solution and use it again.

- *In-Situ Regeneration*—When the ion-exchange column is exhausted, a diluted solution of a sulfuric acid is sent through the resin bed (do not use nitric acid). The acid breaks down the silver-thiosulfate complex being held by the resin. One of the by-products of this decomposition is silver sulfide, an extremely insoluble form of silver, which is precipitated within the resin bead itself. The resin exchange sites are freed for further silver recovery, and the silver sulfide remains in the beads. Each regeneration cycle uses a fresh batch of dilute sulfuric acid (2 percent by volume) as regenerant. Used regenerant should be neutralized before disposal. Eventually, the capacity of the resin starts to drop off, necessitating more frequent regeneration. When the resin needs replacing, remove the used resin and send it for refining. Load a fresh bed of resin in the column and activate it by regeneration.

Ion-exchange systems produce relatively clean water that may, in some cases, be recycled through the system. With some additional features to help reduce biogrowth, and possibly destroy hypo (thiosulfate), ion exchange is the basis of most wash water reuse technologies.

You must take exceptional care to maintain the long-term keeping properties of film when it is processed with recirculated wash water.

## Off-Site Silver Recovery

As an alternative to performing silver recovery in-house, there are companies that provide hauling and centralized silver recovery. The cost of having silver-rich solutions hauled away for recovery typically ranges from $2.00 to $6.00 (U.S. dollars) per gallon. For some facilities that produce small volumes of silver-rich solutions, off-site recovery can be a cost-effective way to reduce silver discharges. For facilities that produce larger volumes, the cost of off-site treatment may be too great. Other considerations for off-site recovery involve collection, safe storage, and handling of silver-rich solutions, as well as compliance with applicable hazardous waste regulations.

**Table 2-17** Summary

- Because of the differences in film technologies, different radiographic film types do not exhibit the same developer component sensitivities.
- Developers and fixers should be mixed to specifications or tolerances set by the manufacturer with an understanding of the sensitivities and needs of the most critical films that will be processed.
- The consequences of incomplete fixing only become apparent after the image has been stored for some time.
- The manufacturer's mixing directions should be closely followed since factors such as mix order, temperature, and stir times are essential to obtaining a good solution.
- A processor quality control program is an effective method for determining if chemistry is mixed properly. Other testing procedures require varying degrees of technical training and may be expensive. Also, results sometimes require careful interpretation.
- Facilities located in the United States should remain aware of local, state, and federal environmental regulations. If necessary, there are several effective means of removing silver from the fixer effluent.
- Facilities located outside the United States should follow the regulations for their country.
- Use recommended replenishment rates.
- If correct mixing and processing procedures are followed and there is adequate ventilation, worker exposure to processing chemistry should not be a health concern. If a potential problem arises, a qualified health and safety professional should be consulted.
- A person who generates, treats, stores, disposes of, or initiates shipment of hazardous waste is responsible for determining if the waste is hazardous, for identifying the waste, and for record-keeping regarding hazardous waste. Such persons are also responsible for providing information to any transporter of the waste.
- It is important to know your effluent's environmental and chemical characteristics.
- It is important to know your sewer use code parameter limitations.
- Determine if a permit to discharge, or notification about such discharge, is required, and if regulations other than sewer codes apply.
- Utilize silver recovery technology that allows you to meet the code limits.

## References


1. Conversion instructions and processing recommendations for KODAK MIN-R 2000 film in KODAK X-OMAT processors, Models 460 RA, 480 RA, 270 RA, 3000 RA, 5000 RA, Multiloader 300, M6B, M6A-N, M6A-W, M7B, M35A-M, M35A, M35 and M35-M, Service Bulletin no. 244, Eastman Kodak Company publication no. 9B8337, catalog no. 191 3664, 1996.
2. Dickerson RE. Co-optimization of film and process chemistry for optimum results. In: Haus AG, ed. Film processing in medical imaging. Madison, WI: Medical Physics Publishing, 1993: 63–78.
3. Fitterman AS, Brayer FC, Cumbo PE. Processing chemistry for medical imaging. Eastman Kodak Company publication no. N–327, 1995.
4. Haist, G. Modern photographic processing. New York: John Wiley and Sons, 1979: Vol. 1.
5. Haus AG. Film processing systems and quality control. In: Gould RG, Boone JM, eds. A categorical course in physics: Technology update and quality improvement of diagnostic x-ray equipment. Oak Brook, IL: Radiological Society of North America, 1996: 49–66.
6. Haus AG. Historical developments in film processing in medical imaging. In: Haus AG, ed. Film processing in medical imaging. Madison, WI: Medical Physics Publishing, 1993: 1–16.
7. Hazard communication for photographic processing facilities, Eastman Kodak Company publication no. J–311, 1996.
8. Introduction to medical radiographic imaging, Eastman Kodak Company publication no. M1–18, 1993.
9. James TH, ed. The theory of the photographic process. 4th Edition. New York: Macmillan, 1977.
10. James TH, Higgens GC. Fundamentals of photographic theory. 2nd Edition. Hastings-on-Hudson, NY: Morgan & Morgan, 1960.
11. Kimme-Smith C, Wuehling P, Kitts EL, Cagnon C, Basic M, Bassett L. Mammography film processor replenishment rate: Bromide level monitoring. Medical Physics 24: 369–372, 1997.
12. Kitts EL. Physics and chemistry of film and processing. RadioGraphics 16: 1467–1479, 1996.
13. Personal protective equipment requirements for photographic processing facilities, Eastman Kodak Company publication no. J–312, 1996.
14. Processing recommendations for KODAK X-OMAT processors for models: M35, M35A, M35A-M, M35-M, M43, M43A, Clinic 1, M7B, M7B-E, 270 RA, 3000 RA, 180 LP, 180 LPS, Multiloader 300, M6A-N, M6AW, M6B, M8, M6RA, 460 RA, 480 RA, & 5000 RA, Service Bulletin no. 30, Eastman Kodak Company publication no. N–923, 1996.
15. The regulation of silver in photographic processing facilities, Eastman Kodak Company publication no. J–214, 1996.
16. The technology of silver recovery for photographic processing facilities, Eastman Kodak Company publication no. J–212, 1995.

# 3

# Processors

- History
- Introduction to Processing
- Developing
- Fixing
- Washing
- Drying
- Transport System
- Replenishment System
- Film Processing Area Layout Considerations
- Processor Installation
- Processing Cycle Time
- Solution Temperatures
- Preventive Maintenance and Cleaning

## History

Before 1900, emulsion-coated glass plates were largely hand-processed in trays. Each plate required special treatment because the amount of emulsion coated on each plate and the amount of exposure it had received were highly variable. Close inspection of the developing image was required. Once the sensitivity of the plates became standardized and the radiation quality and output could be better controlled, it was possible to time development more precisely. Even so, it varied from about 10 to 25 minutes. During development, the plate also had to be agitated constantly to ensure uniform development of the entire image.

As interest in radiography grew, faster and more efficient processing methods became necessary. By 1906, wooden tanks with slotted sides were designed so that plates could easily be inserted in the slots and the development period could be precisely timed. Compartments in the tanks had to be sized for the various plates then in use. By 1910, slotted plate-developing hangers, which had a crossbar at the top for suspending the plate in the solutions, were being manufactured. The hangers could be used to transfer plates to other sections of the tank, thereby eliminating the need for separate compartments for plates of various sizes.

Modified tanks with a developing compartment, rinse area, fixer compartment, and large washing area were soon available. The first tanks were made of wood or wood lined with a lead sheeting; later, they were constructed of soapstone. They produced greater uniformity of results than could be achieved with tray development and also substantially reduced the space required for processing the plates.

A method of standardized tank development was reported in 1929 by F. C. Martin, E. E. Smith, and M. B. Hodgson. These investigators recommended the establishment of a constant time of development for a given temperature based on the rate of exhaustion of the developer. This exhaustion system ensured uniformity of results and made it possible to check exposure time.

Metal hangers were soon used for the hand-processing of medical x-ray films (Figure 3-1). After the exposed film was loaded onto the metal hangers, it was immersed for approximately six minutes in a tank containing developer at 68°F (20°C), then immersed in a stop bath, followed by immersion in a fixer solution. The film was then washed in running water and hung up to drip dry (Figure 3-2). It took approximately one hour to obtain a completely dry and ready-to-read radiograph. The number of films processed per hour depended on the size of the processing tanks, dryer capacity, number of darkroom personnel, etc. Image quality, reproducibility of results, and patient exposure were dependent upon the user's control of processing time, solution temperature and freshness, agitation, and general

**Figure 3-1** 1940s darkroom for processing medical radiographs. A technologist is removing a metal film hanger from a rack of different-sized hangers.

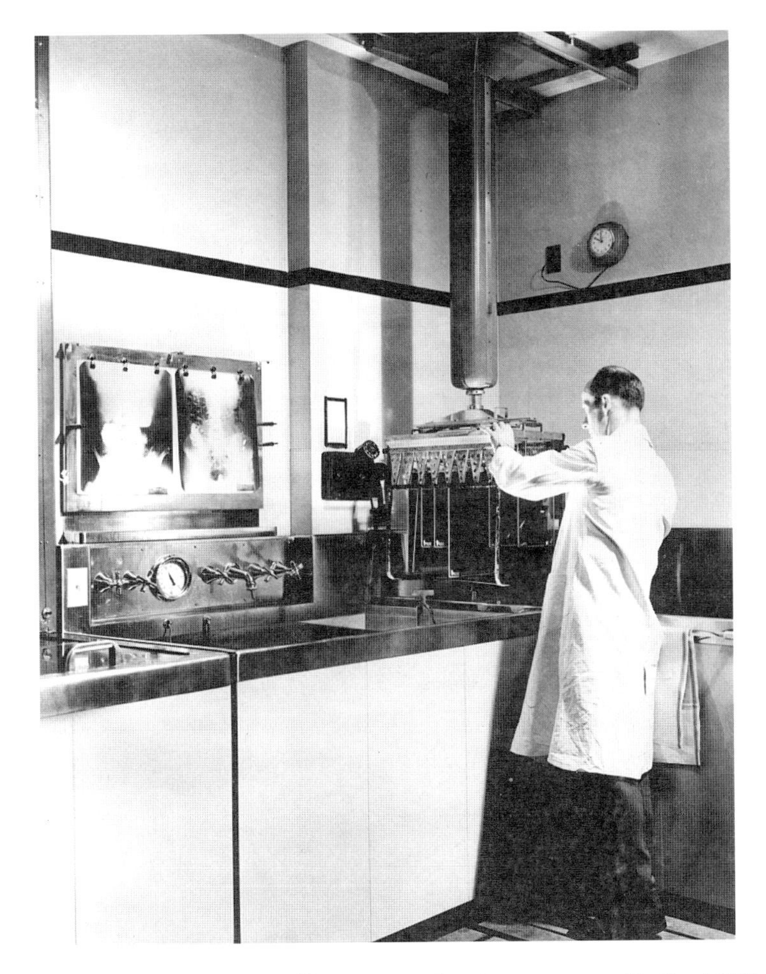

**Figure 3-2** Metal hangers on a ceiling-mounted system were used in the 1940s to move films from solution to solution.

housekeeping. Patient exposure was frequently increased to permit shortened development time for early inspection of the radiograph, a so-called "wet reading." This practice also resulted in reduced contrast.

In 1942, the first prototype automatic x-ray film processor was introduced (Figure 3-3). The first commercially available model processed 120 films per hour using special film hangers and processed one film in approximately 40 minutes.

A significant improvement in automatic x-ray film processing was made in 1956 when the first roller transport processor designed for medical radiographs was introduced (Figure 3-4). This processor accommodated all medical x-ray films designed for exposure with intensifying screens, so the hospitals of the time did not need to change their film inventories in order to switch from hand-processing to automatic processing. The processor was about 10 feet (3 meters) long, weighed nearly three-quarters of a ton, and sold for approximately $33,000 ($180,000 in current U.S. dollars).

Automatic processing resulted in significantly increased productivity, important in busy hospitals. Finished radiographs were now available in six minutes, and variable

**Figure 3-3** The first automatic x-ray film processor was introduced by Pako.

**Figure 3-4** The first roller transport processor for processing medical radiographs was introduced in 1956 by Eastman Kodak Company. (Reprinted with permission from Eastman Kodak Company.)

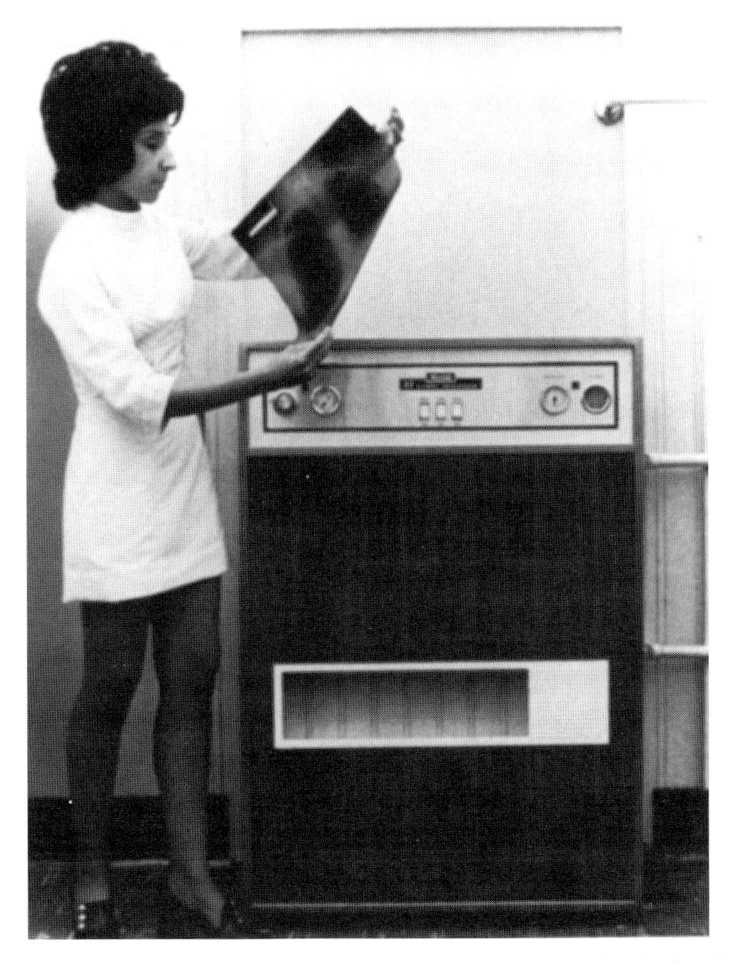

**Figure 3-5** The first processor with the capability of producing a finished radiograph in 90 seconds ("rapid processing") was introduced in 1965 by Eastman Kodak Company. (Reprinted with permission from Eastman Kodak Company.)

image quality caused by the human element was eliminated from processing altogether. Automatic processing made further advances in standardized exposure techniques possible. The number of retakes was reduced as well as the length of time patients needed to wait for radiographs to be read.

In 1965, another significant breakthrough in processing was made when 90-second rapid processing was introduced (Figure 3-5). This advancement combined new chemistry and new film emulsions, an increased development temperature (95°F [35°C]), and the use of a polyester film support for better roller transport—all of which led to faster drying time. The processor, which was 36 inches (91.4 centimeters) long, 30 inches (76.2 centimeters) wide, and 42 inches (106.7 centimeters) high, processed 215 sheets of film per hour in a dry-to-dry time of 90 seconds per film. This type of automatic processing system remains the industry standard as we know it today.

In 1975, the first table-top processor was introduced (Figure 3-6). In 1987, an automatic film processor was introduced that had a processing cycle of approximately

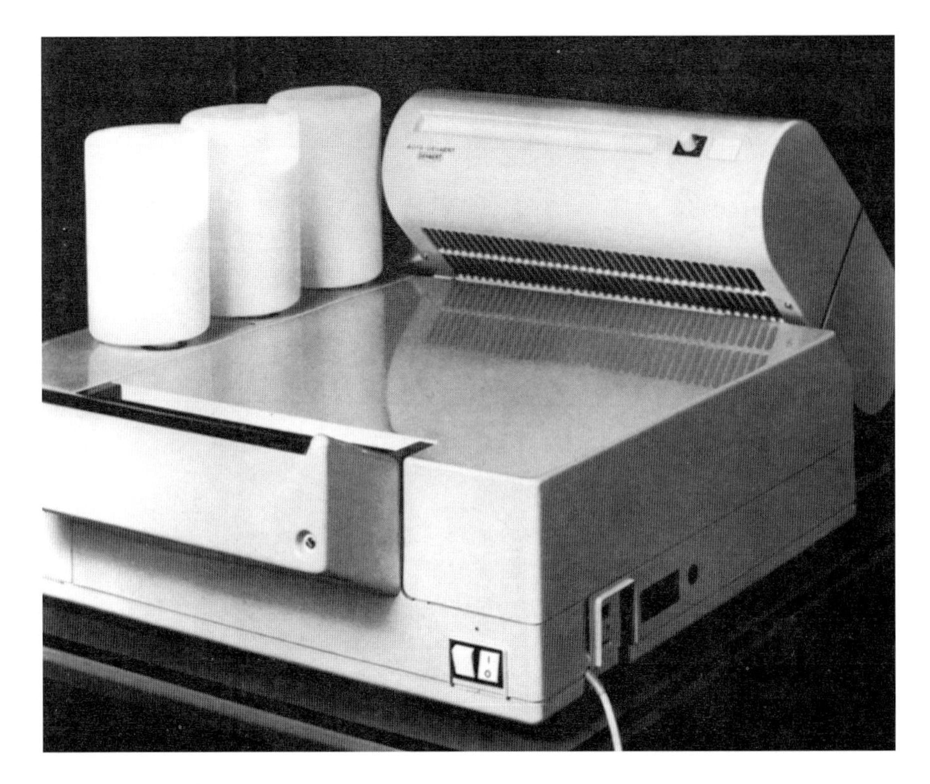

**Figure 3-6** The first table-top processor was introduced in 1975 by Agfa Gevaert. (Reprinted with permission from Agfa Gevaert NV., Belgium.)

45 seconds and required the use of special films (Figure 3-7), and in 1990, an automatic film processor with an approximate 38-second processing cycle was introduced. This processor requires special films and chemicals.

Advancements and improvements in processor technology continue to be made. Now processors may not only be installed in the darkroom or through the darkroom wall but also out in regular room light. Room light systems, discussed later in this chapter, offer easy access to the processing equipment as well as other benefits.

Other important advances related to improved productivity have been made. Early model processors typically had just one processing cycle, but newer models are multicycle capable. Faster cycles provide increased processing capacity and a shorter drop time. Newer models usually also offer extended-cycle processing for some mammographic films.

Typical newer features may include automatic tank fill, area replenishment, and microprocessor control. Improved operator interfaces via liquid crystal displays allow easier monitoring and adjustment of all processing parameters, and display menus are usually programmable in many different languages so that the operator can read the display in his/her native language. Microprocessor control facilitates a change in processor operations and allows upgrades to be made via a laptop computer.

Today, most processor manufacturers offer a broad product line of automatic film processors, ranging in price from approximately $6,000 to $30,000. Due to the

**Figure 3-7** In 1987, Konica Medical Corporation introduced the first processor capable of producing a finished radiograph in just 45 seconds. (Reprinted with permission from Konica Medical Corporation.)

number of choices available, the best processor model for a given application is not always immediately obvious. The best approach when selecting a model is to gather input from everyone, including the technologists, radiologists, business manager/radiology manager, facility manager or site representative, and medical physicist.

## Introduction to Processing

Once the latent image has been formed (Chapter one), film must be processed to produce a visible image. No matter how carefully the patient has been positioned and how carefully technical factors have been controlled, poor processing will always produce poor radiographs. Therefore, it is important to understand some fundamentals regarding this important step in the imaging chain. Note that this book will focus on processing using automatic film processors. Manual processing is described in Appendix B.

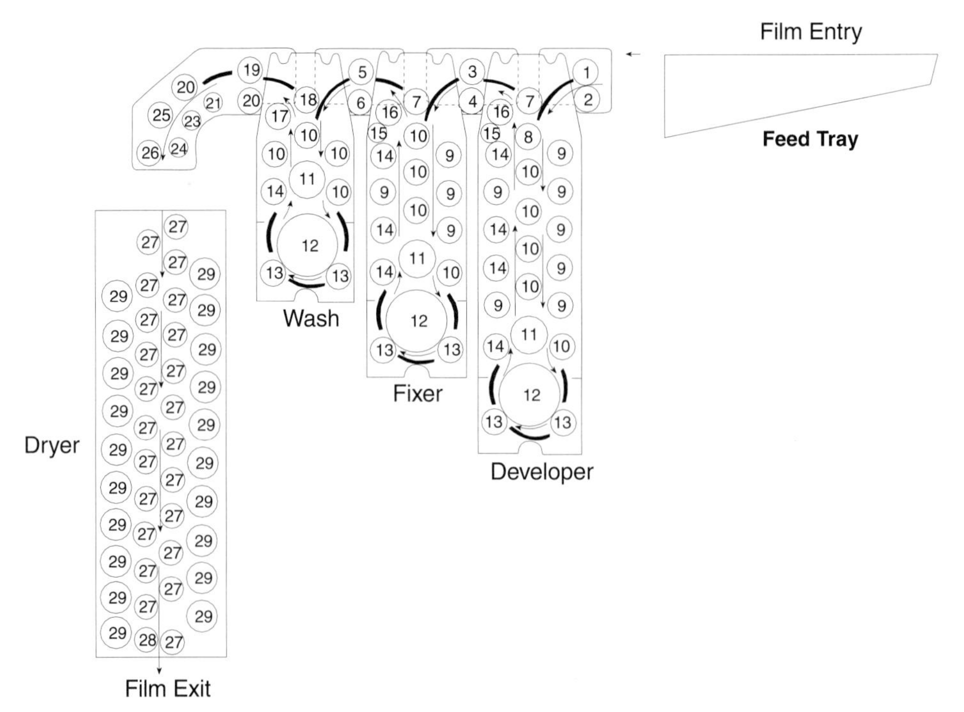


**Figure 3-8** Operation of a typical automatic film processor. Film is manually inserted into the processor transport system from the feed tray (film entry). The film is transported through the developer rack, the fixer rack, the wash rack, and the dryer section, and exits dry and ready to read. The film path is a serpentine-like route. This route enables proper emulsion agitation as well as maximum chemical-to-emulsion "coupling," which produces the optimal development for speed and contrast. Developer makes visible the latent image that was exposed on the film. Fixer essentially stops the development process and makes the resultant image permanent for keeping purposes. Washing removes chemicals to ensure proper and uniform drying and long-term, archival keeping of the radiograph. (Reprinted with permission from Eastman Kodak Company.)

Film processing is a relatively complex series of chemical reactions, many of which occur simultaneously. An automatic film processor transports exposed film through a series of four sections in the machine (Figure 3-8). In each of these sections, a particular set of conditions combines to perform the following four functions:

- *Development:* Converts the invisible latent image into a visible image composed of minute clusters of metallic silver.
- *Fixing:* Removes unexposed, undeveloped silver halide crystals from the emulsion, makes the image permanent, and prepares the film for the remaining steps in the processing cycle.
- *Washing:* Removes excess chemicals from the processed film for long-term stability.
- *Drying:* Prepares the film for handling, viewing, and storage.

## Developing

The developer changes the latent image on the film to a visible image composed of minute clusters of silver. The basic ingredients of a developer are the solvent, developing agents, accelerators (or activators), preservatives, restrainers (or antifoggants), and hardeners.

1. *Solvents:* Water is the solvent in developer. Water dissolves and ionizes the developer chemicals. Film emulsion also absorbs water, causing it to swell; this allows the dissolved developing agents to penetrate the emulsion and to reach all silver halide crystals.
2. *Developing Agents:* A developing agent is a chemical compound capable of converting exposed grains of silver halide into metallic silver. The most common developing agents are hydroquinone and Phenidone.
3. *Accelerators:* Accelerators, or activators, increase the rate of chemical reaction, converting exposed silver halide grains into metallic silver (developing) at a more rapid rate. Accelerators, such as sodium or potassium carbonate and sodium or potassium hydroxide, are chemical bases. Bases have high pH values compared with acids, which have low pH values. The rate of chemical reaction is influenced by the pH and, therefore, the amount of accelerator in the solution.
4. *Preservatives:* An antioxidant preservative, usually sodium or potassium sulfite, retards the oxidation of the high pH developer solution. Preservatives also help maintain the rate of development and prevent staining of the film emulsion layer.
5. *Restrainers:* Restrainers protect unexposed silver halide grains from the action of the developer, minimizing film fog. Potassium bromide and potassium iodide are often used as restrainers or antifoggants. Sometimes additional chemicals supplement the bromide ions in this function.
6. *Hardeners:* Hardeners prevent excessive swelling of the gelatin and damage to the film as it passes through the rollers. They act in the high-temperature, high pH developer solution.

The above summary briefly describes critical active ingredients to which additional chemical components may be added to improve stability and performance. Refer to Chapter two for more information on chemicals.

Some chemical ingredients are consumed during the development process. A replenisher pump transfers fresh chemicals from an external storage container into the developer tank of the processor to restore the developer to its approximate original strength and volume. Replenishment also restores the high pH of the developer solution.

At the end of the development cycle, a considerable amount of developer is retained in the gelatin of the film. Crossover rollers remove excess developer from the film by their "squeegee" action, which prevents the high pH developer from diminishing the effect of the fixer.

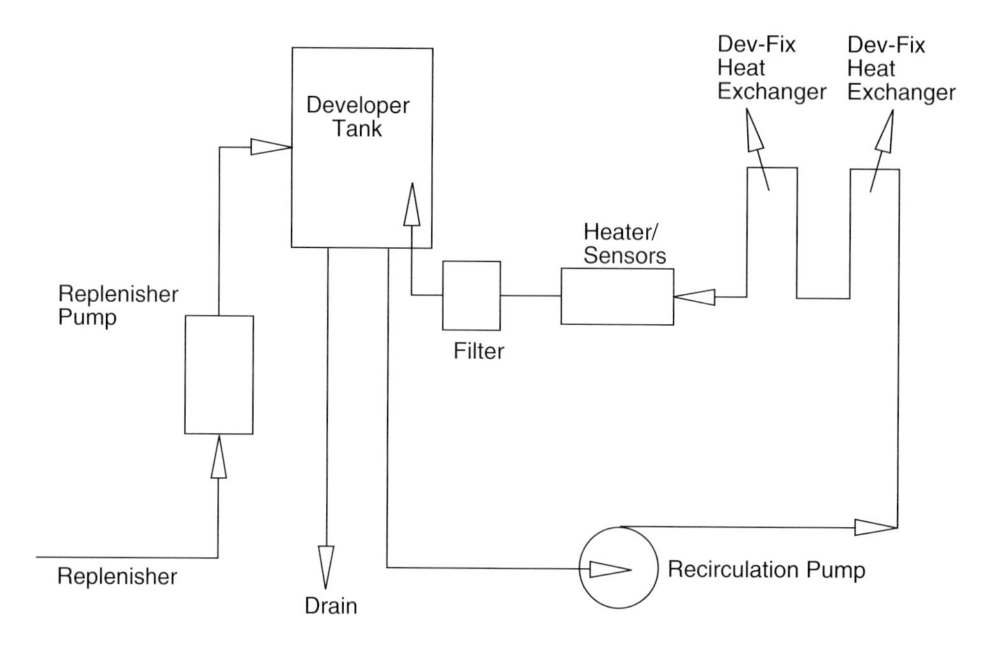


**Figure 3-9** Developer system

## Developer System Components

There are four main sections of the developer system: replenishment, recirculation, temperature control, and drain (Figure 3-9).

### *Replenishment*

The replenisher is the source of fresh chemicals, which may come from replenisher tanks or a chemical mixer. The replenisher pump moves the fresh chemicals from the replenisher supply to the developer processing tank and should be adjustable to allow delivery of different replenishment rates. Typical developer replenishment usage is 60 to 600 milliliters per minute, depending on the productivity of the processor.

### *Recirculation*

The developer tank holds the developer solution and the transport rack. The tank should be easy to clean and maintain. The typical tank volume ranges from 1 to 3 gallons (3.8–11.4 liters). The recirculation pump continuously recirculates the developer for temperature control and mixing of fresh and seasoned chemicals. The recirculation rate typically ranges from 2 to 4 gallons (7.6–15.2 liters) per minute. The filter removes particulate matter from the developer to maintain image quality. Typically, the filter has a porosity of 75 microns; replacement is recommended monthly or after processing 5,000 films. Recirculation also provides developer tank solution agitation, which is critically important for proper development.

Proper functioning of the recirculation pump can be checked by looking for ripples on the surface of the solution in the developer tank. Since the top must be removed from the processor, defeating an interlock switch may be required. This is usually accomplished by placing a small magnet on the microswitch.

Table 3-1 provides an overview of the developer (and fixer) recirculation systems.

### *Temperature Control*

The developer-wash heat exchanger helps maintain the developer temperature. The temperature of the air surrounding the developer tank can rise above that of the developer due to heat from the dryer. This increase in air temperature causes heat flow into the developer tank, thereby raising the developer temperature. The developer-wash heat exchanger cools the developer solution in order to help maintain a constant developer temperature. Typically, the exchanger is a thin-walled, stainless steel tube in the bottom of the wash tank.

The developer-fixer heat exchanger makes use of the developer to heat the fixer solution to its minimum required temperature. This exchanger is usually a thin-walled, stainless steel tube located at the bottom of the fixer tank. If the fixer tank has a heater, the exchanger is not required.

The developer heater and sensors also help maintain the developer temperature. A sensor (a thermostat or a thermistor) monitors the developer temperature and controls the developer heater. Typically, the heaters are rated between 500 and 1,500 watts and are sized according to desired warm-up times and replenisher amounts. The temperature of the developer is usually controlled in the 90 to 100°F range (32.2 to 37.8°C) with a tolerance of ±0.5°F (±0.3°C). Sensors are also added to the design for safety reasons, such as detection of excessive temperature and the presence of solution.

Table 3-2 provides an overview of developer (and fixer) temperature control.

**Table 3-1** Developer and Fixer Recirculation Systems

| | |
|---|---|
| Fluid Recirculation | Needed for temperature control and processing uniformity. |
| Components | High reliability needed due to continuous operation whenever processor is on. |
| Filtration | Needed to keep developer clean and free of particles that may cause film artifacts; generally not needed for fixer. |

**Table 3-2** Developer and Fixer Temperature Control

| | |
|---|---|
| Heater | Sufficient power to heat developer and fixer replenisher and achieve desired warm-up time. |
| Sensors | Control sensors should be located immediately after heater for maximum control and to provide over-temperature protection. |
| Heat Exchanger | Due to the high ambient temperature surrounding the developer tank, cooling of the developer by the water may be needed to limit developer temperature rise; if the fixer system does not include a heater, then a developer-to-fixer heat exchanger may be needed to achieve proper fixer temperature. |

### Drain

As replenisher solution is added, some developer must be expelled through the overflow to a floor drain or collection system. When the developer tank is cleaned, the developer is drained by way of a drain valve. The overflow and drain lines may be combined or separate.

An important consideration for all the components is service life. The typical life goal is 10,000 hours or 5 years. Because these components are in use whenever the processor is on, each component must be carefully selected and tested to achieve the reliability goals and service life set for the processor.

## Fixing

Film leaving the developer via the crossover rollers contains metallic silver grains at the sites where x-ray photons of the aerial image (the pattern of x-ray fluence emerging from the body and incident on the image receptor [film], also referred to as the image in space) exposed the film. With the exception of a small amount of chemical fog, unexposed silver halide crystals are unaffected by the developer. To complete processing, x-ray film must be cleared of undeveloped crystals before washing to prevent discoloration of the film with age or exposure to visible light. The gelatin coatings must also be hardened so that the film will resist abrasion and may be dried with warm air. The basic ingredients of fixer solution are:

1. *Solvent:* The solvent (water) diffuses into the emulsion, carrying with it the other dissolved chemicals of the fixer.
2. *Fixing Agent:* The fixing agent dissolves and removes the undeveloped silver halide grains from the emulsion. This action changes the unexposed areas of the film from a milky-white to a transparent appearance. The two most commonly used fixing agents (also known as hypo) are sodium thiosulfate and ammonium thiosulfate. If a film is improperly fixed, the remaining unexposed silver halide crystals continue to darken upon exposure to visible light after processing has been completed.
3. *Preservative:* Usually sodium sulfite is used as a preservative to prevent decomposition of the fixer.
4. *Hardener:* The hardener is usually an aluminum salt that prevents the gelatin of the emulsion from swelling excessively. It also prevents softening of the emulsion by the wash water or by warm air drying and shortens drying time.
5. *Acidifier:* Acetic acid or other acidic compounds are used to accelerate the action of other chemicals and to neutralize any developer (high pH) that may be carried over from the developer tank.
6. *Buffer:* Buffers are chemical compounds that maintain the solution at a constant pH during the fixing process. When high pH developer neutralizes the low pH acidic solutions of the fixer, buffers act to maintain the pH at the desired level for optimum fixing action.

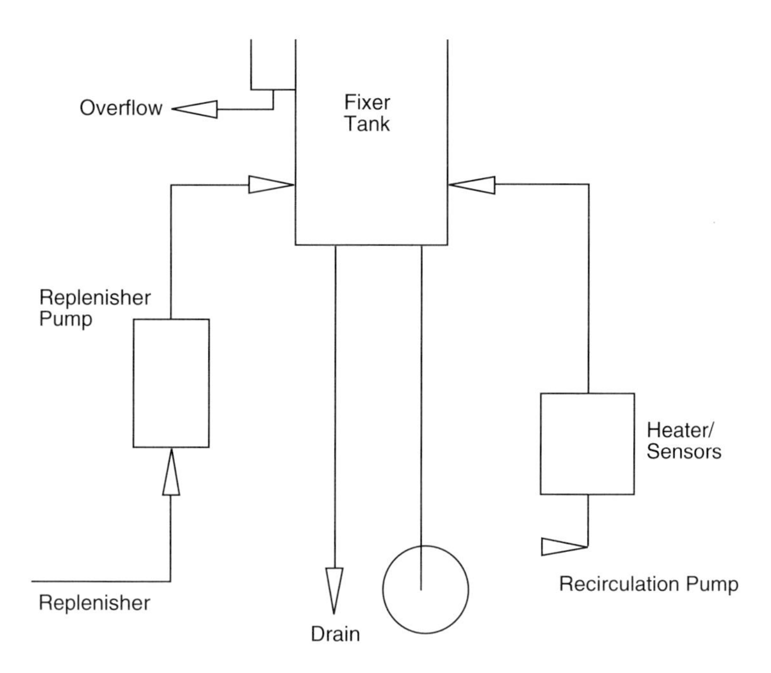


**Figure 3-10** Fixer system

Because some chemical ingredients are consumed during the fixing process, a replenisher pump transfers fresh chemicals from an external storage container into the fixer tank of the processor. This restores the fixer to its approximate original strength and volume.

## Fixer System Components

The fixer system is similar to the developer system but is somewhat simpler. Figure 3-10 shows a typical processor fixer system. It contains the same four sections as the developer: replenishment, recirculation, temperature control, and drain. The function, typical values, and design considerations of each component are described below.

### *Replenishment*

The replenisher pump moves the fresh chemicals from the replenisher supply (a replenisher tank or chemical mixer) to the fixer processing tank. The pump must be adjustable to allow delivery of different replenishment rates. Typical fixer replenishment usage is 80 to 800 milliliters per hour, depending on the productivity of the processor.

### *Recirculation*

The fixer tank contains the fixer solution and transport rack. It should be easy to clean and maintain. The typical volume ranges from 1 to 3 gallons (3.8–11.4 liters). The recirculation pump continuously recirculates the fixer to maintain temperature control and to mix fresh chemicals with seasoned chemicals. The recirculation rate ranges between 2 and 4 gallons (7.6–15.2 liters) per minute. A filter typically is not needed for the fixer solution. Recirculation also provides fixer tank solution agitation, which is critically important for proper fixation.

Proper functioning of the recirculation pump is as important for the fixer tank as it is for the developer tank and can also be verified by the presence of ripples on the surface of the fixer solution in the tank. Since the top must be removed from the processor, defeating an interlock switch may be required. This is usually accomplished by placing a small magnet on the microswitch.

Table 3-1 (page 99) provides an overview of the fixer (and developer) recirculation systems.

### *Temperature Control*

The developer-fixer heat exchanger uses the developer to heat the fixer solution up to its minimum operating temperature. If the fixer has a heater, the exchanger is not required. Typically, the exchanger is a thin-walled, stainless steel tube located at the bottom of the fixer tank. The fixer heater and sensor maintain the fixer temperature. The sensor, which is either a thermostat or a thermistor, monitors the fixer temperature and controls the fixer heater. Typically, the heaters are rated between 500 and 1,500 watts and are sized for the desired warm-up times and replenisher amounts. Fixer temperature is usually controlled or set to minimum points, which may vary between 85 and 95°F (29.4–35°C). The fixer temperature may drift above the set point, but a higher temperature only improves fixing and causes no problems. Sensors are also added to the design for safety reasons, such as detection of excessive temperature and the presence of solution.

Table 3-2 (page 99) provides an overview of fixer (and developer) temperature control.

### *Drain*

As with developer solution, some fixer must be expelled through the overflow to a floor drain or collection system when replenisher solution is added. When the fixer tank is cleaned, the fixer is drained by way of a drain valve. The overflow and drain lines may be combined or separate.

As with the developer system components, an important consideration for the fixer system components is service life and reliability. Typical life is 10,000 hours or 5 years. Selection and testing should be similar to those for the developer system.

## Washing

Washing removes the last traces of processing chemicals and prevents fading or discoloration. One of the important characteristics of a finished radiograph is its archival (long-term) storage capability. Radiographs must be properly washed for long-term image stability.

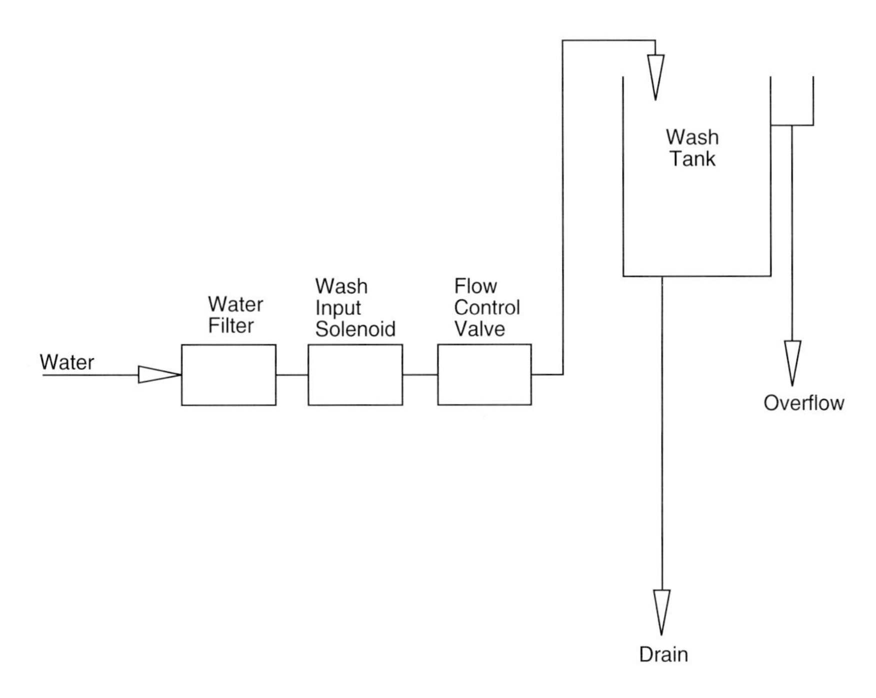


**Figure 3-11** Wash system

## Wash System Components

The wash system is the last of the wet sections. Figure 3-11 shows a diagram of a typical wash system. It consists of the water input and drain sections.

The function, typical values, and design considerations of each component are described below.

### *Water Input*

The filter removes particulate matter from the water to prevent an adverse effect on water flow, as well as preventing the deposit of foreign material (dirt) on the films. Typically, the filter has a porosity of 50 microns; replacement is recommended every 1 to 3 months.

### *Input Solenoid*

The solenoid controls water usage by allowing the water to be turned on and off as needed. Water flow is turned off during standby periods, except when the developer solution needs to be cooled. It is increasingly important to use as little water as possible for conservation reasons.

### *Flow Control Valve*

The flow control valve regulates the flow of water into the processor when fluctuations occur in water line pressure. The typical flow rate of the incoming water is from 0.25 to1.5 gallons (0.95 to 5.7 liters) per minute.

**Table 3-3** Wash System

| | |
|---|---|
| Flow Rate | Wash water must be replenished at the appropriate rate to ensure adequate washing of the film for long-term keeping. |
| Solenoid Valve | Used to reduce water consumption when processor is in standby mode. |
| Temperature | General goal is to accept water temperatures from 40°F (4.4°C) up to the developer temperature. |
| Filtration | Water should be filtered to eliminate particles that may reduce water flow. |
| Regulations | Anti-siphoning regulations must be met. |

### *Temperature Control*

Water helps stabilize the temperature of the processing solutions. Some processors operate using both hot and cold water; the temperature is controlled by a mixing valve. Other processors require only a cold water supply and control temperature with internal heating elements. Supply water must be clean and free of contaminants, sludge, and other deposits that can interfere with optimum processing.

Most processors do not control water temperature within the processor as is necessary for the developer and fixer solutions. There are, however, temperature limits on incoming water. Because water is sometimes used to cool the developer, an upper limit of 5–10°F (2.8–5.6°C) below the developer temperature is desirable. The typical low limit for incoming water is 40°F (4.4°C).

### *Drain*

As water is replaced by fresh incoming water, excess water is expelled through the overflow. When the tank needs cleaning, a drain valve is opened to drain the tank. When turned off, most processors now automatically drain the wash tank in order to reduce algae growth.

An important regulatory concern centers on the proper management of spent wash waters containing photochemicals. The particular manner that is employed to dispose of spent wash waters must be carefully evaluated with respect to national and local regulatory requirements.

Table 3-3 provides an overview of the wash system.

## Drying

The last step in processing an x-ray film is drying. A blower supplies heated air to the dryer section of the film processor. Most of the warm air is recirculated; the rest is vented to prevent buildup of excessive humidity in the dryer. The dryer temperature should be set as low as possible while still being consistent with good drying. Use of a lower setting will also result in energy savings for processor operation. The dryer temperature should never exceed the film manufacturer's recommendations.

Many users tend to overdry films, which may cause surface-pattern artifacts. Severe drying artifacts that are visible on transmitted light can adversely affect the diagnostic interpretability of films, but drying artifacts that are visible only on reflected light are considered normal and acceptable.

## Dryer System Components

Figure 3-12 shows a typical dryer system. It consists of air movement and temperature-control sections. The function, values, and design considerations of each dryer system component are described below.

### *Air Movement*

The blower moves heated air to the film surface to promote drying. Typical air-flow capacity of the blower is 100–300 cubic feet (3–9 cubic meters) per minute. Every effort should be made to reduce the noise of the blower in order to lessen irritation in the work area. Blowers that operate with 50- or 60-hertz power sources must be used so that the unit will operate efficiently on the world's power sources.

Air tubes are the critical part of the air movement system. They uniformly provide air turbulence at the film's surface to accelerate moisture evaporation. Higher air flow typically results in greater heat transfer and, therefore, quicker drying.

As film is dried, the moisture content of the air increases. The air exhaust allows some of the humid air to be ducted away from the dryer to stabilize the humidity in the dryer at an acceptable level. Typically, relative humidity values for exhaust range between 25 and 100 percent.

### *Temperature Control*

As with the developer and fixer systems, the dryer has a heater and related control and safety sensors. Typically, the power of the heater ranges from 1,500 to 3,000 watts. The heater is sized to control air temperatures between 100 and 160°F (37.8 and 71.2°C) and for rapid warm-up times. The control sensor is either a thermostat

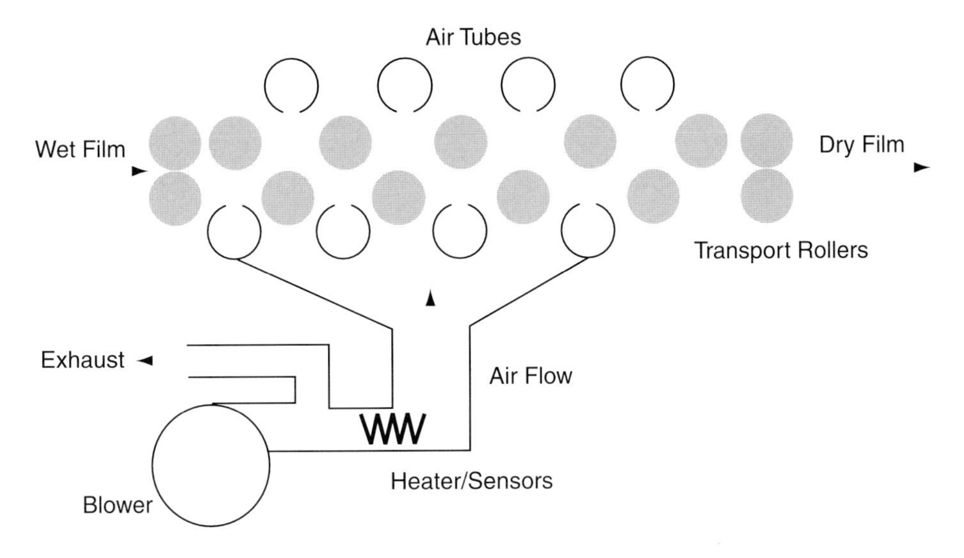


**Figure 3-12** Dryer system

**Table 3-4** Dryer System

| | |
|---|---|
| Blower | Provides air movement to film; airflow removes moisture from the surrounding air as evaporation from film takes place. |
| Heater | Provides heat to air; higher temperature air increases evaporation from film, thereby increasing drying rate. |
| Sensors | Provide temperature control and safety protection (over-temperature, airflow). |
| Air Tubes | Provide uniform air distribution of the heated air directly to the film surface. |

or a thermistor, and safety sensors include an over-temperature sensor and sometimes an air-flow switch. The air-flow switch creates air flow across the heater before the heater is turned on.

Table 3-4 provides an overview of the dryer system.

# Transport System

The transport system moves the film through the developer, fixer, washing, and drying sections of the processor. The diagram in Figure 3-8 (page 96) shows a typical transport system.

Optimum processing demands that each film spends exactly specified amounts of time in each section of the processor. Film moves through the processor by a system of rollers driven by a constant speed motor. This motion also produces the required amount of chemical agitation at the film surfaces, which improves uniformity of processing.

Some newer model processors are available that function at several different speeds, according to the processing cycle specifications of the manufacturer. For each speed, the time of transport is carefully coordinated with temperature and chemical concentrations to achieve the desired processing results and film characteristics.

The transport system has many design considerations that must be balanced. Film transport must be very reliable because as many as 50,000 to 100,000 films of various sizes and types will typically be processed over its life. If film is damaged by a faulty transport system, patients may need to have radiographs repeated, increasing radiation dose. Processor components are usually tested to ensure compatibility with chemicals, to avoid photographic effects, and to reduce component failure.

## Transport System Components

### *Rollers*

Various manufacturers of processors use different methods for film transport, but all generally include rollers that move the film on either an opposed or staggered path. Processor rollers are composed of different types of materials depending upon function. These materials include, but are not limited to, phenolic, plastic, silicon, and stainless steel.

Phenolic rollers are used wherever uniform wetting of the film is required. Plastic rollers are used to drive film through the rack. They exhibit enough "tooth" to drive the film but are smooth enough to prevent wet pressure or other processing artifacts. Silicon rollers are used where a delicate surface is required in order to facilitate the passage of film without doing any damage to the emulsion surface. Stainless steel rollers are used to drive film against the squeegee rollers. The function of certain rollers is also critical to image quality. Squeegee rollers remove as much of the chemical (or water) as possible when the film is leaving one solution and entering the next. Proper squeegee roller function is dependent upon the integrity of the rollers and upon the springs that hold them tightly together.

Many processor rollers have 1-inch (2.54 centimeter) diameters. There are also 2-inch (5 centimeter) diameter rollers, called "master" rollers, at the bottom of the turnaround assemblies of the developer, fixer, and wash racks.

Generally, rougher rollers hold more solution than smoother rollers, e.g., the knurled roller with a deep diamond-scored pattern in the fixer rack is designed to hold a lot of fixer in order to stop development quickly.

During preventive maintenance, it is essential that all rollers and other components be closely inspected for the first visible signs of wear so they may be changed before they begin to affect image quality. Additionally, any replacement parts should be those specified by the manufacturer in order to minimize artifacts and maintain processor reliability.

### *Crossover Assemblies*

Crossover assemblies move film from one section of the processor to the next. Processors contain three crossover assemblies: the entrance detector crossover assembly, the developer-to-fix crossover assembly, and the fix-to-wash crossover assembly.

The entrance detector crossover assembly transports the film into the developer and, in most processors, causes the activation of the replenisher pumps in order to deliver fresh developer and fixer replenisher while the passage of film separates the stepped rollers of the assembly. The developer-to-fix crossover assembly removes excess developer from the film before it enters the fixer tank. The fix-to-wash crossover assembly removes excess fixer from the film before it enters the wash tank. The developer-to-fix and fix-to-wash crossover assemblies should be thoroughly cleaned on a daily basis, especially if the processor is used for mammography film.

### *Turnaround Assemblies*

Turnaround assemblies are located at the bottom of the developer, fixer, and wash racks. They change the direction of the film 180 degrees.

### *Guide Shoes*

Guide shoes are located throughout the processor to help direct film through the crossover, turnaround, and rack assemblies of the processor. The guide shoes change the travel direction of the film in areas where rollers would not be adequate. They are usually manufactured from a special stainless steel and polished to a very smooth and bright finish by an electro-polish process.

**Table 3-5** Transport System

| | |
|---|---|
| Transport | Must be highly reliable to ensure no loss of patient films. |
| Film Artifacts | All transport components must be smooth and free of sharp edges, abrasion points, and high stress points to ensure highest possible film quality. |
| Cleanliness | Transport system must be maintained (cleaned, adjusted, checked for wear, and parts replaced as specified) to ensure highest reliability and film quality. |

Guide shoes coming into contact with film often leave small (1/4 inch [6.35 millimeters] in length) plus-density artifacts on the leading or trailing edge of films. Guide shoe marks of this length are inherent in many processors and are considered normal and acceptable.

Guide shoe marks are frequently noted along the chest wall of mammography films that have been manually processed with the long dimension of the film as the leading edge. In order to move these marks so as not to interfere with the clinical information, all mammography films should be placed on the film feed tray so that the short dimension edge of the film is the leading edge. This will position these marks along the short dimension edge where a minimal amount of breast tissue is imaged.

Refer to Appendix C for a discussion of processing mammography film.

Table 3-5 provides an overview of the transport system.

## Replenishment System

Film processing consumes some of the chemical components. Therefore, accurate replenishment of developer and fixer solutions is necessary to maintain consistent, high-quality film processing. The amount of replenishment required depends on the amount of film processed. Replenishment rates are determined by either total surface area or total length of film processed. Both over- and under-replenishment can lead to suboptimal radiographs.

Chapter two contains additional information on replenishment.

## Film Processing Area Layout Considerations

Following a comprehensive planning process is very important when designing a new darkroom and film processing area or the renovation of an existing film processing area. Proper planning will prevent a number of problems at the time of the actual construction and equipment installation. Examples of avoidable problems are not allowing for adequate space for the equipment, failing to provide the necessary utilities, or installing the equipment incorrectly. Some of these problems are not only very costly to correct but also may result in frustrating delays in the completion of a very vital part of a radiology department. Thorough planning will also ensure that the floor space is used efficiently. This is very important because space is limited in most hospitals.

All individuals involved in such planning should be qualified and thoroughly familiar with the operation and needs of medical imaging processing.

The planning process should include the following six steps: (1) determining the requirements of the area, (2) evaluating the current area, (3) defining the redesign or design needs, (4) obtaining equipment information, (5) developing a conceptual plan, and (6) reviewing plans with an architect.

## Determining the Requirements of the Area

The first step is to make the major decisions regarding the requirements for the area. Consider the functions that will be performed in this space and decide what type of processing area (room light or darkroom) is most appropriate. A significant decision is whether or not a darkroom for processing films is adequate, or if space for viewing and sorting films, performing quality control procedures, or doing any other work is required. The minimum amount of space that is required for a darkroom with one processor is approximately five feet by eight feet (1.5 meters by 2.4 meters). More space is required for two processors. Additional space might also be required, depending on what functions will be performed in the area.

There are advantages to both room light processing areas and traditional darkrooms.

The advantages of room light processing areas include:

- Automated film handling.
- Easy access to equipment.
- Fast turnaround time.
- Effective use of people.
- Combination of different functions in one area.

The advantages of traditional darkrooms include:

- More flexibility in processing certain specialty films.
- Lower investment in equipment.

One of the major benefits of a room light setup is easy access to the equipment. Technologists may walk just a few steps from the x-ray control area to the processing equipment. This is especially important in trauma or emergency room areas where it is essential to stay with the patient as much as possible. Emergency and trauma rooms are also frequently staffed around the clock, with technologists on the off-shifts responsible for processing their own films. Convenience is extremely important in this type of situation.

After a cassette is fed into the room light equipment, it is automatically opened and unloaded, the exposed film is transported into the processor, and the cassette is reloaded with an unexposed film. With a darkroom, the technologist must enter the darkroom, manually unload and reload the cassette, feed the film into the processor, exit the darkroom, and retrieve the processed films.

Room light equipment offers fast turnaround. Cassettes are reloaded and returned in just a few seconds. Room light equipment is also easy to incorporate into an open, multifunctional work area. Therefore, processing, sorting, and quality control related activities can easily be combined into one very efficient space. With a darkroom, the space is dedicated only to processing. With the introduction of room light equipment, operating costs may be reduced because labor is used more effectively. Operating costs must be weighed against the initial cost of the equipment.

Room light areas are not always ideal, however. There are cases in which traditional darkrooms are more appropriate. They are more flexible, especially for certain specialty films, such as orthopedic films, and require a lower investment in equipment. Even with room light equipment, a small darkroom is generally needed to load film magazines. Therefore, the darkroom cannot be completely eliminated. It may be possible, however, to share a darkroom with another area.

The next decision concerns equipment needs. Determine the expected workload and film volume for the rooms that will be using the room light equipment and/or darkroom. Then determine the desired film access time and the number and models of processors that will be required. The equipment manufacturer can provide information on equipment capacities, as well as guidelines for appropriate uses. It is also important to give serious consideration to the need for a backup processor. Replenisher tanks and silver recovery units also need to be selected. As with the processor or processors, the size of these units depends on the film volume and the required film access time.

Centralized systems for replenishment and silver recovery are also an option. The centralized systems remove the chemicals from the immediate vicinity of the processor and offer advantages of scale. Other pieces of equipment to consider include duplicators, identification cameras, passboxes, and film bins.

## Evaluating the Current Area

The second step in the planning process is evaluating the current area. The following information is especially useful when renovating an existing space, but many of the points also apply to new construction. The only difference is that new construction offers more flexibility with space. In an existing facility, it is common to have limitations on the amount of space that is available. It is important to compare the amount of space that is available with the requirements determined in the first step to determine the changes that need to be made.

When evaluating the current space, be sure to note the location of existing utilities such as power, water, and drains because they are required for processor operation. Processors also require exhaust venting. Additionally, locate any load-bearing walls or columns because these cannot be changed.

## Defining the Redesign Needs

The third step in the planning process is to determine the redesign needs. This involves comparing and reconciling the information obtained in the previous steps and making major decisions as to whether any walls need to be moved. Examples of very common renovations are removing walls to change a darkroom into a large work area, reducing a darkroom in size to accommodate room light equipment outside, or adding walls to create a darkroom out of an open space.

One should be especially cautious if considering the conversion of existing small rooms, such as closets or bathrooms, into darkrooms, particularly if the darkroom is to be used for special applications, such as mammography. Proper ventilation, light-tightness, and cleanliness are critical to high-quality mammographic images.

Another consideration is determining where utilities need to be provided. Whenever possible, use the current utilities because it is most cost-efficient. When this is not possible, utilities will need to be installed in the area; this will add to the cost of the redesign. Moving or adding a drain is one of the most difficult changes to make, especially on a ground floor concrete slab.

## Obtaining Equipment Information

The fourth step is to contact the manufacturer or manufacturers to request more detailed equipment information. It is helpful to obtain specific information on all equipment that is being considered. When contacting the manufacturer(s), ask for site specifications, which provide detailed information on equipment dimensions and weight, the required clearances for proper use, utility requirements, and installation information. It is important to obtain this information early in the planning process in order to avoid costly changes later.

## Developing a Conceptual Plan

After all of this information has been gathered and most of the fundamental decisions have been made about the processing area, it is time to develop the conceptual design. This may be something as simple as a sketch if the changes are minor, such as removing one processor and replacing it with a new or different model. However, it is important to develop a detailed drawing if any major changes are under consideration. Consider obtaining assistance from an architect or the equipment manufacturer when creating the drawings, especially if limited space is available. One of the most important considerations in developing a conceptual design is to plan for the optimal placement of equipment for good work flow. It is very important to allow for the clearances recommended for service and ease of use. In many cases, using only the minimal clearances may increase service time or inefficiency. If any structural changes are to be made, an architect should make construction blueprints.

There are some specific guidelines to consider when designing a darkroom. The first is to determine the most appropriate size. The proper size depends upon the functions to be performed in the darkroom and the number of personnel who will need to use it. For example, if a darkroom technician processes all films, a small room would be adequate; if all technical staff routinely process their own films, a larger room would be necessary. Consideration should also be given to the work counters and storage cabinets to ensure adequate work and storage space.

Another consideration is whether or not a revolving door is required or if a hinged door is adequate. A revolving door allows access to the darkroom at any time without light leak dangers, but it is more expensive and requires more floor space.

The work flow must be understood in order to plan properly for room light equipment. Observe and ask questions of those working in an existing processing area to understand the work flow. The objective is to locate the equipment for easy access and good traffic flow, resulting in a very functional area. If possible, place the feed end of the equipment close to the control areas of the adjacent x-ray rooms. Be sure to allow for space around the equipment for good traffic flow, ensuring that one person standing by the equipment will not block the flow of traffic through the area. If the equipment is located off a main corridor, try to set the equipment back a few feet from the corridor so that a person using the equipment will not block traffic flow in the corridor. Providing the recommended clearances around the equipment will result in an efficient work area and easier service. When space is limited, consider requesting assistance from the manufacturer for best utilization of space. Some manufacturers offer assistance with facilities planning involving their equipment.

There are also some general layout considerations that apply to darkrooms and room light areas. For instance, it is most convenient to orient the processor's receiving bin toward the viewing counter. This facilitates film retrieval and review. Consider locating the processing area close to the film sorting, reading, and work areas. As previously mentioned, be sure to plan for the necessary utilities and allow the proper clearances for service. Using less than the recommended clearances will increase service time, thereby increasing downtime. Plans must also be made for cleaning processor racks and other components. A large sink should ideally be located right in the processing area; a janitor's closet nearby might also be used.

The last item to think about is the best location for replenisher tanks and silver recovery units. They need to be in a location that is easy to access and close to the processor. Consider locating them under a hinged or removable counter so they are out of the way, yet close to the equipment. This results in a cleaner processing area.

## Reviewing Plans with the Architect

Once the conceptual design has been developed, it should be reviewed with an architect. This step is absolutely necessary if any structural changes are being made because a conceptual design is not suitable for construction purposes. The architect will make sure that all of the proper construction plans have been made and will develop blueprints.

## Typical Layouts

Two typical layouts are included to illustrate some of the important points (Figures 3-13 and 3-14). In Figure 3-13, the current layout shows four x-ray rooms around a darkroom and viewing area. There is a corridor between two of the rooms to provide access to the darkroom. There is access to the darkroom from the other rooms by exiting into the corridor and walking through the main work area and into the darkroom. Passboxes are located in each of the x-ray rooms.

The proposed layout illustrates some of the points that have been discussed regarding room light equipment. The current darkroom was eliminated completely and the area was opened up to create a large, open work space. Since the control areas

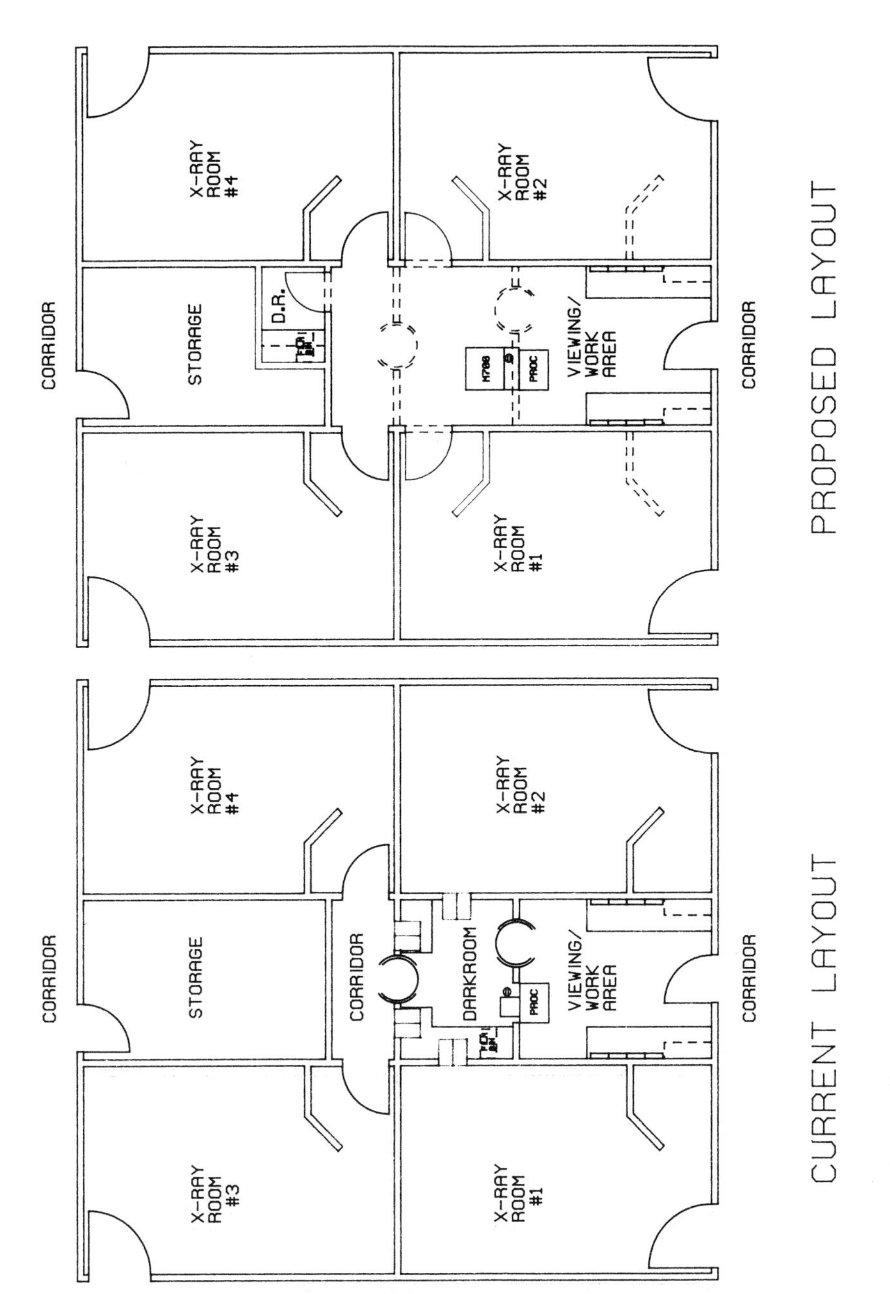


**Figure 3-13** Current and proposed layout for darkroom, viewing, and work area for a radiographic suite

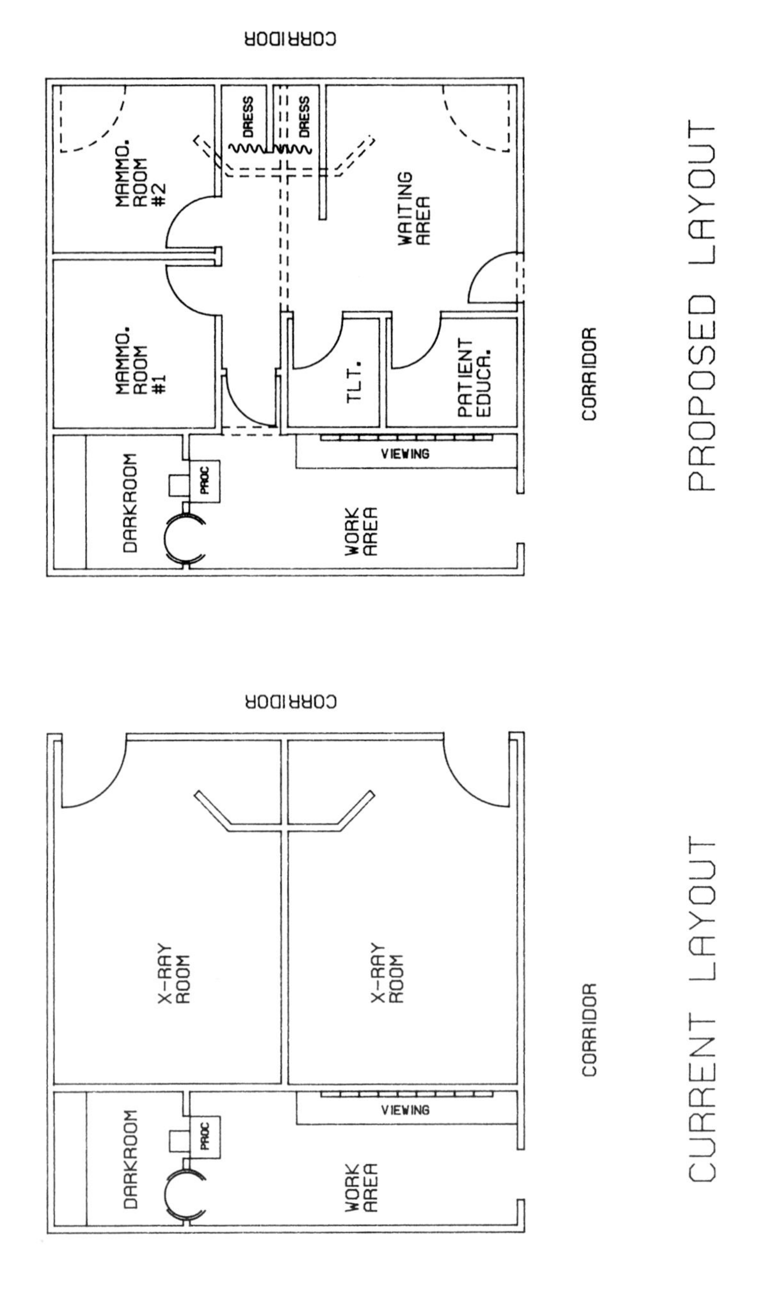


**Figure 3-14** Current and proposed layout for darkroom, work area, waiting area, etc., for a mammography suite

for two of the rooms were not close enough to the processing equipment for convenience, they were relocated. Now there are four doors leading from the control areas that open up close to the feed end of the processing equipment. This is a very convenient layout for the user, and service clearances are adequate for easy service. Plenty of work space is provided at the busy feed end of the equipment.

The processor remains in its current location, which minimizes the cost because the existing utilities can be used. However, moving the control areas does result in additional construction costs. A small darkroom has been added in the former storage area. If there is no other darkroom with a processor relatively close by, consider adding a processor in the darkroom as a backup.

Figure 3-14 illustrates the renovation of an existing area into a new mammography suite. The two previous x-ray rooms have been replaced with two new mammography rooms.

The suite has been designed for good traffic flow. A waiting area with a separate room for patient education is located inside the entrance from the corridor. Just beyond the waiting area are the dressing rooms. The restroom facility is located across from the dressing area, and the two mammography rooms are located just beyond the dressing area.

Since the area was already adjacent to a darkroom, it was not necessary to add a new one. Direct access to the processing area has been provided by adding a new walkway from the mammography suite, with a door for privacy. This allows for convenience, but also screens this area from patients.

## Processor Installation

For a processor to operate properly, it must be installed according to the recommendations in site specification and installation documents. These documents are available from all processor manufacturers, and they provide critical information concerning room layout and service requirements. The processor user should ensure that all specifications listed by the manufacturer of the processor are thoroughly met. Failure to meet these specifications could affect processor warranties.

Radiographic film and equipment dealer service personnel frequently receive comprehensive training from the manufacturer of the processor; processor installation should be done only by qualified individuals.

### Room Layout

#### *Recommended Service Clearances*

Figure 3-15 shows a typical floor plan for a single automatic film processor. Work space must be provided for operation, maintenance, and service. Figure 3-16 shows typical ideal recommended service clearance distances; note that 36 inches (91.4 centimeters) between the device and any wall is ideal. A clearance of less than this distance may impede service. Figure 3-17 shows the minimum distances required. This information should be requested from the processor manufacturer while still in the planning stage to ensure that adequate distances are maintained.

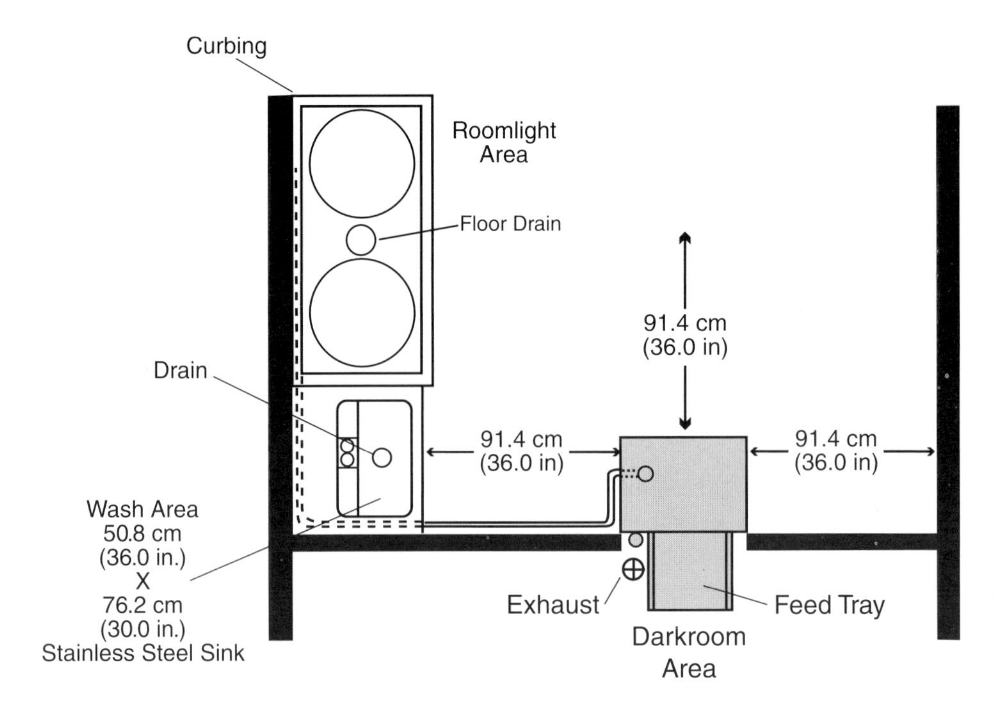


**Figure 3-15** Typical floor plan for a darkroom and work area for a single automatic film processor. (Reprinted with permission from Eastman Kodak Company.)

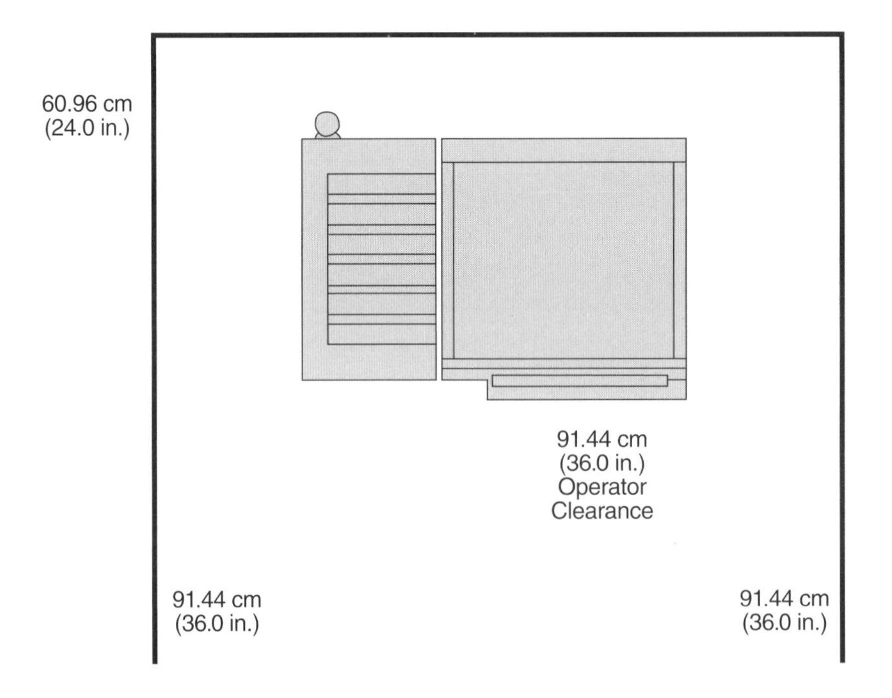


**Figure 3-16** Ideal recommended service clearance distances. (Reprinted with permission from Eastman Kodak Company.)

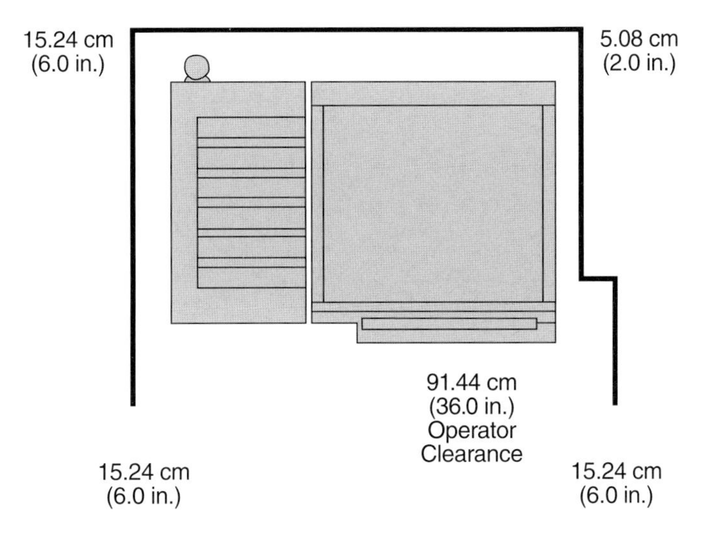


**Figure 3-17** Minimum clearance distances. (Reprinted with permission from Eastman Kodak Company.)

## *Dryer Exhaust*

The purpose of the dryer exhaust is to help avoid corrosion to the sensitive electronic components of the processor and any equipment docked to the processor, as well as to remove moisture from the dryer. The air exhaust hose should have as smooth a surface as possible, and each processor should have its own vent. Acceptable duct materials include flexible, thin-walled pipes made of schedule 40 or 20 PVC plastic, aluminum, or galvanized steel.

An adjustable air gap should be used inline with the dryer exhaust and should not exceed two inches. This helps prevent positive static pressure from entering back into the processor.

The dryer exhaust should be connected to the building exhaust system. For best processing results, the building exhaust system should be operational 24 hours per day/7 days per week. The disposal of effluent air must comply with prevailing environmental codes.

## *Processor Ventilation*

Negative static pressure of the building exhaust system should be measured and should comply with the recommendations of the processor manufacturer. Table 3-6 shows typical recommendations for two different diameter size exhaust ducts, 3 inches (76 millimeters) or 4 inches (102 millimeters). Negative static pressure should be measured by a qualified service engineer with an air-flow meter or equivalent tool. Note that the power to the processor should be off and the exhaust should be disconnected from the processor when measuring negative static pressure; the state of the building exhaust system is what is of interest. Measurements are taken by inserting the "L"-shaped metal tube provided with the meter through a small hole

**Table 3-6** Duct Diameter and Negative Static Pressure

| *Duct Diameter* | *Negative Static Pressure, Water Head* | |
|---|---|---|
| | Minimum | Maximum |
| 3 inches (76 mm) | 0.03 inch (0.76 mm) | 0.04 inch (1.02 mm) |
| 4 inches (102 mm) | 0.01 inch (0.25 mm) | 0.02 inch (0.51 mm) |

(Reprinted with permission from Eastman Kodak Company.)

created in the duct material so that the tube tip opening is pointed in the direction of airflow away from the processor.

Negative static measurements should be taken during normal operating hours. Negative airflow should remain constant when the power to the processor is on and the processor is in either the "run" or "standby" modes, and also when the power to the processor has been turned off ("shutdown" mode). It is advantageous to repeat measurements periodically throughout the year to ensure proper processor operation, to prolong the life of processor components, and to minimize processing artifacts. Additionally, they should be repeated after any changes in venting, such as adjustments to the size of the air gap. If ventilation is found to be inadequate or marginal, auxiliary ventilation fan kits are available.

### *Darkroom Ventilation and Air Exchanges*

It is recommended that there be a minimum of 10 complete changes of air per hour inside the darkroom and two per hour outside the darkroom. Air exchanges are necessary for high-quality processing and safe handling and storage of film and chemicals. Clean fresh air also provides a better environment for personnel.

All incoming air should be clean or filtered. This is especially important for darkrooms used for processing mammography film. If dust is a problem, consider installing an air filtering unit into the air circulation system to clean the incoming air. Electrostatic air cleaning systems can also be added to improve the air quality. Filters and air filtering equipment should be serviced at the intervals recommended by the manufacturer.

Appendix D contains information on the features of a mammographic darkroom and the practices designed to help maximize its cleanliness.

Darkroom air pressure must be greater than the surrounding room air pressure. This helps ensure that air is moving in the direction of film travel across the processor's film feed tray, not against it. It also helps prevent unfiltered air from entering the darkroom. To relieve changes in air pressure caused by opening and closing doors and to provide ventilation, a lightproof vent should be installed in the darkroom door or in a wall if a light lock configuration entrance is used.

### *Environmental Conditions*

Temperature for film use and storage should be between 50 to 70°F (10 to 21°C). Chemicals should be used and stored between 40 and 85°F (5 and 30°C). Relative

**Figure 3-18** A floating lid should always be used on top of the developer replenisher to reduce evaporation and chemical oxidation. (Reprinted with permission from Eastman Kodak Company.)

humidity should be between 30 and 50 percent. These values are compatible with photographic work and provide comfortable working conditions for most personnel.

Temperature and relative humidity specifications for processing and other equipment have a much wider range than those required for film and chemicals. The recommended temperature range for proper processor operation is 59 to 86°F (15 to 30°C); relative humidity should be maintained between 15 and 76 percent. Automated film handling equipment may have different temperature and relative humidity specifications; these can be found in the site specifications documents provided by the equipment manufacturer.

Dust covers and floating lids should always be used; they will minimize the evaporation and oxidation of chemicals (Figure 3-18). A stainless steel catch pan should sit directly under chemical storage tanks and processing equipment to prevent excessive damage from solution leaks or spills.

### *Replenishment Lines*

Each processor should have its own dedicated replenishment lines. Using lines with "T" or "Y" fittings increases the risk of replenishment starvation. Not all replenishment pumps are alike. If two pumps are activated at the same time on processors using common lines, under-replenishment could occur.

Heavy-walled tubing of 3/8-inch (0.95-centimeter) inside diameter is recommended to help reduce the possibility of kinked lines. Kinked lines could also result in under-replenishment.

### *Sink*

A sink made of noncorrosive material such as stainless steel, fiberglass, or inert plastic should be located conveniently to the processor. The sink should be large enough to accommodate all processor components and facilitate their cleaning. The daily cleaning of the developer-to-fix and fix-to-wash crossover assemblies is of particular importance.

## Service Requirements

### *Electrical*

Voltage varies with each processor; therefore, each processor must have a dedicated line. Table 3-7 shows the typical voltage requirements worldwide.

### *Water*

Characteristics of water used in processors:

*Temperature:* 40 to 85°F (4 to 30°C). An optional mixing valve can be used that has the ability to mix cold and hot water so a constant temperature is maintained.

*Pressure:* 138 to 448 KPa (20 to 65 PSI). Two pressure gauges may be placed inline, if desired; one goes before the water filter and one immediately after it. With this arrangement, line pressure may be monitored along with the pressure exiting the filter, indicating when the filter must be cleaned or changed. If only one gauge is used, it should be placed between the processor and filter.

*Filtration:* 50-micron filter suggested.

*Check Valve:* New processors have a water gap of approximately 0.8 inches (20 millimeters). Local codes should be checked and followed. It may be necessary to install a water gap with some older model processors.

*Flow:* Flow control valves are used to vary the water flow between approximately one quart (0.946 liter) to six quarts (5.676 liters) per minute. It is essential to ensure adequate water flow to the processor. If the flow rate is too low, improper washing of the films may occur.

Many processors now use garden-hose-type connections. These connections are easy to obtain and install; they may be standard equipment with many processors.

### *Drain*

Brass, copper, and galvanized steel should not be used for processor drains because these materials can corrode and deteriorate when exposed to processing chemicals. Noncorrosive material, PVC or equivalent, is recommended. Materials used for drains must comply with local codes.

**Table 3-7** Typical Voltage Requirements Worldwide

| | |
|---|---|
| United States | 120 V AC |
| | 208 V AC |
| | 240 V AC |
| Europe | 220 V AC |
| | 230 V AC |
| | 240 V AC |
| Japan | 100 V AC |
| | 200 V AC |
| Frequency | 50/60 Hz |
| Current | Varies from 15 amperes to 35 amperes |
| Connection | Three or four wire, plus earth ground |
| Phase | Single or three phase |

Solid connections between hoses and the drain pipe should not be made. The drain should be located as close to the processor as possible, not exceeding 60 inches (1.5 meters).

Drain capacity may vary from processor to processor. The maximum for modern processors is four gallons (15.14 liters) per minute of any kind of solution—water, developer, or fixer. Use, at minimum, three-inch pipe with proper pitch and free of obstructions.

## Processing Cycle Time

Processing cycle time is usually defined as the time it takes for the leading edge of the film to enter and exit the processor, or the leading edge of the film to enter and the trailing edge of the film to exit the processor. The latter definition will be used in this book. This is also referred to as "drop time."

Processing cycles on processor models available today range from approximately 30 to 210 seconds. The so-called conventional, or standard, processing cycle for general-purpose processing in medical imaging is between 90 and 150 seconds (Table 3-8). Many automatic processors operate at 90 seconds for standard-cycle processing. Developer temperature and replenishment rates are determined by the processing cycle and set by film and chemical manufacturers in order to achieve the desired sensitometric characteristics (speed, contrast, and base plus fog) for the type of film being processed.

Film processors with shorter processing cycles of approximately 30–45 seconds have quicker access time and greater per-hour film volume capacity. They will generally be referred to in this text as "very rapid." They may prove to be especially useful for angiography, special procedures, trauma, emergency rooms, and surgery applications where fast access time is important. Film processors with very rapid

Table 3-8 Film Processing Recommendations for Kodak Medical X-Ray Films

| | Kodak X-Omat Processors | | | | | |
|---|---|---|---|---|---|---|
| *Processing Cycle* | *M35 M35A[2]* | *M7 M7A M7B[3] (2 minute)* | *M6A-N M6AW M6B[3]* | *M8[3]* | *270 RA 3000 RA[3]* | *480 RA 5000 RA[3]* |
| Standard Processing Cycle[1] (seconds) | 150 | 120 | 90 | 90 | 111 | 95 |
| Kwik/RA Cycle[4] (seconds) | - | - | - | - | 62 | 45 |
| Development Time (seconds) | 33 | 27 | 25 | 21 | 27<br>14.5 | 24 standard<br>11.0 Kwik/RA |
| Developer Temperature | 92°F (33.5°C) | 94°F (34.4°C) | 95°F (35°C) | 96°F (35.6°C) | 94°F standard (34.4°C)<br>96°F Kwik/RA (35.6°C) | 95°F standard (35°C)<br>98°F Kwik/RA (36.5°C) |

[1]Kodak RP X-Omat chemicals.
[2]Based on 43 centimeters of film travel, leading edge entering the processor to the trailing edge exiting the processor.
[3]Based on 35 centimeters of film travel, leading edge entering the processor to the trailing edge exiting the processor.
[4]Kodak X-Omat RA/30 chemicals recommended.
(Reprinted with permission from Eastman Kodak Company.)

processing cycles must be used together with specially designed film and chemicals to achieve sensitometric characteristics (speed, contrast, and base plus fog) that are comparable to the sensitometric characteristics obtained with standard processing cycles, film, and chemicals.

For some single-emulsion films, specifically mammographic film, film contrast and speed are increased by extended-cycle processing without significantly affecting film fog (Figure 3-19). This method increases the length of time the film remains in the developer while keeping the developer temperature essentially unchanged (Table 3-9). Extended-cycle processing is not recommended for double-emulsion film because film contrast and speed are not altered significantly.

Close inspection of Table 3-9 reveals that the processing time (drop time) for extended-cycle processing is always longer than the time for standard processing for the same model processor. Note, however, that by design, some processors with cycle times in excess of 90 seconds deliver "standard cycle" performance and not "extended-cycle" performance.

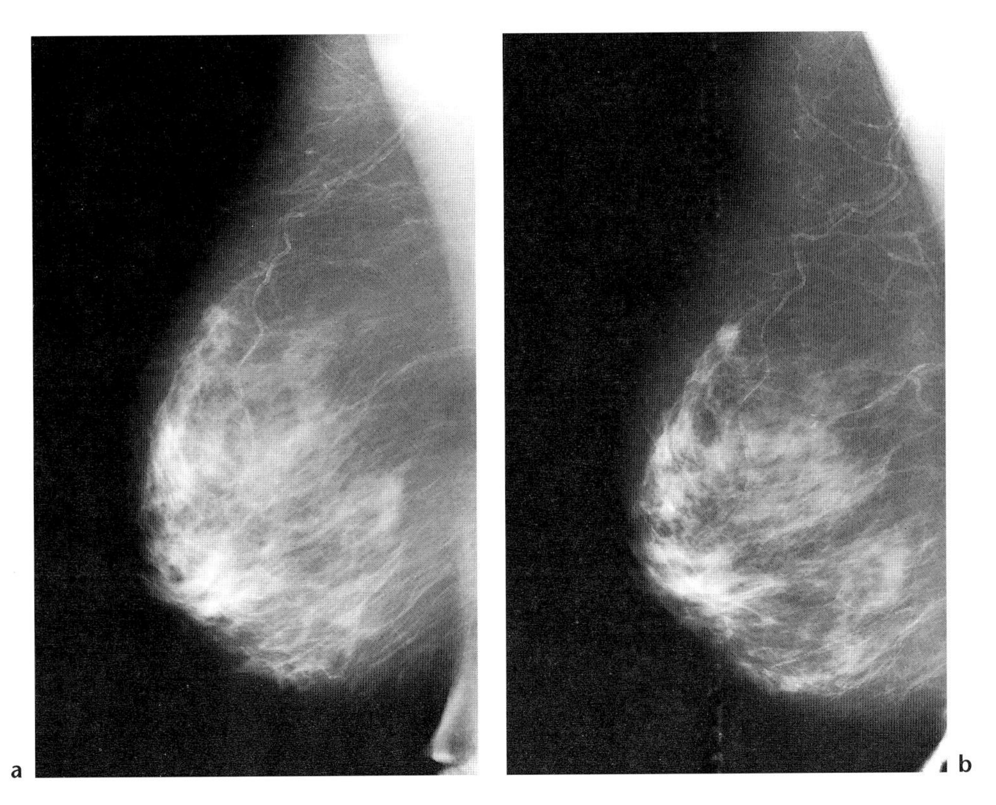


**Figure 3-19** Mediolateral oblique projection mammograms taken of the same woman using the same type of film. (a) Processed in a standard cycle. (b) Processed in an extended cycle. The image processed in the extended cycle is higher in contrast, allowing better visualization of details. (Reprinted with permission from László Tabár, M.D.)

Extended-cycle processing is available on multicycle-capable processors. Older model processors may be converted to extended-cycle processing by the installation of kits designed for that purpose.

Whether using standard, extended, or very rapid processing cycles, be sure to consult and adhere to the film manufacturer's recommendations for time and temperature in order to achieve optimum processing.

## Solution Temperatures

Three solutions—developer, fixer, and water—are used in automatic film processors. Each of these solutions has specific temperature requirements.

### Developer Temperature

Developer temperatures in automatic film processors range from 92 to 101°F (33.5 to 38.3°C) for all film types. The specific developer temperature depends on film type, processing cycle, and the manufacturer's recommendations. Figure 3-20 illustrates

Table 3-9 Processing Conditions for Kodak Mammography Films[1]

| *Film and Processing Cycle* | *Kodak X-Omat Processors* | | | | | |
|---|---|---|---|---|---|---|
| | *M35 M35A M35A-M M20* | *M6B M6-N M6A-N M6AW* | *M7 M7A M7B* | *M8* | *270 RA[2] 3000 RA* | *480 RA[3] 5000 RA* |
| Min-R H, Min-R M, or Min-R 2000, Standard Processing Cycle | | | | | | |
| Processing Time(s)[4] | 140 | 90 | 122[6] | 90 | 100 | 88 |
| Developer Time(s)[5] | 32 | 24 | 27 | 21.5 | 26.6 | 23.8 |
| Developer Temperature | 92°F (33.5°C) | 95°F (35°C) | 94°F (34.4°C) | 96°F (35.6°C) | 94°F (34.4°C) | 95°F (35°C) |
| Min-R E, Extended Processing Cycle | | | | | | |
| Processing Time(s)[4] | 203 | 172 | 188 | NA | 209 | 172 |
| Developer Time(s)[5] | 49.5 | 47 | 43 | | 52 | 47 |
| Developer Temperature | 95°F (35°C) | 95°F (35°C) | 96°F (35.6°C) | | 94°F (34.4°C) | 95°F (35°C) |

[1]Kodak RP X-Omat chemicals recommended.

[2]Listed processing times are associated with a front film exit. Top film exit processing times are 214 seconds for extended cycle processing and 107 seconds for standard cycle processing.

[3]Some variation may occur in the listed times due to adjustable turnarounds.

[4]Processing time is based on 24 centimeters of film travel, leading edge entering the processor to trailing edge exiting the processor.

[5]Developer time is based on the leading edge of the film into the developer to the leading edge of the film into the fixer.

[6]Some M7 and M7A processors may have a longer standard processing time (140 seconds rather than 122 seconds). The temperature recommended for standard processing of all Kodak mammography films in this case is 92°F (33.5°C).

NA - not applicable.

(Reprinted with permission from Eastman Kodak Company.)

the effect of developer temperature on sensitometric characteristics. Similar variations in results are to be expected with changes in development time.

Note that when the developer temperature is lower than the manufacturer's recommendation, film speed is reduced. As a result, an unnecessary increase in radiation dose will be required to produce radiographs of comparable optical density. Film contrast is also reduced when developer temperature is lowered. Conversely, if the developer temperature is higher or the development time longer (extended-cycle processing) than the film manufacturer's recommendation, film speed increases. This may permit a reduction in radiation dose and result in an increase in film contrast.

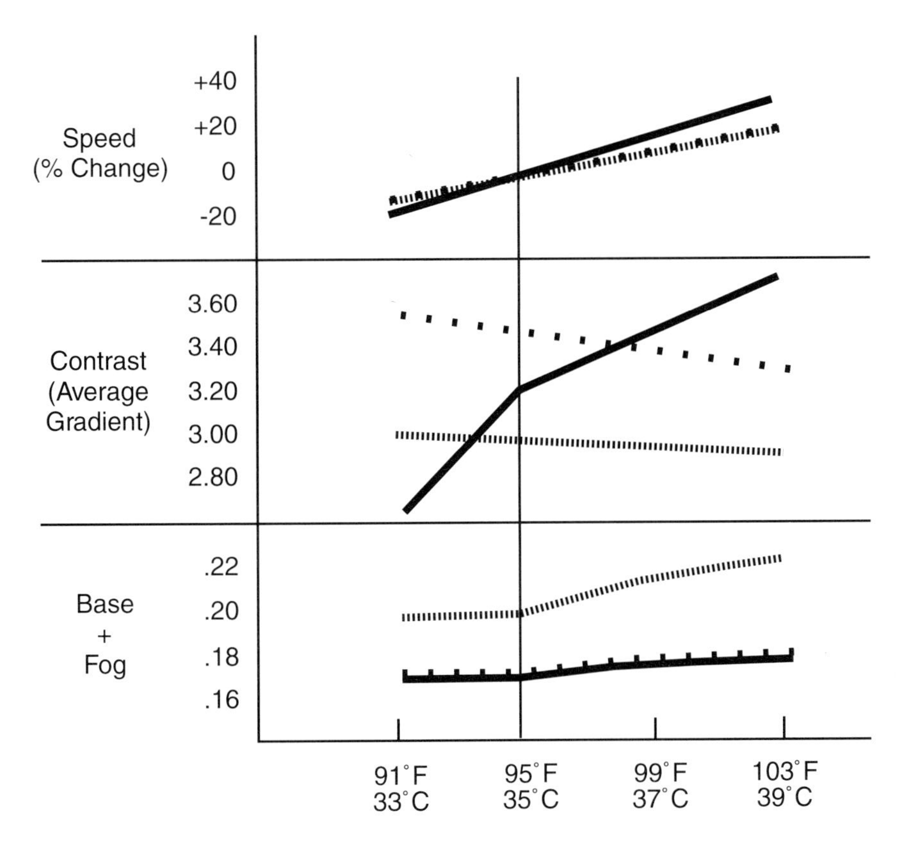


**Figure 3-20** Effect of developer temperature on film speed, contrast, and base plus fog for single-emulsion mammography film (3D grains: solid line; cubic grains: dotted line) and double-emulsion film (tabular grains: broken line). The processor and chemicals recommended by the film manufacturer were used. The vertical line represents the recommendations for a standard processing cycle. (Reprinted with permission from Eastman Kodak Company. EKC publication no. N-314.)

However, these changes can usually be expected to increase radiographic noise, which is discussed in Chapter four. Increased developer temperature may also increase film fog. Finally, use of developer temperatures higher than those recommended by the manufacturer may result in developer instability (increased rate of oxidation) and increased susceptibility to processing artifacts.

The National Council on Radiation Protection and Measurement (NCRP) in NCRP Report No. 99 and the American College of Radiology (ACR) in *Mammography Quality Control* indicate that the developer temperature should be within ±0.5°F (±0.3°C) of that recommended by the manufacturer for the specific film-developer combination being used. The accuracy, precision, and repeatability of measurements with the thermometer are most important. The thermometer used to measure developer temperature should have an accuracy at least equal to, or better than, the variability recommended in the NCRP and ACR documents.

In radiology or medical imaging departments, a variety of thermometers are used to measure developer temperature. These thermometers vary in accuracy, precision, ease of reading, and cost. Clinical digital thermometers, generally available in pharmacies and supermarkets, are inexpensive but accurate devices for measuring temperature of the developer solution. It is also recommended that the thermometers used to measure developer temperature be evaluated against a thermometer that has a calibration traceable to the National Institute of Standards and Technology.

## Fixer Temperature

Fixer temperature is not as critical as developer temperature. Maintenance of the proper developer and water temperatures should ensure acceptable fixer temperature. The fixer temperature generally varies between 85 and 95°F (29.4 and 35°C). Fixer at higher temperatures only improves fixing and causes no problems. Cold fixer may cause decreased fixer activity, retained hypo on film, and failure to pass a fixer (hypo) retention test. Long-term image stability will also be affected.

## Water Temperature

Water temperature should be maintained between 40 and 85°F (4.5 and 29.5°C) for most processors. Extremes in water temperature can cause biological growth, improper developer and/or fixer temperatures, improper washing and fixing, artifacts, and transport problems.

Water that is too hot or too cold may require the use of chillers or heaters for proper processor function. Water warmer than the recommended range may cause increased developer temperature and sensitometric changes.

## General Thermometer Information

A clinical digital thermometer, as discussed above, is an excellent choice for measuring developer temperature (Figure 3-21). Such thermometers have a reading range of approximately 90 to 108°F (32 to 42.2°C). Note that these thermometers will not register temperatures colder than the lower limit; a thermometer with a wider temperature range may be needed to check fixer and water temperatures.

Thermometers that contain mercury should never be used. Also, thermometers made of glass, whether they contain alcohol or mercury, should never be used due to the possibility of breakage. Digital thermometers, which provide a reading to the nearest tenth of a degree, are preferable to nondigital (scale) thermometers, which require an estimation of temperature.

For the purposes of the daily processor quality control procedure, the probe of the thermometer (with known accuracy) should be placed in the same location of the developer tank each time, according to the recommendations of the processor manufacturer. This is usually on the nondrive side of the processor between the side plate of the developer rack and the rack support. As developer temperature readings may vary by as much as 1°F (0.7°C) if the thermometer probe is placed anywhere in the developer tank, the probe should always be placed in a consistent location. A consistent location for measuring fixer and water temperatures should also be chosen.

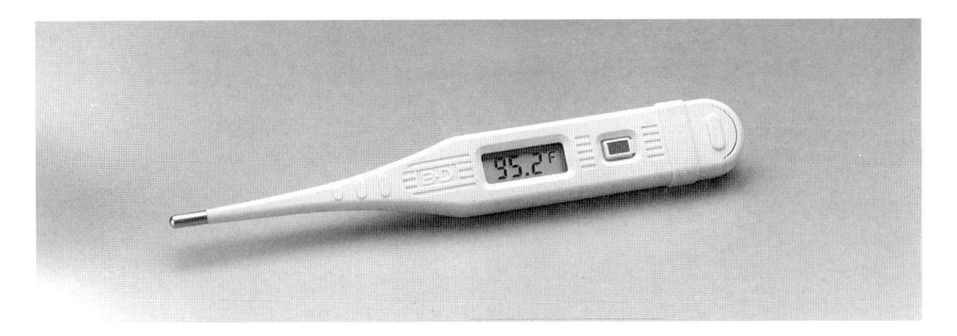


**Figure 3-21** Clinical digital thermometers are inexpensive but accurate devices for measuring the temperature of the developer solution.

The probe should be rinsed with warm water and dried thoroughly before recapping and storing to prevent corrosion and to prolong the useful life of the thermometer.

# Preventive Maintenance and Cleaning

Automatic film processors are complex machines, combining chemical, mechanical, thermal, electrical, and often microcomputer-controlled systems. To maintain optimum processor performance, all of these systems must be operating within tolerances specified by the processor manufacturer. Even deviations that seem insignificant, such as a small nonuniformity on a roller, can affect image quality.

In order to maintain processor performance, the schedule of periodic preventive maintenance and cleaning recommended by the manufacturer must be followed. Lack of proper maintenance can cause component failure and processing artifacts. Only qualified individuals who have been thoroughly trained in correct procedures for cleaning, testing, and repair should perform processor maintenance procedures; facility personnel should also be trained for those procedures for which they have responsibility.

Processor manuals generally specify daily, weekly, monthly, and periodic maintenance and cleaning procedures as well as other important procedures such as startup and shutdown.

Immediate action should be taken at the advent of sudden changes in film quality.

## Daily Maintenance and Cleaning Procedures

The daily maintenance and cleaning procedure should be done by facility personnel at the end of the day along with the shutdown procedure. Cleaning the crossover assemblies on a daily basis is particularly important to reduce processing artifacts, especially on mammographic films (Figure 3-22).

Many facilities fail to perform the daily maintenance and cleaning procedure, or perform it haphazardly or incorrectly.

The steps in the daily maintenance and cleaning procedure are:

1. Turn off the main power switch. (The main power to the processor should always be turned off before servicing the unit.)

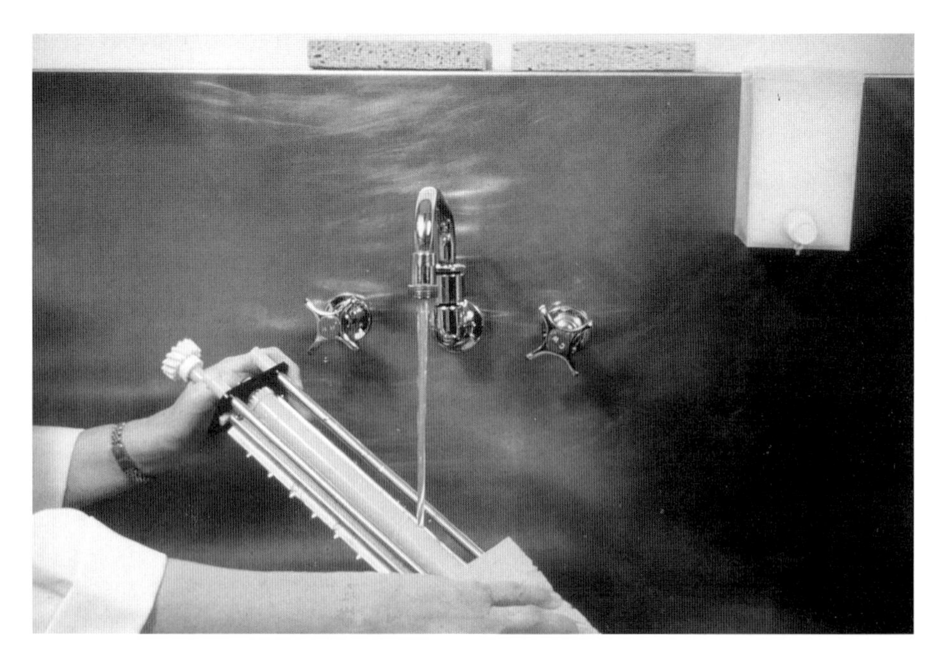

**Figure 3-22** Cleaning the crossover assemblies (developer-to-fix and fix-to-wash) on a daily basis is particularly important to reduce processing artifacts, especially on mammographic films. (Reprinted with permission from Eastman Kodak Company.)

2. Disconnect the main power.
3. Remove the evaporation covers and the developer-to-fix and fix-to-wash crossover assemblies and take to the sink for cleaning. (Leave the entrance detector crossover assembly in place; it may be wiped occasionally with a barely damp nonabrasive cloth but should not be rinsed under running water.)
4. Rinse the evaporation covers.
5. Clean the crossover assemblies, using care not to twist them out of square or disturb the guide shoes. (Damaged guide shoes can cause artifacts.)
6. Rinse the crossover assemblies in warm water using a damp cloth or synthetic sponge. The use of different colored sponges (i.e., one for cleaning developer components, one for fixer components, and one for general cleaning around the processor and darkroom) is particularly helpful. Do not use any abrasive materials on the racks, crossover assemblies, or squeegee rollers. (The aim is only to remove dirt that is removable; attempts to remove permanent chemical stains will result in permanent roller damage.)
7. Rotate and clean the rollers.
8. Wipe off all chemical deposits above solution levels in the processor.
9. Clean off debris or buildup from the dryer area with a warm, damp sponge.
10. Reinstall the crossover assemblies and evaporation covers in the processor. The crossover assemblies and evaporation covers must be returned to their respective locations in order to avoid causing processing artifacts. Some processor manufacturers use a washer located on the drive side of each assembly to identify the developer-to-fix (D/F) and fix-to-wash (F/W) crossover

assemblies (Figure 3-23). A permanent black ink pen may be used to designate the particular location for each evaporation cover.

11. Install the top cover.
12. Keep the top open overnight for venting purposes. The evaporation covers may be used to position the cover so that it remains ajar (Figure 3-24).

## Weekly, Monthly, and Periodic Maintenance and Cleaning Procedures

Weekly, monthly, and periodic maintenance and cleaning procedures generally involve the removal of all racks from the processor, inspection of all components for proper alignment and function, changing the developer filter, changing the chemicals, etc. All manufacturers of processors provide product-specific publications with the details of these steps. These procedures are usually performed by the processor service company.

Two precautions should be taken whenever racks are removed from a processor containing developer and fixer solutions: (1) use a rack drip tray when removing or replacing racks, and (2) use splash guards between the developer and fixer tanks to prevent contamination of the developer.

In all cases, the main power to the processor should always be turned off before performing any maintenance or cleaning procedures; the manufacturer's recommendations should be closely followed; all rollers and other components should be closely

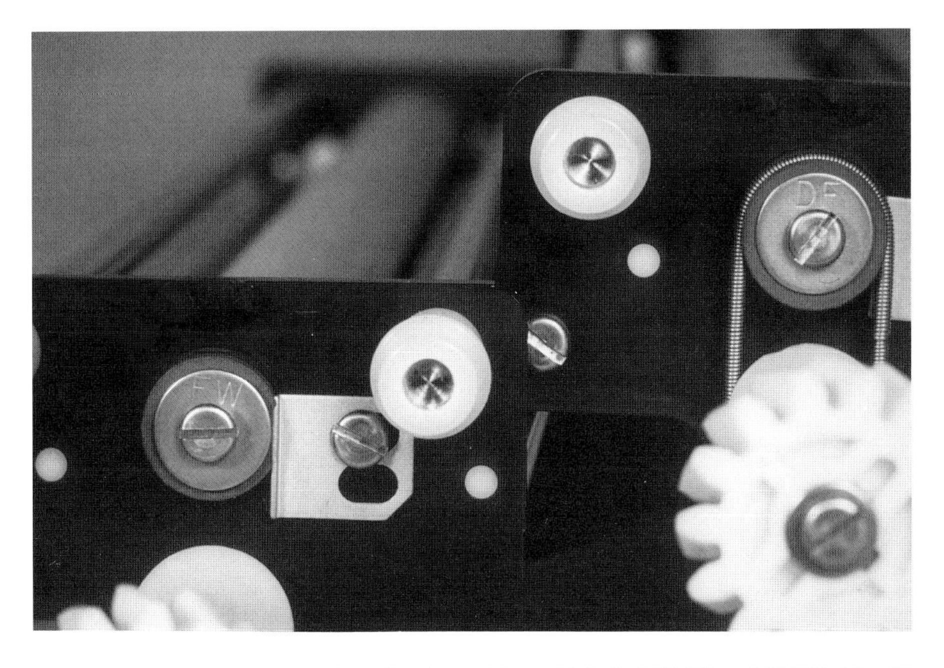


**Figure 3-23** A washer located on the drive side and labeled "D/F" and "F/W" identifies the developer-to-fix and fix-to-wash crossover assemblies, respectively; component identification should be checked to ensure correct replacement in the processor after cleaning or service. (Reprinted with permission from Eastman Kodak Company.)

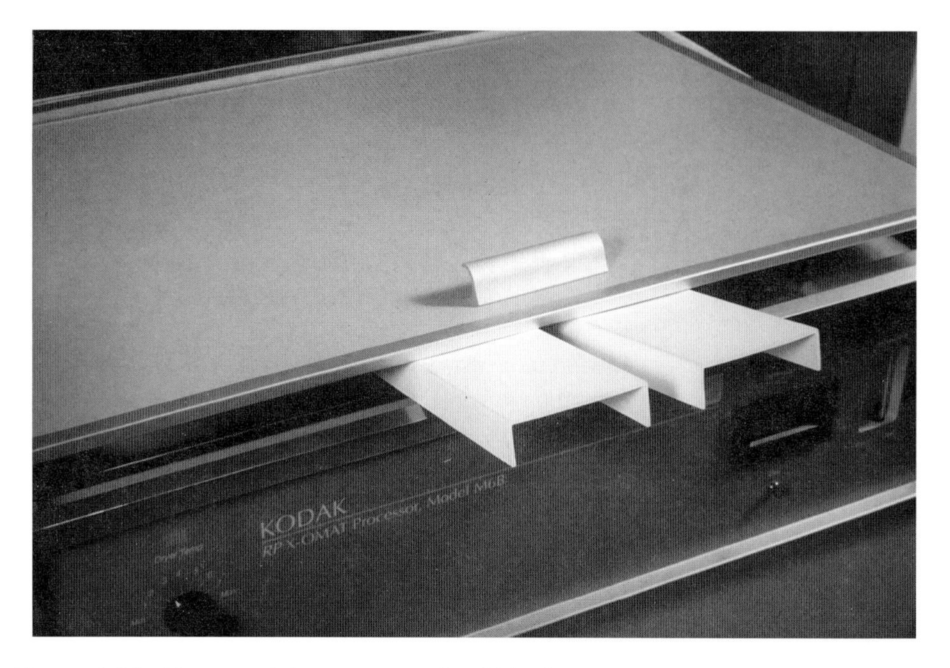


**Figure 3-24** The top of the processor should be kept open for venting purposes. The evaporation covers may be used to position the cover so it remains ajar.

inspected for the first visible signs of wear so they may be changed before affecting image quality; and replacement parts should be those specified by the manufacturer to minimize artifacts and maintain processor reliability.

## *System Cleaner*

System cleaner, sold as separate chemicals for cleaning the developer and fixer tanks and racks, may be used periodically (i.e., at least every three months, or more frequently, if necessary). Whenever system cleaner is used, instructions for use should be followed carefully and appropriate personal protective equipment (Chapter Two) (e.g., rubber gloves and goggles or safety glasses with side shields) should be used. Use with adequate ventilation. Wash hands thoroughly after handling.

There are two types of developer system cleaners available, and their use differs as follows:

1. Developer system cleaner, containing sodium dichronate, should be used sparingly and only when absolutely necessary. Developer racks should never be submerged in this chemical. After removing a developer rack from the tank for cleaning, flush the rack with warm water, and clean the rack with a non-abrasive sponge and a small amount of developer system cleaner poured onto the sponge. Avoid any spraying of developer system cleaner containing sodium dichronate. The used cleaning solution must be disposed of according to local environmental regulations (it may not be poured down the drain).
2. Developer system cleaner (with neutralizer), containing potassium permanganate, may also be used to clean the developer tank and rack. The developer

rack may be completely submerged. Check with local authorities on the disposal of the used cleaning solution.

In addition, system cleaner for the fixer and wash systems may be used according to the manufacturer's instructions.

Material Safety Data Sheets (MSDSs) should be obtained for all types of system cleaners used and read before using. Also note that attempts to remove permanent discolorations on processor rollers from developer and fixer will result in roller damage.

## Startup Procedure

The purpose of the startup procedure is to make sure everything in the processor is operating correctly. As the startup procedure is generally performed by facility personnel, the steps are described below:

1. Turn on the water supply.
2. Turn on the main power switch on the processor.
3. Remove the top cover.
4. Check that the developer rack, fixer rack, wash rack, fix-to-wash crossover assembly, developer-to-fix crossover assembly, entrance detector crossover assembly, and dryer assembly are seated correctly.
5. Check that the developer, fixer, and wash solutions are at the overflow level of the weirs.
6. Check that water is flowing from the water inlet tube of the wash tank.
7. Check that the evaporation covers are installed.
8. Install the top cover.
9. Press the run/standby switch.
10. Feed one or two sheets of cleanup film into the processor when the digital temperature display indicates the correct developer temperature for the processing mode selected has been reached.
11. Set the dryer temperature to the lowest setting to dry the film. (The correct temperature for the dryer depends on the condition of the air and on the amount and type of film fed into the processor; processors frequently are set with too hot a dryer temperature, resulting in artifacts.)
12. Perform daily processor quality control.

## Shutdown Procedure

Shut the processor down if it will not be used overnight. It is not desirable to leave a processor on without processing films as chemical oxidation will occur, negatively impacting processor quality control results.

Follow these steps:

1. Turn off the main power switch and water supply.
2. Disconnect the main power.
3. Remove the top cover.
4. Remove the evaporation covers.
5. Perform the daily maintenance and cleanup procedure.
6. Install the top cover with a 2-inch (5 centimeter) opening.

## References


1. American College of Radiology (ACR). Mammography quality control manual for radiologists, medical physicists and technologists. Reston, VA: American College of Radiology, 1994.
2. American College of Radiology (ACR). Recommended specifications for new mammography equipment: Screen-film x-ray systems, image receptors, and film processors. Reston, VA: American College of Radiology, 1995.
3. Dryer venting requirements for all KODAK X-OMAT processors, Service Bulletin no. 101, Eastman Kodak Company publication no. N-763, 1990.
4. Haus AG. Film processing systems and quality control. In: Gould RG ,Boone JM, eds. A categorical course in physics: Technology update and quality improvement of diagnostic x-ray equipment. Oak Brook, IL: Radiological Society of North America, 1996: 49–66.
5. Haus AG. Historical developments in film processing in medical imaging. In: Haus AG, ed. Film processing in medical imaging. Madison, WI: Medical Physics Publishing, 1993: 1–16.
6. Kimme-Smith K, Rothchild PA, Bassett LW, Gould RH, Moler C. Mammographic film processor developer temperature, development time, and chemistry: Effect on dose, contrast and noise. AJR 152: 35–40, 1989.
7. KODAK auxiliary ventilation fan kit/110V, Service Bulletin no. 158, Eastman Kodak Company publication no. N-768, 1993.
8. National Council on Radiation Protection and Measurements (NCRP). Report 99. Quality assurance for diagnostic imaging equipment. Bethesda, MD: National Council on Radiation Protection and Measurements, 1988.
9. Oemcke KW. Design basics and component functions of automatic film processors. In: Haus AG, ed. Film processing in medical imaging. Madison, WI: Medical Physics Publishing, 1993: 79–88.
10. Schmidt RA, Doi K, Sekiya M, Xu XW, Giger ML, Lu CT, Mojtahedi S, MacMahon H. Evaluation of radiographs developed by a new ultra-rapid film processing system. AJR 154: 1107–1110, 1990.
11. Suleiman OH, Slayton RJ, Conway BJ, Rueter FG. Effects of temperature, chemistry, and immersion time on x-ray film. Radiology 177: 132, 1990.
12. Tabár L, Haus AG. Processing mammographic films: Technical and clinical considerations. Radiology 173: 65–69, 1989.
13. Wilson WB, Haus AG, Nierman C, Lillie R, Batz BA. Evaluation of a clinical thermometer for measuring developer temperature in automatic film processors. Medical Physics 20: 823–824, 1993.

# 4

# IMAGE QUALITY

- Background
- Exposure and Density
- The Characteristic Curve
- Film Speed
- Radiographic Contrast
- Latitude
- Image Blur (Unsharpness)
- Radiographic Noise
- Artifacts
- Viewing Conditions

## Background

The choice of a screen-film combination, combined with film-processing conditions, substantially affects radiographic image quality (contrast, blur, and noise) and radiation dose. Film type (single- or double-emulsion, silver halide content, grain morphology, and spectral sensitivity), processing conditions (chemicals, temperature, time, and agitation), fog level (storage, safelight, light leaks), and optical density level affect film contrast. Film contrast characteristics (gradient) determine how the x-ray intensity pattern will be related to the optical density pattern in the radiograph. The type of screen (phosphor layer thickness, light-absorbing dyes and pigments, phosphor particle size), speed of the screen-film processing system (sensitivity), film granularity, screen uniformity, and film contrast affect radiographic noise. Figure 4-1 illustrates the importance of film processing and, in particular, film contrast on image quality.

Image quality in screen-film radiography is affected by numerous technical variables. As outlined in Table 4-1, image quality refers to the aggregate effect of these elements on the appearance of the radiographic image. The radiographic image is the result of the trade-offs among the many components involved. Some factors can be controlled and even optimized so that radiographs with good image quality can be obtained at a low radiation dose to the patient.

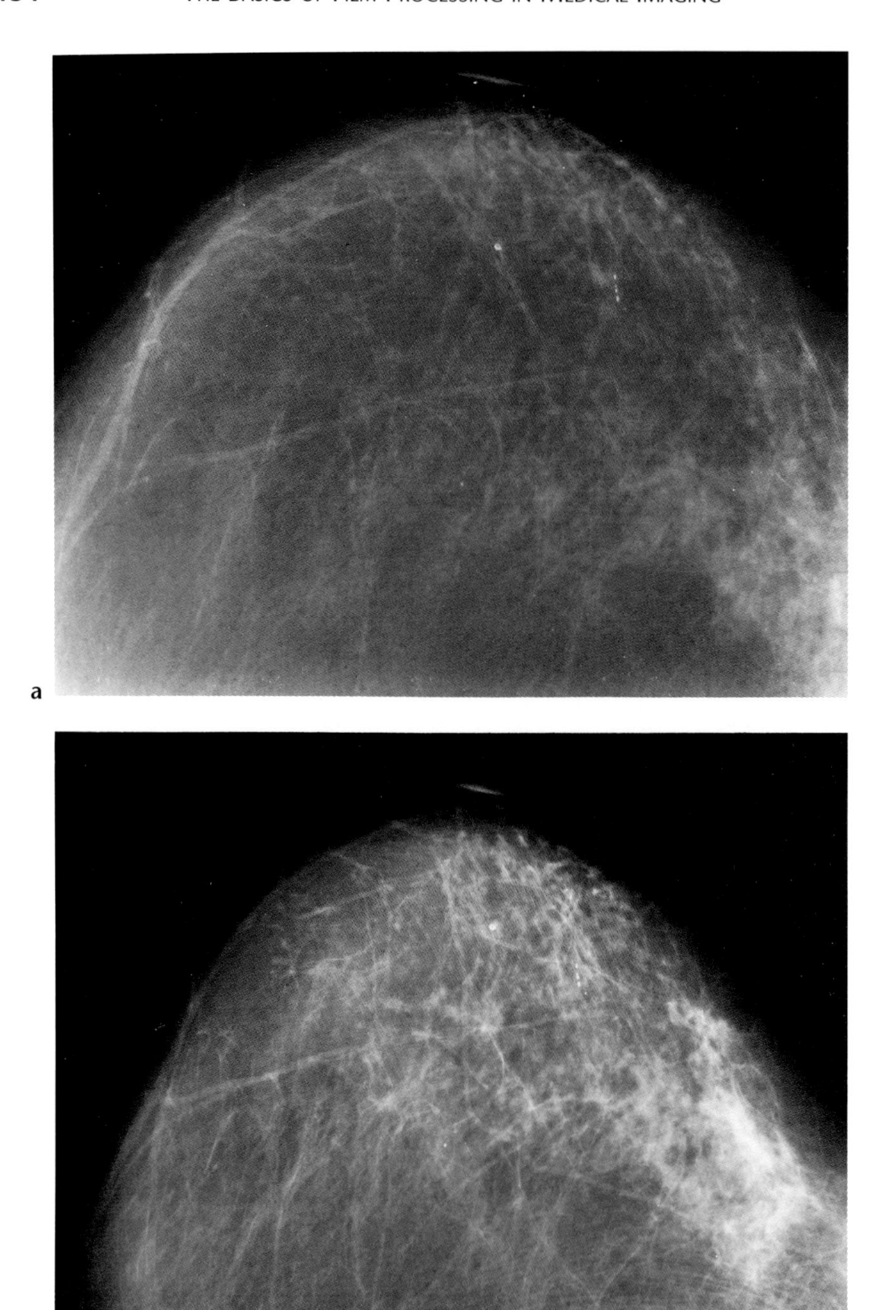


**Figure 4-1** Mammograms demonstrating the importance of film processing and contrast. In this comparison of the same breast, all factors are similar, including film type, except that the mammogram in (a) was processed less-than-optimally by today's standards. The mammogram in (b) was processed in a dedicated processor that was properly optimized according to the manufacturer's recommendations. (Images courtesy of Wende Logan Young, M.D.)

**Table 4-1** Factors Affecting Mammographic Image Quality

| *RADIOGRAPHIC SHARPNESS* | | *RADIOGRAPHIC NOISE* | |
|---|---|---|---|
| *Radiographic Contrast* | *Radiographic Blur (Unsharpness)* | *Radiographic Mottle* | *Radiographic Artifacts* |
| **Subject Contrast** | **Motion Blur** | **Receptor Graininess** | **Handling** |
| Absorption differences | Breast immobilization (compression) | | Crimp marks |
| Thickness | Exposure time | **Quantum Mottle** | Finger marks |
| Density | | Film speed | Scratches |
| Atomic number | **Geometric Blur** | Film contrast | Static |
| Radiation quality | Focal spot size | Screen absorption | Exposure |
| Target material (e.g., Mo, W, Rh)* | Focal spot-object distance | Screen conversion efficiency | Fog |
| Kilovoltage (kVp) | Object-image receptor distance | Light diffusion | **Processing** |
| Filtration (e.g., Mo, Rh)* | | Radiation quality | Streaks |
| Scatter radiation | **Screen-Film Blur** | | Spots |
| Beam collimation | Phosphor thickness | **Structure Mottle** | Scratches |
| Compression | Light-absorbing dyes and pigments | | Dirt |
| Air gap | Phosphor particle size | | Stains |
| Grid | Screen-film contact | | |
| **Film Contrast** | | | |
| Film type | | | |
| Processing | | | |
| Chemistry | | | |
| Temperature | | | |
| Time | | | |
| Agitation | | | |
| Photographic density | | | |
| Fog | | | |
| Storage | | | |
| Safelight | | | |
| Light leaks | | | |

*MO-Molybdenum, W-Tungsten, Rh-Rhodium.

For this discussion, radiographic image quality will be limited to measurable factors in the imaging process affected by the screen-film and film-processing systems—not factors related to the interpretation of the image. There is no single "optimum" or "necessary" image quality applicable to all types of radiographic examinations. For example, the image characteristics accepted as necessary to visualize a lesion in the chest can differ from those needed to visualize a bone fracture. There is no single standard for diagnostic image quality, and there are no generally accepted objective methods for predicting clinical performance (or diagnostic image quality) based on measurements of technical factors affecting image quality.

## Exposure and Density

Efforts to quantify the relationship between exposure and the appearance of the resulting image are as old as photography itself. The first success in this search was the work of Hurter and Driffield, published in 1890. The curve resulting from a plot of optical density versus log exposure, which they proposed, continues to be useful today and is known as the characteristic curve, the sensitometric curve, the D log E curve, or the H&D curve (after the originators) (Figure 4-2). The study and measurement of these relationships are known as sensitometry.

For the evaluation of film contrasts and relative speeds of radiographic film or screen-film combinations, it is important to know the density increments resulting from known exposure increments (i.e., to describe the sensitometric properties of a given film or screen-film combination).

The term "exposure" may be used to designate the radiographic conditions—that is, kVp, mA and time, or mAs—used for a certain examination. It may also be used in the somewhat more restricted sense of the actual tube current (mA) and time (seconds), or mAs used to produce a radiograph. The term "radiation exposure" may

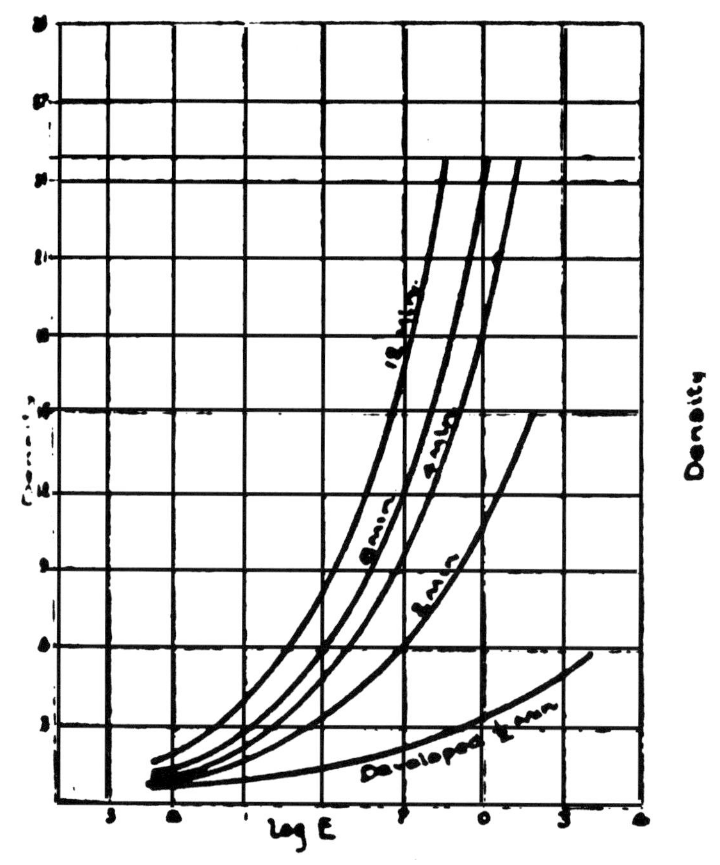


**Figure 4-2** Characteristic curves, published in 1917 by Hodgson, compared films processed at different developer temperatures. (Reproduced from Hodgson MB. The sensitometry of x-ray materials. British Journal of Photography 64: 654–657, 1917.)

refer specifically to the patient exposure in milliroentgens (mR). Exposure may also be used to refer to the amount of radiation, either light or x-ray, reaching a certain area of the film; it will be used in this last sense for the remainder of this discussion. Remember, however, that film blackening may be caused by direct x-ray exposure without the use of intensifying screens or by both x-ray and light exposure when screens are used.

Exposure, then, will be defined as the amount of x-ray or light energy reaching a particular area of the film. This energy is responsible for producing a certain optical density on the processed film.

Exposure may be specified in absolute units using ergs per square centimeter for light or x-rays. The Roentgen (R) is a measure of the exposure in air that produces a specified amount of ionization in an ion chamber. One thousand mR equals one R. In most cases, using relative exposure (ratio of absolute exposures) is more convenient and just as functional as using absolute units.

The exposure received by the film is related to the kVp, mAs, focal-film distance (FFD), and other technical factors.

Optical density is a quantitative measure of film blackening. In many instances, the term "optical density" is simply shortened to density. For example, the lung fields in a chest radiograph demonstrate a greater film-blackening effect and, therefore, have a higher (optical) density than the lighter mediastinal segment of the radiograph. It is important not to confuse this term with physical density (or mass density), which is defined as mass per unit volume. For example, mediastinal structures have greater (physical) density and absorb more x-radiation than the surrounding lungs.

The image in a radiograph is built up of innumerable tiny masses of metallic silver distributed throughout the emulsion layer(s) of the film. This image is viewed by the transmitted light from an illuminator or viewbox. The relative transparency (the ability of a portion of the radiograph to transmit light from the viewbox) of the various areas in the radiograph depends upon the distribution of the developed black silver masses. The degree to which the silver attenuates (absorbs), or interferes with, the transmission of light through a small area of a radiograph is related to the amount of silver in that area and its physical structure. It is this variation in the quantity of transmitted light that makes up the image seen by the eye.

The heavier the deposit of black silver, the greater the absorption of light, and the darker the area appears. This film blackening is defined as optical density. Regions of lower or higher optical density correspond to the amount of light transmitted by the radiograph. The higher the optical density, the lower the amount (intensity) of light transmitted. A device called a densitometer is used to measure density (Figure 4-3).

The densitometer compares the relative amounts of light entering a particular area on one side of the processed film with the amount of light emerging from the film. The measured amount of light is defined in terms of luminance, a quantitative measure of its brightness. For a small area of film, the ratio of the incident light to the transmitted light is called the opacity of the film. Frequently, film has opacities with large numerical values, such as 100, 1,000, or 10,000. To simplify mathematical operations with these large numbers, logarithms are used:

$$\text{optical density} = \log(\text{opacity}) = \log(1/\text{transmittance})$$

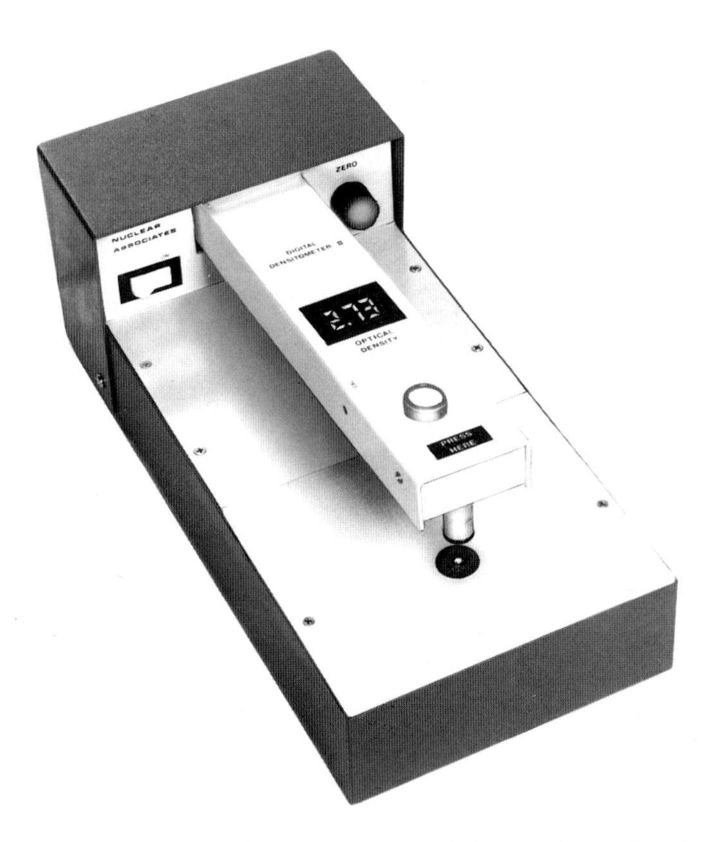


**Figure 4-3** A densitometer used to measure optical density. (Printed with permission from Nuclear Associates.)

Medical radiographs contain many different optical densities in the various areas that constitute the image. These optical densities range from about 0.30 in the relatively clear areas to more than 3.00 in the very blackest areas (Figure 4-4). In other words, an area of the film with an optical density of 0.30 contains enough silver deposits to allow half of the light produced by the viewbox to pass through the radiograph to the observer. The darkest portions of the radiograph, with an optical density of 3.00, allow only 1/1,000 of the viewbox light to be transmitted. It is important to note that mammography images may have optical densities of greater than 4.0.

## The Characteristic Curve

It is often convenient to be able to describe the characteristics of a film-process combination by relating the optical density of a radiograph to the radiation exposure required to produce that density. If a series of known exposures is given to different areas of a film and the densities produced after processing are measured, a graph can be plotted showing the density resulting from each exposure. For most applications, it is more useful to plot the relationship between density and the logarithm of relative exposure (Figure 4-5). In radiography, the ratios of exposure are usually more significant than the absolute values of the exposures themselves.

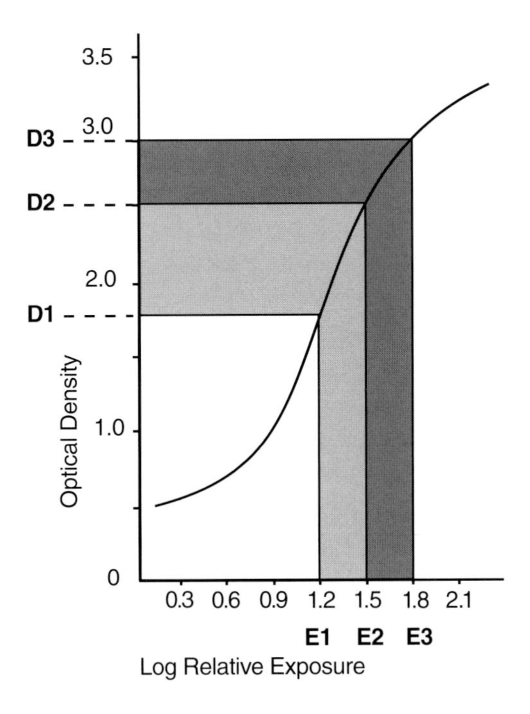


**Figure 4-4** Log relative exposure and optical density. A screen-film system characteristic curve shows the output (optical density) on the y-axis resulting from input conditions (x-ray exposure) as log relative exposure on the x-axis. Doubling the input x-ray exposure is shown as an increase in log relative exposure of 0.3. Note that doubling the exposure from E1 to E2 produces a difference in optical density from D1 to D2. This density difference is greater than D2 to D3, which is produced when doubling x-ray exposure from E2 to E3.

Therefore, it is usually sufficient and more convenient to express the exposures to the film using a scale based on relative exposures, that is, what happens when film is exposed to twice as much x-radiation.

There are several advantages to using a logarithmic scale for the horizontal exposure axis of the characteristic curve. The most significant advantage is that exposures with a constant ratio will always be separated by the same distance on the logarithmic scale, no matter what their absolute values. Therefore, two exposures, one of which is twice the other, will always be separated by 0.3 on the logarithmic exposure scale. For example, two exposures of 2 and 4 mR would be separated by 0.3 on the log relative exposure axis.

A sensitometer is a device that produces a series of light exposures on film with luminances having a constant ratio to one another. When the data points in Figure 4-5 are connected by a solid line, the resulting curve is called the characteristic curve. Characteristic curves are defined for each film-process combination and can be used to provide information about the relative sensitivity (speed) and contrast of a particular film or screen-film system.

## Introduction to Logarithms

A logarithm, or log, is an exponent. Logs are often used to represent and compare numbers that cover a wide range of values on the same axis of a graph, like comparing 1 and 1,000. There are three key items about logs that are important to understanding logarithmic plots, such as the characteristic curve:

1. Numbers are often expressed using exponents. When you see the word log think of the phrase "the power on 10 producing..." For example,

| Number | | | Log |
|---|---|---|---|
| 1 | = | $1 \times 10^0$ | 0 |
| 10 | = | $1 \times 10^1$ | 1 |
| 100 | = | $1 \times 10^2$ | 2 |

2. Because logs are exponents, the log of the product of two numbers is equal to the sum of their logs. That is,

$$\text{Log}\,(A \times B) = \text{Log}\,A + \text{Log}\,B$$
$$10^2 \times 10^1 = 10^3$$
$$\text{Log}\,(100 \times 10) = \text{Log}\,100 + \text{Log}\,10 = 3$$

3. The log of 2 is 0.3. This is a somewhat less intuitive concept than the others because it is not easy to visualize a fractional exponent. Nonetheless, the following is true:

$$\text{Log}\,2 = 0.3$$
$$\text{Log}\,20 = \text{Log}\,(10 \times 2) = 1 + 0.3 = 1.3$$

In principle, one can obtain the series of relative exposures necessary to cover the full density range of the film either by varying the exposure time and keeping the x-ray intensity constant (time-scale sensitometry) or by varying the x-ray intensity and keeping the exposure time constant (intensity-scale sensitometry). The measurement method used depends on the system whose sensitometric properties are to be measured and on the manner in which the system is to be used. In the case of radiographic film exposed to direct x-rays (without intensifying screens), either method can be used and will give the same results.

When radiographic film is used with intensifying screens, the film is primarily exposed by light from the screens. Time-scale and intensity-scale sensitometry will result in different characteristic curves, in general, because of reciprocity law failure when light emitted from the screen exposes the film. The effect of reciprocity law failure can be important in screen-film mammography, in which long exposure times are used because additional exposure may be required to compensate for reduced film speed. The definition given for exposure (Exposure = Intensity x Time) states that the response of the film to radiation of a given quality will be unchanged if the product of intensity and time remains the same. It is implied that this relationship

**Introduction to Logarithms *(Continued)***

It is often useful to know the effect of doubling the exposure to a screen-film combination. Since the horizontal axis (abscissa) of the characteristic curve is log relative exposure, doubling the amount of exposure means adding 0.3 to the log.

For example, using Figure 4-4, point E1 represents a given amount of exposure and the optical density that it produces on film. To know the optical density that is produced by twice as much radiation exposure to the screen-film combination, add 0.3 to the x-axis value and find point E2. One can see the effect on optical density from doubling the exposure.

The effect of doubling the exposure again can be seen by adding another 0.3 to the x-axis value, point E3. It can be seen that the optical density did not increase as much as from E1 to E2. This is because the screen-film combination has been overexposed, that is, point E3 is beyond the straight-line portion of the characteristic curve.

## Examples

Consider the measurement of optical density in these examples. When the silver in a radiograph allows 1/10 of the light to pass through, the opacity is the ratio of the original incident light to the transmitted light, that is, 10/1 or simply 10. The logarithm of 10 is 1 and, therefore, the area is said to have an optical density of 1.00. Again, if the silver allows only 1 percent or 1/100 of the light to pass through, the ratio is 100/1. The logarithm of 100 is 2 and, therefore, the area has an optical density of 2.00.

remains constant, regardless of whether long or short exposure times are used, provided that the time changes are compensated for by a proportional change in intensity. This relationship, also known as the reciprocity law, applies to direct exposure of film by x-rays; however, the law fails for exposure by screen-produced light.

Therefore, the sensitometric technique that most nearly simulates the actual conditions under which the screen-film combination is exposed should be used. During a typical radiographic exposure, the exposure time does not vary over the area of the film, but rather the radiation intensity and x-ray spectrum are modulated by the x-ray absorption in different parts of the subject. Intensity-scale sensitometry is generally the method of choice for exposure of screen-film combinations. One method of varying the radiation intensity transmitted to the image receptor during the sensitometric exposure is to make use of the decrease in intensity as the image receptor is moved away from the radiation source (inverse square law) (Figure 4-6).

The slope or steepness of the characteristic curve is related to film contrast. Consider the curves in Figure 4-7. Assume that both films were exposed together and processed together so that the differences between them are attributable to differences in film type. Superimposed on the curves are steps to show how density differences between steps change with exposure.

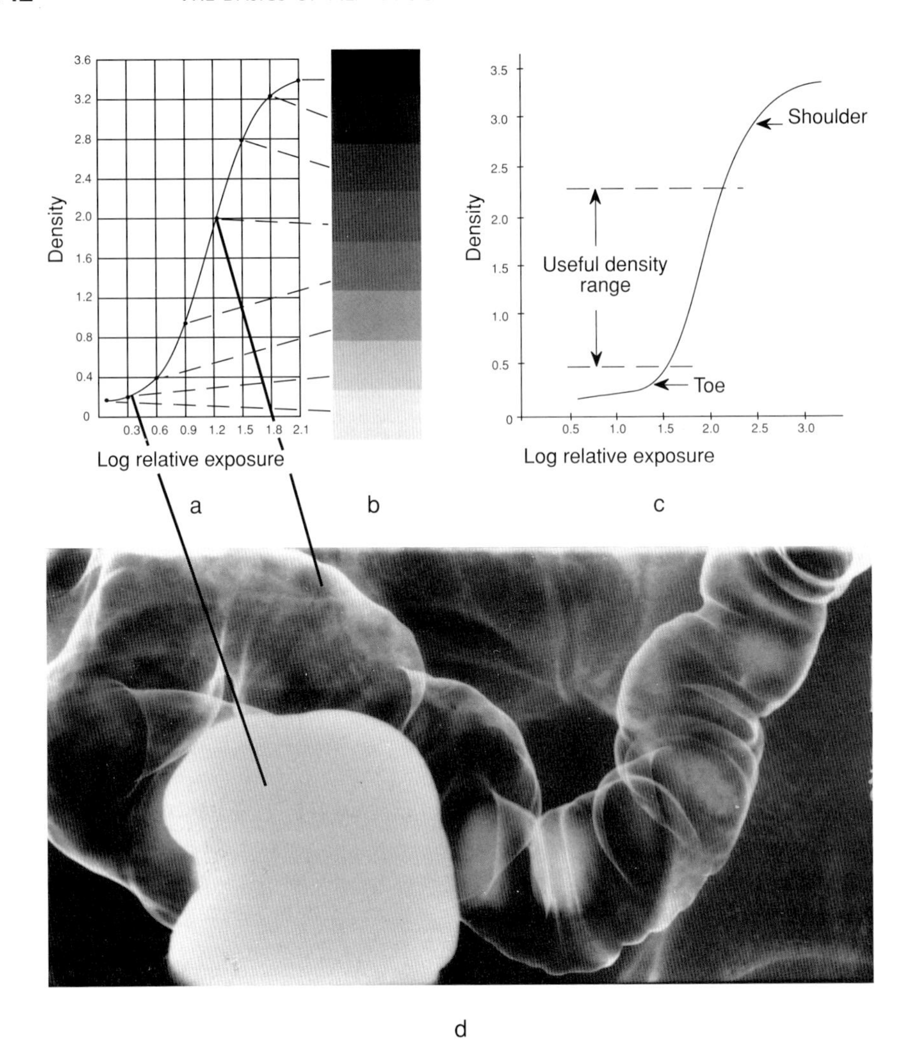


**Figure 4-5** The characteristic curve of a typical medical x-ray film. (a) The curve expresses the relationship between the logarithm of relative exposure (horizontal axis) and the resulting optical density (vertical axis). (b) In the center, a processed strip of film that was given a carefully controlled series of exposures is shown. Each step of the strip was given an exposure twice as great as that of the step below it. (a and b) Each step is connected by a dotted line to a corresponding point on the graph, which shows the density of that step and the log relative exposure that caused it. When the points corresponding to the various steps are connected, the resulting curve is called the characteristic curve, the sensitometric curve, the D log E curve, or the H&D curve for the screen-film combination. (c) The useful density range in medical radiography is shown. (d) A barium-filled rectum is imaged near the toe of the curve; the aerated colon is imaged within the straight-line portion of the curve. Various shades of gray correspond to the sensitometric steps in the useful density range (c).

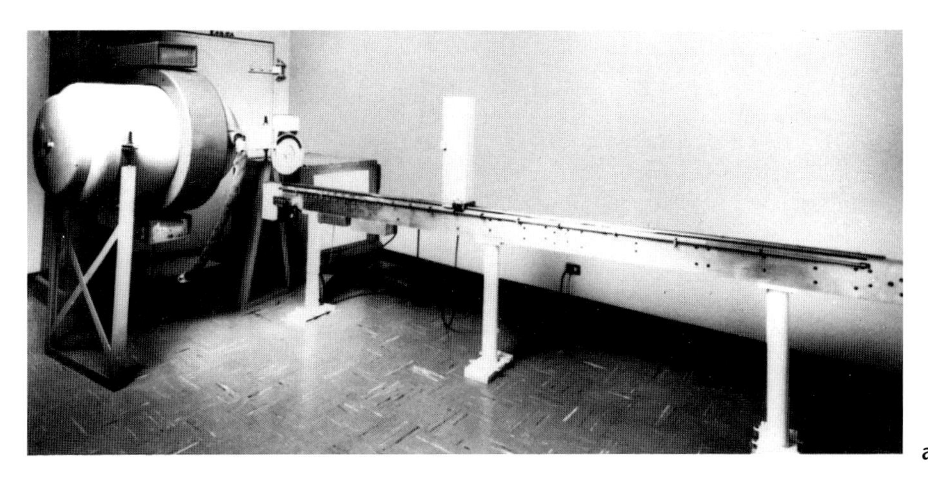

a

| Density | Log relative exposure | Relative exposure | Focal spot-film distance (inches) |
|---|---|---|---|
| .21 | 0.0 | 1.00 | 126.0 |
| .22 | 0.1 | 1.26 | 112.0 |
| .23 | 0.2 | 1.59 | 100.0 |
| .25 | 0.3 | 2.00 | 89.0 |
| .28 | 0.4 | 2.51 | 79.3 |
| .31 | 0.5 | 3.16 | 70.6 |
| .37 | 0.6 | 3.98 | 63.0 |
| .47 | 0.7 | 5.02 | 56.0 |
| .59 | 0.8 | 6.31 | 50.0 |
| .75 | 0.9 | 7.95 | 44.6 |
| .97 | 1.0 | 10.0 | 39.8 |
| 1.26 | 1.1 | 12.6 | 35.5 |
| 1.57 | 1.2 | 15.9 | 31.6 |
| 1.91 | 1.3 | 20.0 | 28.2 |
| 2.27 | 1.4 | 25.1 | 25.1 |
| 2.57 | 1.5 | 31.6 | 22.4 |
| 2.80 | 1.6 | 39.8 | 20.0 |
| 2.98 | 1.7 | 50.2 | 17.8 |
| 3.11 | 1.8 | 63.1 | 15.8 |
| 3.21 | 1.9 | 79.5 | 14.1 |
| 3.25 | 2.0 | 100.0 | 12.6 |
| 3.29 | 2.1 | 125.8 | 11.2 |
| 3.32 | 2.2 | 158.5 | 10.0 |

b

**Figure 4-6** (a) An intensity-scale x-ray sensitometer positioned with the image receptor placed at a distance from the x-ray source (according to the inverse square law) to measure the radiation intensity. (b) The film strip produced with the sensitometer, with the corresponding measurements for film density, log relative exposure, relative exposure, and focal spot-to-film distance.

Notice how much steeper the slope is in the middle of the curve for Film B relative to the corresponding slope for Film A. Film B has a steeper slope because it produces a larger density difference for a given exposure ratio (or equal log E) than does Film A. Film B also has higher contrast than Film A. Therefore, the film with a characteristic curve that rises more rapidly, that is, a steeper curve, has higher contrast than a film with a characteristic curve that has a more gradual slope.

There are several ways of measuring the steepness (film contrast) of the characteristic curve, including average gradient and contrast index, also referred to as contrast, density difference, or DD (Chapter five).

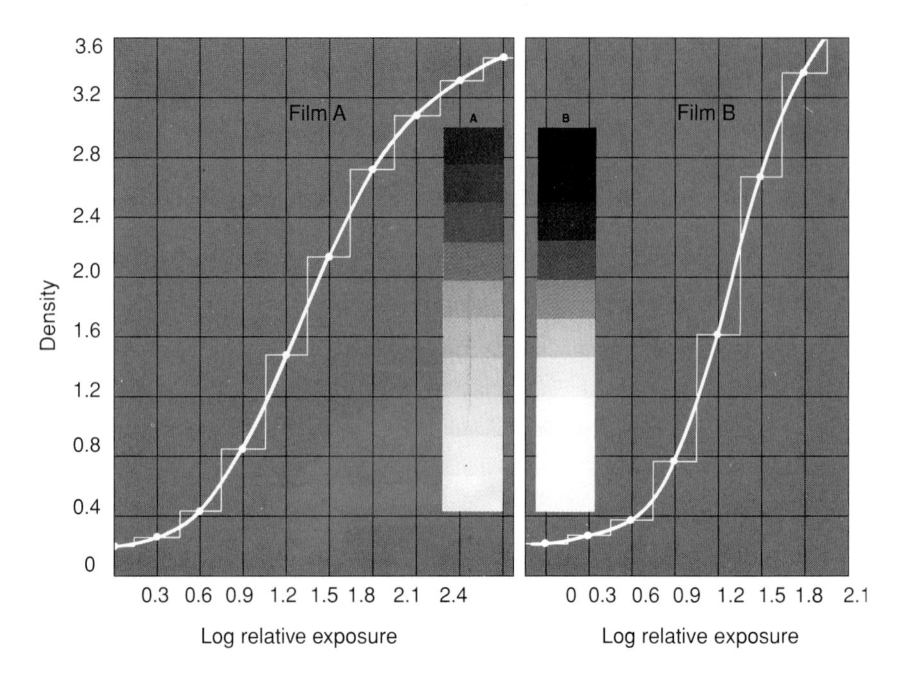


**Figure 4-7** Film contrast differences. Radiographic contrast is the difference in optical density between two areas in a radiograph. Radiographs of a stepped-wedge taken under identical exposure conditions with two screen-film combinations, A and B, and their associated characteristic curves are shown above. Film contrast is related to the slope or steepness of the characteristic curves. The density differences between steps in the middle-density range of curve B are larger than those for curve A. The slope of characteristic curve B is higher or steeper than the slope of curve A. Since exposure conditions are the same for both strips, the higher radiographic contrast of B results from higher film contrast. This difference may be attributable to differences in the composition of the two films, or if the film is the same, to differences in processing conditions. Note that film contrast depends on the density level. Density differences between steps are smaller in the low-density region at the bottom (toe) and the high-density region at the top (shoulder) of the characteristic curve than they are in the middle-density region. Contrast is lower in the toe and shoulder than in the middle of the curve.

The exposure range over which the steps are well differentiated is an indicator of film latitude. The low-contrast characteristic curve A produces easily distinguishable steps over a wider range of exposures on the log relative exposure axis than does curve B.

The average gradient describes the average film contrast over a selected usable range of film density and is defined as the slope of a straight line between two points of specified densities. The density range of 0.25–2.0 above the base plus fog is used to determine the average gradient of medical screen-film systems and x-ray films (Figure 4-8).

The average gradient is defined as change in density divided by change in the log relative exposure, or $1.75/\log E_2 - \log E_1$, where $E_1$ is the exposure corresponding to the net density of 0.25 above the base plus fog density, and $E_2$ is the exposure corresponding to the net density of 2.00 above the base plus fog density. In the example of

Figure 4-8, the film base plus fog is 0.20. The average gradient is (2.00 + 0.20) – (0.25 + 0.20)/1.12 – 0.50 = 1.75/0.62 = 2.8. Typical average gradients of screen-film combinations range from approximately 2.2 to 3.5 (Table 4-2). Examples of clinical images with different average gradients are shown in Figure 4-9.

Figure 4-10 shows the characteristic curves of several film and screen-film systems, illustrating the range of contrast and relative speed differences of commercially available recording systems for medical imaging.

Examples of the use of sensitometric data to evaluate film contrast for screen-film combinations designed for thoracic imaging are shown in Figure 4-11.

## Film Speed

The speed, or sensitivity, of a radiographic material is inversely related to the exposure required to produce a given effect. Therefore, if in Figure 4-12, Film A requires only half the exposure needed by Film B to produce a given density, Film A is said to have twice the speed or to be twice as fast as Film B. (A factor of two in exposure is represented by a log relative exposure difference of 0.3.) Speeds of radiographic films are often determined from the exposures required to produce a density of 1.00 above the base plus fog of the films (base plus fog is the optical density of the film base plus the density of the emulsion layers in areas that have not been intentionally exposed). Values of base plus fog differ, in part, owing to inherent differences in the film base and emulsions used for different film types.

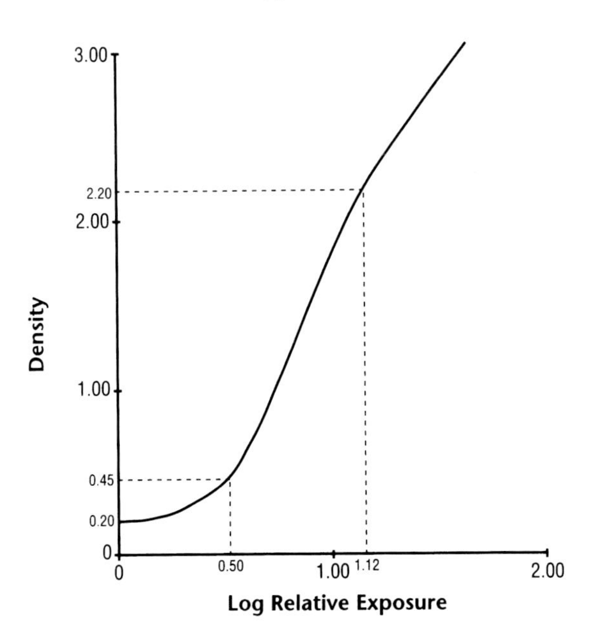


**Figure 4-8** Average gradient. The density range of 0.25 to 2.00 above the base plus fog density is used to determine the average gradient.

**Table 4-2** Typical Average Gradients Determined from Characteristic (H&D) Curves for Several Medical Imaging Modalities

| *Modality* | *Average Gradient* |
|---|---|
| CT scan, MRI, etc. | |
| Helium-neon laser film (HeNe) | 1.9$^{+}$ |
| Infrared laser film (IR) | 2.5$^{+}$ |
| Chest | 2.2$^{++}$ |
| General purpose (head and neck, extremities, etc.) | 2.8$^{++}$ |
| Mammography | 3.3$^{++}$ |
| Radiation therapy portal localization | 5.7$^{++}$ |

$^{+}$ Determined from specially designed laser sensitometers.
$^{++}$ Determined from an intensity scale x-ray sensitometer.

The speed or sensitivity of a film is indicated by the location of its characteristic curve along the log relative exposure axis. For example, in Figure 4-12, the curve for the faster film (A) will lie toward the left of the curve for a slower system. Therefore, compared with a slower system, a faster system requires less exposure to produce the same optical density. Conversely, the curve for a slower film (B) will lie toward the right side of the diagram in the region where the relative exposure, or its logarithm, is larger.

Absolute film speed is often defined as the reciprocal of the exposure required to produce an optical density of 1.0 above base plus fog. The approximate film sensitivity values are sometimes based on the exposures required to produce a given optical density on a radiograph of a phantom or test object at specified exposure conditions (peak kilovoltage, filtration, exposure time, processing conditions, etc.). Factors that influence screen-film speed (or the exposure required to produce a given density on the film) include: (1) film type, (2) screen type, (3) film-processing conditions, (4) film storage conditions, and (5) reciprocity law failure. Sensitivities for several screen-film combinations are shown in Table 4-3.

The following factors influence the speed or exposure required to produce a given density on a film:

1. *Film Type.* The composition of the film and the way it is manufactured affect film speed. The ingredients used in making the emulsion, the manner in which they are treated and combined, and the technique by which they are coated on the film base all play a part in determining the sensitivity of the film.
2. *Exposure Type.* Significantly more exposure is required to produce a given density when the film is exposed to x-rays alone than when intensifying screens are used.

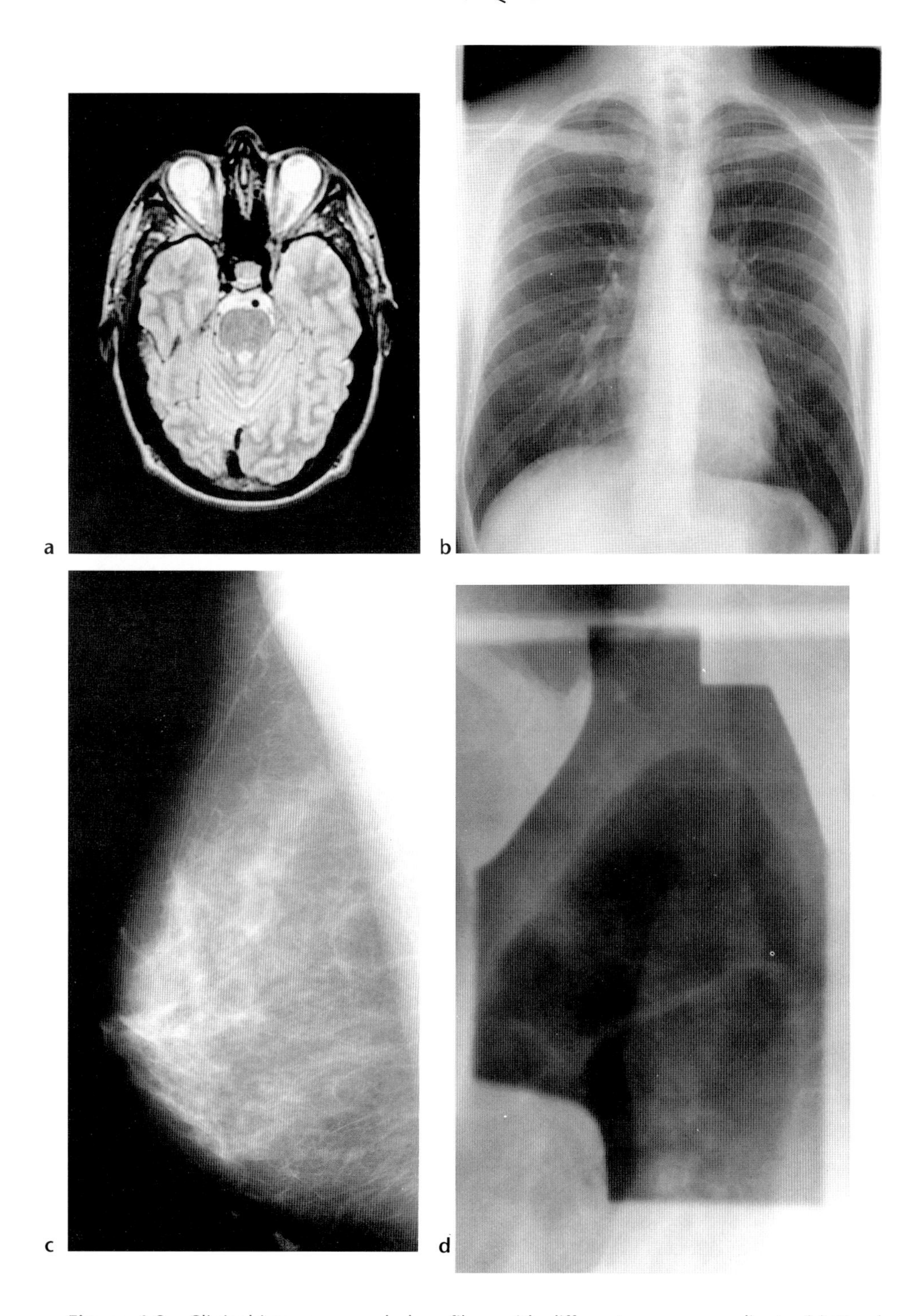


**Figure 4-9** Clinical images recorded on films with different average gradients. (a) Head CT image on helium-neon laser film: average gradient 1.9. (b) Chest: average gradient 2.20. (c) Mammogram: average gradient 3.2. (d) Radiation therapy portal localization radiograph of the chest: average gradient 5.7.

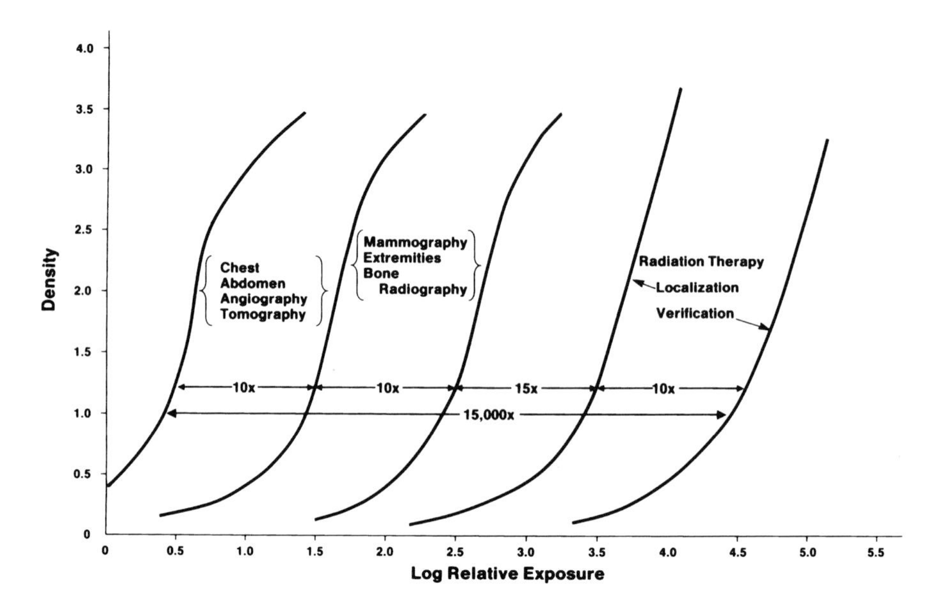


**Figure 4-10** Characteristic curves of several film and screen-film combinations illustrate the range of relative speed and film contrast differences among commercially available recording systems used in medical imaging.

**Figure 4-11 (facing page)** (a) Characteristic curves of a screen-film combination designed for thoracic imaging (dashed line) and a conventional high-contrast screen-film combination (solid line). These characteristic curves are matched at density = 2.0, corresponding to a typical maximum viewable lung field density. These data show that the optical density of the high-contrast film is inadequate to capture and display the full range of exposures required for chest radiography. Approximate log exposure values for the maximum lung field density of 2.0 and for a typical mediastinum log exposure value are shown as vertical reference lines. (b) Slopes of the characteristic curves in (a) are calculated at each log exposure value. Because the log E values exiting a given patient are the same, independent of the screen-film contrast, the contrast of these two systems can easily be compared at the indicated reference mediastinum exposure level or at the other exposure levels of interest. (c) Slopes of the characteristic curves in (a) are calculated as a function of density. Although it is relatively easy to think of contrast as a function of density for understanding the behavior of a single system, it is more difficult to compare the contrast of two or more systems by using these data, because the density levels corresponding to any specific anatomic structure will always differ. In this chest radiography example, for the high-contrast conventional combination, the mediastinum is imaged at a very different density level than is the same structure for the wide-latitude thoracic combination. Thus, to compare the contrast of the mediastinum as imaged with these two systems by using contrast-versus-density graphs, one must first determine the density at which it will be radiographed for each system and then locate these densities on the graphs, as shown here. This two-step process is much less convenient and less understandable than directly using the contrast-versus-log E data in (b). Also, one must repeat this procedure for other structures of interest, instead of directly and naturally using the data in (b).

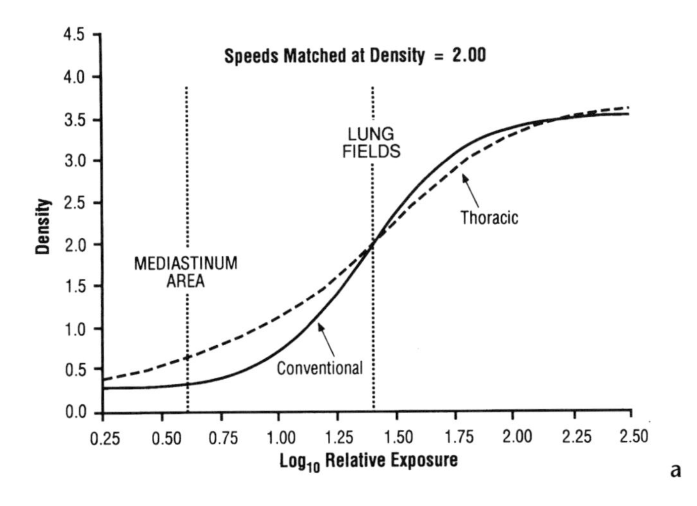
Speeds Matched at Density = 2.00
LUNG FIELDS
MEDIASTINUM AREA
Thoracic
Conventional
Density
Log10 Relative Exposure
4.5 4.0 3.5 3.0 2.5 2.0 1.5 1.0 0.5 0.0
0.25 0.50 0.75 1.00 1.25 1.50 1.75 2.00 2.25 2.50
a

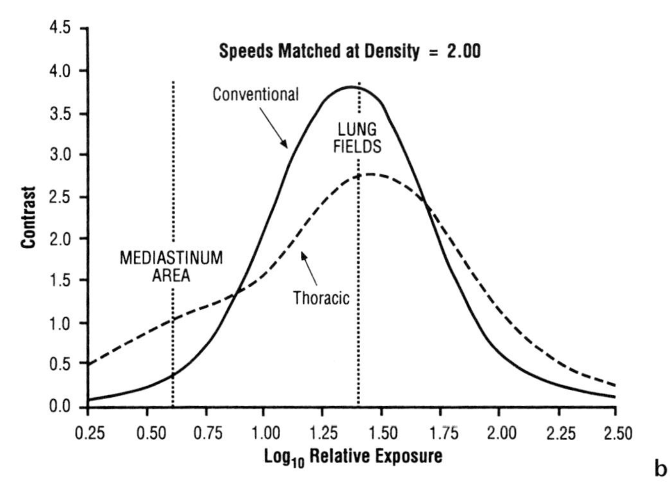
Speeds Matched at Density = 2.00
Conventional
LUNG FIELDS
MEDIASTINUM AREA
Thoracic
Contrast
Log10 Relative Exposure
4.5 4.0 3.5 3.0 2.5 2.0 1.5 1.0 0.5 0.0
0.25 0.50 0.75 1.00 1.25 1.50 1.75 2.00 2.25 2.50
b

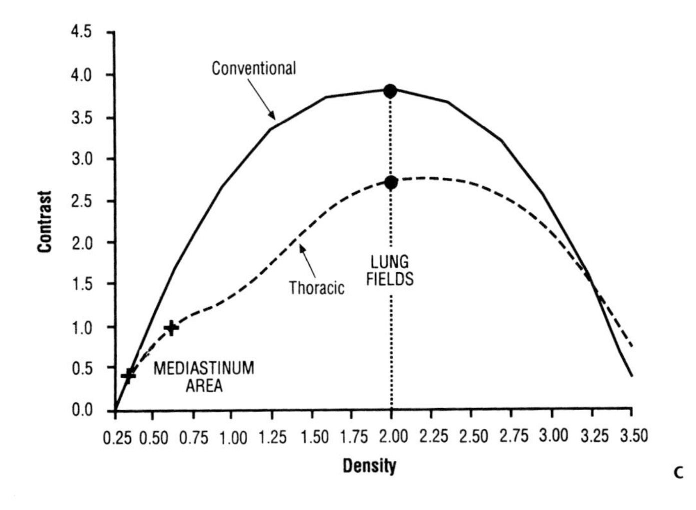
Conventional
Thoracic
LUNG FIELDS
MEDIASTINUM AREA
Contrast
Density
4.5 4.0 3.5 3.0 2.5 2.0 1.5 1.0 0.5 0.0
0.25 0.50 0.75 1.00 1.25 1.50 1.75 2.00 2.25 2.50 2.75 3.00 3.25 3.50
c

**Figure 4-11**

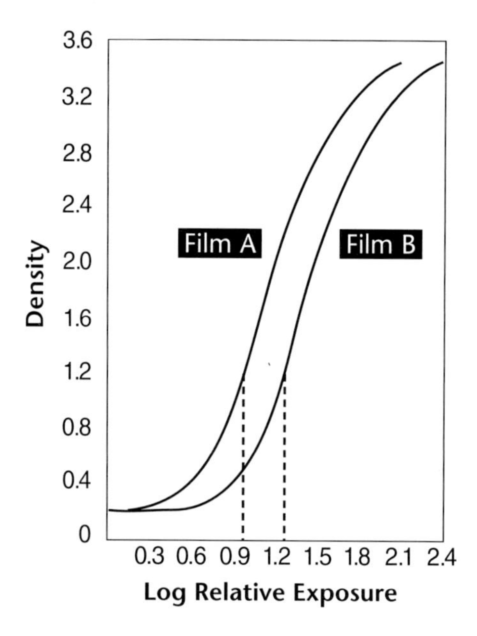


**Figure 4-12** Characteristic curves for films differing by a factor of 2 in speed. Because photographic speed is inversely related to exposure, the faster or more sensitive film (A) will require less exposure than the slower film (B) to produce a density of 1.20 (a density of 1.00 and a base plus fog density of 0.20). Film A requires 10 units of relative exposure (log 10 equals 1.0), and Film B requires 20 units of relative exposure (log 20 equals 1.3) to produce the same optical density of 1.20 on each film. Note that the characteristic curves of both films have the same shape. Since their slope or steepness is the same, film contrast is also the same.

**Table 4-3** Approximate Relative Speeds and Receptor Sensitivities for Several Screen-Film Combinations

| *Screen Pairs* | *Film* | *Relative Speed** | *Approximate Receptor Sensitivity (mR)+* |
|---|---|---|---|
| 1 | 1 | 100 | 1.28 |
| 2 | 1 | 300 | 0.43 |
| 3 | 1 | 400 | 0.32 |
| 4 | 3 | 600 | 0.21 |
| 2 | 2 | 500 | 0.26 |
| 3 | 2 | 800 | 0.16 |
| 4 | 2 | 1,200 | 0.11 |

* Relative speeds based on average values of radiographs of chest phantoms at about 80 kVp with scatter and at about 125 kVp without scatter and a pelvic phantom at about 70 kVp without scatter. Screen number 1 and film number 1 are arbitrarily assigned a relative speed of 100.

+ Receptor sensitivity is defined as the exposure (mR) required to produce an optical density of 1.0. Approximate receptor sensitivity values are based on radiographs of an extremity phantom exposed at 60 kVp with scatter.

In the case of screen light exposures, the sensitivity, or speed of the film, depends on the wavelength, or color, of the exposing light. The manner in which the film responds to radiation of different wavelengths is called its spectral sensitivity. A graph that shows the relationship of the film's sensitivity to the wavelength of exposing radiation is called a spectral sensitivity curve.

3. *Reciprocity Law Failure.* The definition given for exposure implies that the response (optical density) of film to radiation of a given spectral quality will be unchanged if technical factors (mA, time, distance, grid) are adjusted so that the mR remains constant. The reciprocity law implies that this is the case regardless of whether long or short exposure times are used. In other words, at a given kVp, 100 mAs would be expected to produce the same optical density on film whether the exposure was made at 1000 mA for 0.1 seconds or 25 mA for 4 seconds. This example of the reciprocity law is true for direct x-ray exposures, as in industrial radiography. However, when film is exposed to visible light, as from an intensifying screen, reciprocity law failure may occur.

   For exposure times used in most medical radiography, the effect of reciprocity law failure is usually insignificant for most screen-film images. Long exposure times (in excess of one second) may produce clinically meaningful effects. In mammographic imaging, for example, exposure times requiring 2 seconds need about 15 percent more mAs to produce the same optical density as for exposures of under 1 second.

4. *Processing Conditions.* Among the most important factors affecting film speed are the conditions used in processing the film after exposure. The chemical formulation of the solutions used, the way in which they are mixed and replenished, their temperatures, the manner in which they are agitated, the duration of developer and fixer immersion, and washing and drying conditions all contribute to the film's speed, contrast, and appearance.

   The chemical action of the processing solutions amplifies the effect of exposure by a factor of many millions. It follows, then, that changes in processing conditions can have a significant effect on film's response. For this reason, optimal radiographic results can only be guaranteed when the film manufacturer's recommendations for screen-film combination, processing chemicals, and automatic processing conditions are followed precisely. In addition, processing conditions should be monitored systematically as part of a complete quality assurance program.

5. *Ambient Conditions.* Film sensitivity may also be affected by such factors as temperature, humidity, age, chemical fumes, and storage conditions. Temperature extremes, high humidity, and atmospheres containing chemical contaminants should be avoided. Protection must also be provided from light leaks, excessive safelight exposure, x-radiation, and gamma radiation in order to avoid fogging film.

   If storage, handling, and processing conditions are not carefully controlled, a high fog density may be produced that adversely affects the contrast of the radiographic image. Natural background radiation can, over time, fog film. Even if film is protected from man-made sources of radiation (e.g., that produced by x-ray machines and radioactive materials used in nuclear medicine), it is being constantly exposed to natural background radiation. This source of film fog is sometimes confused with chemical fog and mistakenly attributed to chemical changes in the emulsion as it ages.

Natural background radiation consists of high-energy cosmic rays from outer space that penetrate the earth's atmosphere and emissions from naturally occurring radioactive materials in the environment. Natural background radiation levels are typically low, but may accumulate over time to a level that will produce noticeable film fog. In some instances, this natural background radiation exposure may be responsible for a major portion of fog arising during storage of x-ray film. Since it is difficult to control natural background radiation, the best protection is to rotate film stock so that the oldest film is used first.

After a radiographic film has been exposed to light, it is more sensitive to subsequent exposures to light, such as from safelights or a minimal light leak, than unexposed film. For this reason, there will be a greater increase in film density from excessive safelight exposure if it occurs after normal patient exposure. The color of light transmitted by safelight filters is selected to provide only wavelengths for which the film has minimal sensitivity. Periodic testing for unwanted sources of light in the darkroom is an essential component of quality control (Appendix D).

## Radiographic Contrast

Radiographic contrast is the term used to describe the difference in optical density between two areas of interest in a radiograph. Large differences in optical density indicate high radiographic contrast, also known as short-scale contrast. An image with high radiographic contrast is one in which regions on the film corresponding to areas of high x-ray absorption in the patient appear very white, while regions corresponding to areas of low x-ray absorption are very dark.

The components of radiographic contrast are subject contrast and film contrast.

Subject contrast is the ratio of x-ray fluences emerging from two adjacent regions of the subject. It is a function of the pattern of x-ray fluences in the aerial image that expose the image receptor. Subject contrast depends upon factors that affect x-ray absorption in the subject (x-radiation beam spectrum and the nature of the subject) and upon scattered radiation (Table 4-4).

Film contrast is a measure of how differences in x-ray fluence, incident on the image receptor, will be converted into differences in optical density on a radiograph. Film contrast is the slope of the (film) characteristic curve. It is affected by film type, processing conditions, density level in the radiograph, fog level of the film, and type of exposure (whether screen light or direct x-ray) (Table 4-5).

In radiographic imaging (Figure 4-13a), the input consists of the different x-ray fluences within the aerial image that strike the image receptor, that is, the subject

**Table 4-4** Factors Affecting Subject Contrast

| *Factor* | *Change* | *Subject Contrast* |
|---|---|---|
| *Beam spectrum* | | |
| kVp | ↑ | ↓ |
| Generator percent ripple | ↑ | ↑ |
| Target material | W to Mo* | ↑ |
| *Scatter* | | |
| kVp | ↑ | ↓ |
| Beam collimation (field size) | ↑ | ↓ |
| Grid or air-gap technique | Yes | ↑ |
| Compression | Yes | ↑ |
| *Nature of subject* | | |
| Differences in x-ray absorption | ↑ | ↑ |
| Patient thickness | ↑ | ↓ |

Arrows indicate increase (↑) or decrease (↓).
*W-Tungsten, Mo-Molybdenum.

**Table 4-5** Factors Affecting Film Contrast

| *Factor* | *Change* | *Film Contrast* |
|---|---|---|
| *Film type* | | |
| High contrast (short-scale) | Yes | ↑ |
| Wide latitude (long-scale) | Yes | ↓ |
| *Position on film characteristic curve (density level)* | | |
| Toe | | ↓ |
| Linear position | | ↑ |
| Shoulder | | ↓ |
| *Fog level* | ↑ | ↓ |
| *Film processing* | | |
| Suboptimal | Yes | ↓ |

Arrows indicate increase (↑) or decrease (↓).

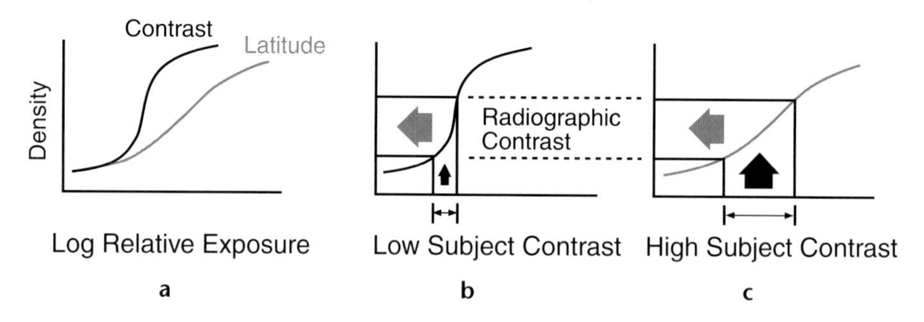


**Figure 4-13** Radiographic contrast. Radiographic contrast is the term used to describe the difference in optical density between two areas of interest in a radiograph. The slope of the characteristic curve is a measure of film contrast. (a) A wide-latitude film can be used for long-scale imaging, such as chest radiography. (b) Low subject contrast in the aerial image may be imaged with a high-contrast film to produce good radiographic contrast. (c) Similar radiographic contrast may be obtained with high subject contrast in the aerial image and a latitude film.

contrast along the log E axis. The characteristic curves represent the response of two screen-film systems, A and B, the slope of which is the film contrast. The output of the radiographic system is the optical density produced on the film. In a given screen-film system, the difference in optical density is the radiographic contrast.

Consider the effects of different subject and film contrast on radiographic contrast, as shown in Figure 4-13. In Figure 4-13b, the input or subject contrast is low, due to the high kVp needed on a portable exam (nongrid) to penetrate a thick body part. A high-contrast, screen-film system may be used to provide high radiographic contrast (output). In Figure 4-13c, the subject contrast is increased, but film contrast is reduced. This could have been caused by performing the same exam at a lower kVp in the imaging department with a grid (increased subject contrast) and with the inadvertent use of a chest cassette loaded with wide-latitude film (decreased film contrast, long scale). The result is comparable output, that is, radiographic contrast. In the first exam, the lower subject contrast input has been somewhat compensated for by the use of a high-contrast, screen-film system.

Factors that affect film contrast include film type (Chapter one), processing conditions (Chapters two and three), optical density, and film fog. Film contrast changes with density (Figure 4-14). Note that in the middle density part of the curve (generally between densities of about 1.00 and 2.00), the gradient (contrast) is higher than in regions of greater or lesser density.

Exposure conditions should be selected so that the structures of diagnostic interest in the radiograph will appear in the range of optical densities where contrast is highest. If the radiograph is under-exposed, these structures will fall on the bottom, or toe, of the characteristic curve (Figure 4-14). If the radiograph is over-exposed, the structures of interest will fall within the shoulder of the curve (Figure 4-14), corresponding to the higher optical densities. In either case, radiographic contrast will be lower, making it more difficult (or perhaps impossible) to distinguish

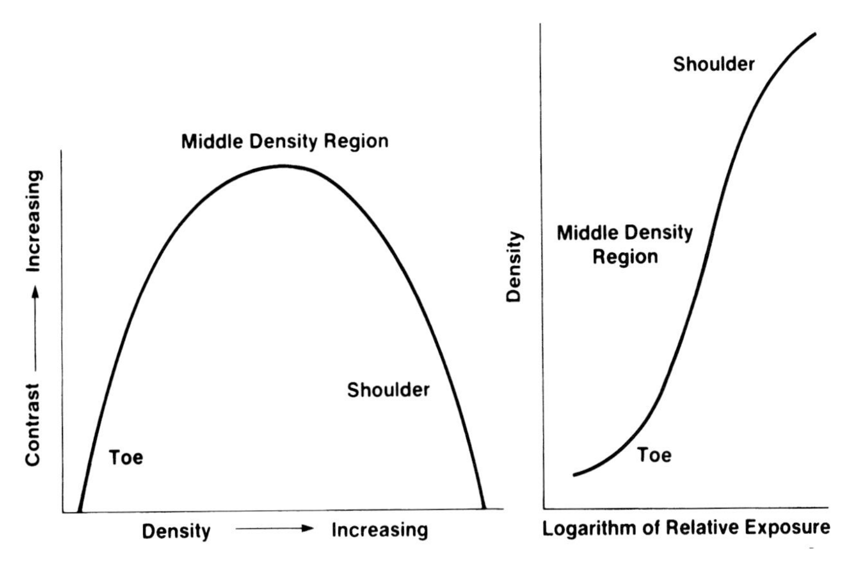


**Figure 4-14** Film contrast, as determined from the slope of the characteristic curve for a typical medical x-ray film, increases as photographic density increases until a maximum is reached, in the middle-density region, and then declines. The maximum contrast attained and the density range over which it extends are functions of the film type and processing conditions. If a radiograph is under-exposed, structures of interest will have low contrast because their density is low—placing them on the toe of the characteristic curve. On the other hand, if the radiograph is over-exposed, structures of interest will have low contrast because their density is high—placing them on the shoulder of the characteristic curve. Visibility will be greatest if the radiograph is exposed so that structures of interest appear in the middle-density region, where film contrast is highest. Quantum mottle also will be most apparent in the middle-density region, where contrast is high.

these structures than if the structures of interest had been exposed to fall in the middle-density region.

It is important to keep the fog level of film as low as possible because the radiographic effect of fog reduces film contrast. Fog density that is added to the normal fog level of a film—whether it arises from unwanted exposure, undesirable ambient conditions, or improper processing—increases density the most in the lower part of the characteristic curve. This is illustrated in Figure 4-15, showing the effect of improper processing that produced a "pulled-out" toe on the characteristic curve, resulting in a lower slope (reduced film contrast).

## Latitude

Latitude describes the range of structures in the subject that can be satisfactorily imaged. Latitude can be subdivided into two categories: exposure latitude, which is associated with subject contrast, and film latitude, which is associated with film contrast. Film contrast and film latitude are reciprocally related. In other words, as film contrast increases, film latitude decreases, and vice versa.

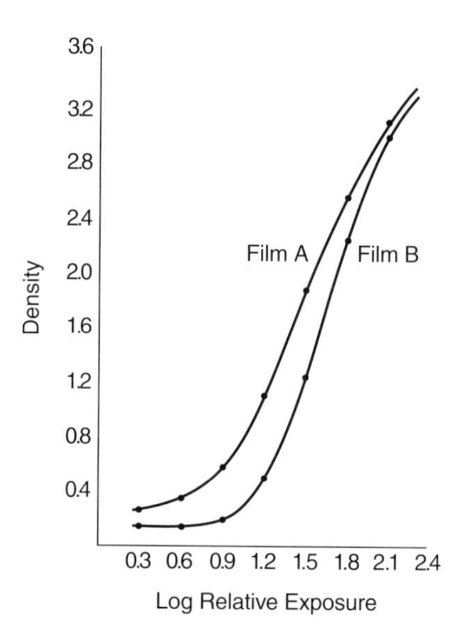


**Figure 4-15** The effect of fog on radiographic film. The characteristic curve of a typical medical x-ray film is represented by A. Characteristic curve B represents an x-ray film with added fog density caused by improper processing. This "pulls out" the toe of the curve, resulting in the lower slope and reduced film contrast seen in B.

Figure 4-16 illustrates the effect of kVp on subject contrast. Three strips of film were exposed under an aluminum stepped-wedge at 40, 70, and 100 kVp. The strips were cut from the same sheet of x-ray film before exposure and processed together, so the differences in appearance among them are due to differences in subject contrast. Because the penetrating power of the x-radiation changes as kVp changes, the ratio of transmitted x-ray fluences, or subject contrast, between steps 7 and 12 of the aluminum wedge is greater for the 40 kVp exposure than for the 100 kVp exposure.

If it is essential that both steps 7 and 12 be imaged on a single strip, one can see that there is little latitude or margin for error in the 40 kVp exposure. If the strip had been somewhat over-exposed so that the density of step 14 extended down to step 12, step 12 would then be indistinguishable from the steps on either side of it. On the other hand, if the 40 kVp strip had been under-exposed so that the lower density of step 5 shifted to step 7, step 7 would be indistinguishable from the steps on either side of it.

Next, consider steps 7 and 12 on the 100 kVp strip. Here, subject contrast is lower than for the 40 kVp strip and exposure latitude is greater. In other words, the densities of steps 7 and 12 could be shifted several steps in either direction by a change in exposure, and the images of these steps would still be distinct from the adjacent steps. Therefore, the lower subject contrast of the 100 kVp exposure produces greater exposure latitude (long-scale contrast) than the 40 kVp exposure. On the other hand, the differences between steps are more readily apparent in the higher contrast 40 kVp exposure (short-scale contrast).

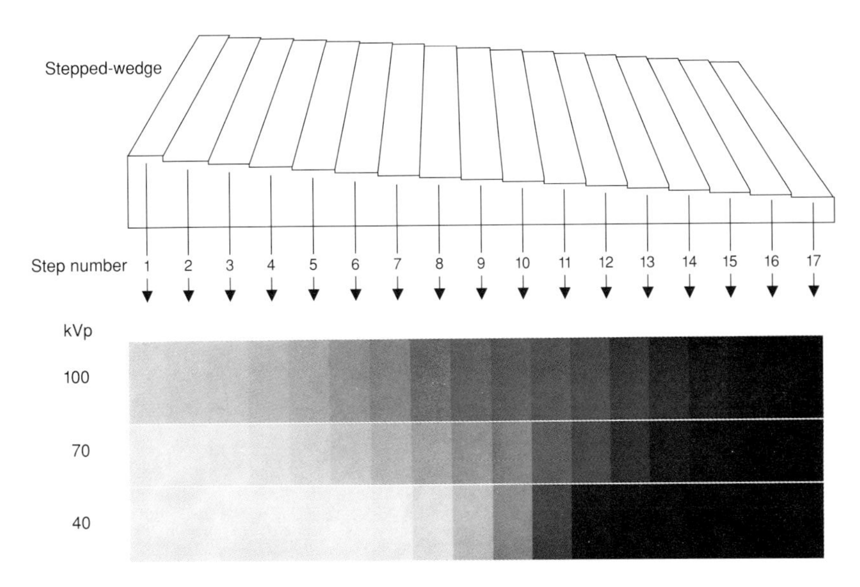


**Figure 4-16** Latitude versus contrast. Latitude is related to the range of structures in a subject that can be imaged satisfactorily. In the strip labeled 100 kVp, more steps can be seen than in the strip labeled 40 kVp. Therefore, strip 100 has greater latitude than strip 40. Strip 100 also has lower radiographic contrast than strip 40. As contrast decreases, latitude increases, and vice versa. The difference in radiographic contrast between strip 40 and 100 could be the result of either a difference in subject contrast or difference in film contrast.

To illustrate film latitude, consider the characteristic curves of two films with the same speed screens, as shown in Figure 4-7 (page 144). Film A is a wide-latitude film; Film B is a high-contrast film. The characteristic curve of Film B is steeper than Film A, owing to the higher contrast of Film B. Latitude is the range on the log relative exposure axis within which a satisfactory radiograph is produced. From this perspective, Film A has a wider latitude than Film B. As one can see from these curves, as film contrast increases, film latitude decreases.

Wide-latitude film is advantageous in radiography of the chest, where the subject contrast inherent in the aerial image is high. The use of latitude film prevents the lung fields from appearing completely black, which would obscure subtle vascular markings. Similarly, latitude film demonstrates small differences in x-ray absorption within the denser mediastinal region.

High-contrast film is used to visualize both osseous structures and body parts enhanced by the use of iodinated contrast media or barium. High-contrast film is also important in mammography. Therefore, different procedures require different types of film (latitude or contrast) to achieve specific imaging objectives.

## Image Blur (Unsharpness)

Blur, or unsharpness, results from three causes: motion, geometric blur, and receptor blur. For screen-film radiography, light diffusion (i.e., spreading of the light emitted by the intensifying screen before it is recorded by the film) causes blurring. Factors involved include: (1) phosphor layer thickness in the screen, (2) phosphor particle size, (3) light-absorbing dyes and pigments in the screen, (4) reflective screen support, (5) light crossover, and (6) screen-film contact.

The contribution of blur to the overall impression of sharpness of the edge of an object in an image is illustrated in Figure 4-17. The magnitude of the density change across the boundary, or the density difference between the structure of interest and its surroundings, is the contrast. Blur is the width of the transition region in the radiograph, where the density changes between the structure of interest and its surroundings.

The components of radiographic contrast are subject contrast and film contrast. Similarly, those factors in Tables 4-4 and 4-5 (page 153) that cause an increase in subject and film contrast (radiographic contrast) also cause perceptibility of the boundary to improve. That is, they improve radiographic sharpness. In other words, as the density difference between the steps increases, visibility of the edge improves and sharpness increases, if other factors remain unchanged.

The three radiographs of a bone specimen shown in Figure 4-18 also demonstrate the effects of radiographic contrast and radiographic blur on the image. Figure 4-18a was made with a screen, which produces considerable image blur because of its thickness. Even though it has high contrast, this radiograph exhibits increased image blur because of the lateral spreading of the light within the screen-film combination. Figure 4-18b, made without screens, shows minimal image blur. However, the film

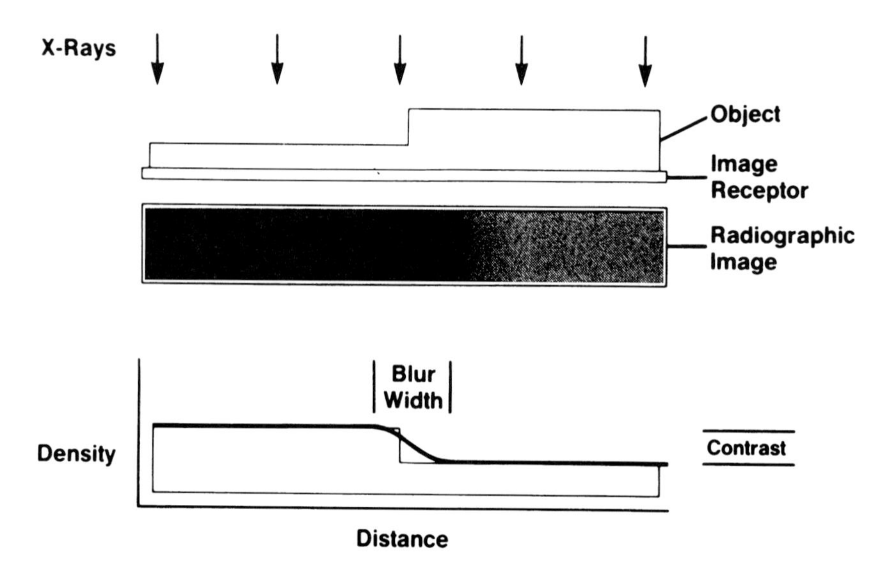


**Figure 4-17** This diagram depicts an object with a sharp vertical boundary as it is being exposed to x-rays, illustrating concepts of contrast and blur.

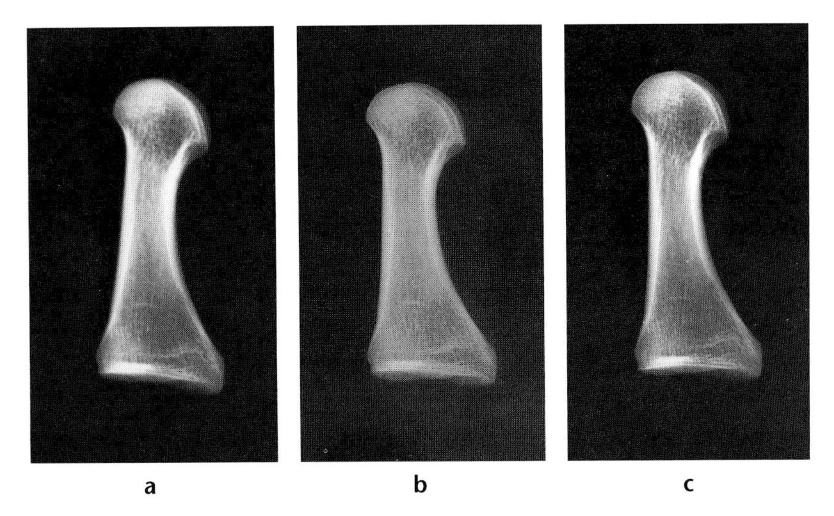


**Figure 4-18** Radiographic contrast and image blur. (a) Radiograph of a bone specimen made with a screen-film combination that produces considerable image blur. Note that even though the radiographic contrast is high, the image is not sharp. (b) A direct exposure radiograph made without intensifying screens exhibits minimal blur. The low film contrast impairs the perceptibility of detail in the image. (c) A radiograph made with thin screens containing a light-absorbing dye to reduce light diffusion is combined with a high-contrast film. This reduction in image blur improves the perceptibility of detail when compared with (a) and (b).

contrast is low, which tends to offset the gain from reduced image blur. The result is that sharpness is less than optimal. Figure 4-18c was made with screens that had a thin phosphor layer containing a light-absorbing dye, thereby reducing image blur from light diffusion. This effect, combined with high film contrast, produces an image with boundaries that are more distinct (reduced image blur) than the other two radiographs.

There are several methods used for measuring and evaluating image blur. The methods include square-wave response function (SWRF), modulation transfer function (MTF), line-spread function (LSF), edge response function, and bar pattern resolution assessment.

Image blur is not significantly affected by film processing variables and, therefore, will not be discussed in more detail here.

## Radiographic Noise

Unwanted variations in the optical density across a radiograph impair the ability to distinguish objects. When these variations in density are random, they are collectively described as radiographic noise.

## Radiographic Mottle

Radiographic mottle is the density variation in a radiograph made with intensifying screens that have been exposed to uniform x-radiation fluence. Three components of radiographic mottle are structure mottle, film graininess, and quantum mottle.

Structure mottle is the density fluctuation resulting from nonuniformity in the structure of intensifying screens. Inhomogeneities in the phosphor layer, such as clumping of the phosphor crystals or coating variations, would result in varying amounts of screen light output and, therefore, film density fluctuations. However, in most commercially available screens, structure mottle is not a practical problem. The combined effect of structure mottle and quantum mottle (described later in this chapter) is sometimes referred to as screen mottle.

Film graininess is the visual impression of the density variations (granularity) in a film exposed to a uniform source of light. In this case, the exposing light does not arise from intensifying screens because exposure by intensifying screens produces quantum mottle, which usually overwhelms any film graininess, except in mammography. Film granularity is caused by the random distribution of deposits of developed silver. Figure 4-19b shows the pattern produced on x-ray film by uniform exposure to light alone. Notice how much smoother it appears than the pattern of radiographic mottle shown in Figure 4-19a, which was made with intensifying screens. The density fluctuations in Figure 4-19a are due almost exclusively to quantum mottle, since the other two components—structure mottle and film graininess—generally make a less important contribution to this pattern.

The following discussion about the insignificance of film graininess relative to quantum mottle is generally true for radiographs as normally viewed on a viewbox. However, in cases where a portion of the radiograph is magnified by a lens or by projection on a screen, film graininess can be seen, particularly for low-density areas.

## Quantum Mottle

The effect of a limited number of x-ray photons (or quanta) on the appearance of the radiograph is called quantum mottle and is the major contributor to image noise. This mottled image is due to insufficient x-ray photons. This effect is similar to the

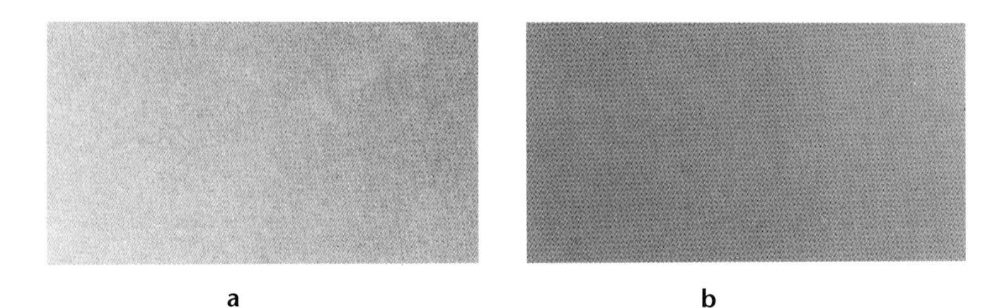


**Figure 4-19** Radiograph of mottle versus film granularity. (a) This image was produced using a screen-film combination that was given a uniform x-ray exposure. The irregular pattern of optical densities is due primarily to quantum mottle. (b) A second image was produced by uniform exposure of the film to light alone (without screens). The noise in this image is due to film granularity.

phenomenon we experience when we try to see under low-light conditions (insufficient light photons). If you turn off the room lights and allow a few moments to adapt to the dark, you will be able to see some image detail, but will notice a mottled appearance to the visual image. This effect is analogous to that observed in an image produced by insufficient x-ray photons in a radiograph.

Quantum mottle is seen as the variation in density of a radiograph exposed to a uniform fluence of x-ray photons, which results from the random spatial distribution of the x-ray quanta absorbed in the screen. For a given intensifying screen and film contrast level, as the number of x-ray absorption events in the screen decreases, the quantum mottle (noise) increases.

Every screen-film image contains a certain degree of quantum mottle. When quantum mottle is very visible, it interferes with the ability to interpret radiographs, particularly small, low-contrast objects in the image. The negative impact of quantum mottle in an image is affected by both the amount and visibility of the mottle, as influenced by the following factors:

1. *Absorption.* Changing the thickness of the phosphor layer or changing to a phosphor with different absorption characteristics alters the absorption of x-ray quanta in the screen. However, we know as screen thickness increases, screen blur also increases. The combination of these effects is such that it is possible for a thicker, faster screen to produce no noticeable increase in quantum mottle. While the appearance (texture) of quantum mottle in the radiograph may not be appreciably altered, in this case, the overall image blur would increase, thereby reducing image quality. Minimal quantum mottle in a fast screen-film combination is indicative of increased image blur introduced by the screen.

   Conversely, if x-ray absorption is increased by changing to a phosphor having greater absorption without a change in screen thickness, then image blur and quantum mottle might be the same as for the less absorbing screen if patient exposure were reduced to provide radiographs of equivalent optical density. Remember that it is the number of absorbed x-ray quanta, not incident quanta, that controls the amount of quantum mottle.
2. *Screen Conversion Efficiency.* As screen conversion efficiency increases, quantum mottle increases, other factors being equal. Screen conversion efficiency is the ratio of light energy emitted by the screen to the x-ray energy absorbed in it. A more efficient phosphor produces more light for each x-ray quantum absorbed; therefore, fewer x-ray quanta are needed to produce a given density in the radiograph than that required for a less efficient phosphor. As the number of quanta used to produce a given density decreases, quantum mottle increases.
3. *Radiation Quality.* When the x-ray beam is made "harder," that is, when the average energy of the photons in the beam is increased by raising kVp, the average energy of the x-ray quanta absorbed by the screen is increased. The greater the energy of each photon, the brighter the screen light (the greater the number of light photons) emitted when the x-ray photon is absorbed.

The brighter screen light emission implies that fewer x-ray photons will be required to produce a given film density. For this reason, the use of higher kVp increases quantum mottle somewhat.

4. *Film Speed.* If film speed is increased, fewer x-ray quanta are needed to produce a given density in the radiograph, and quantum mottle increases. Conversely, a slower film offers one way of reducing quantum mottle. The relation between film speed and quantum mottle is illustrated in Figure 4-20. This concept is demonstrated radiographically in Figure 4-21.

   Film speed or sensitivity depends on the ingredients in the emulsion and the manner in which they are treated, combined, and coated. Additional factors that influence film speed are spectral sensitivity (i.e., relative sensitivity to the color of the exposing light), the length of the exposure (i.e., reciprocity failure), ambient conditions (i.e., temperature, humidity, and safelighting), and processing conditions.

   In the foregoing discussion, the interdependence of the actions of the screens and the film on their combined speed have been discussed. Because of the important influence processing conditions have on the speed of the system, there are those who believe that terminology should be amended to refer to the "screen-film-process combination" rather than to the "screen-film combination" alone.

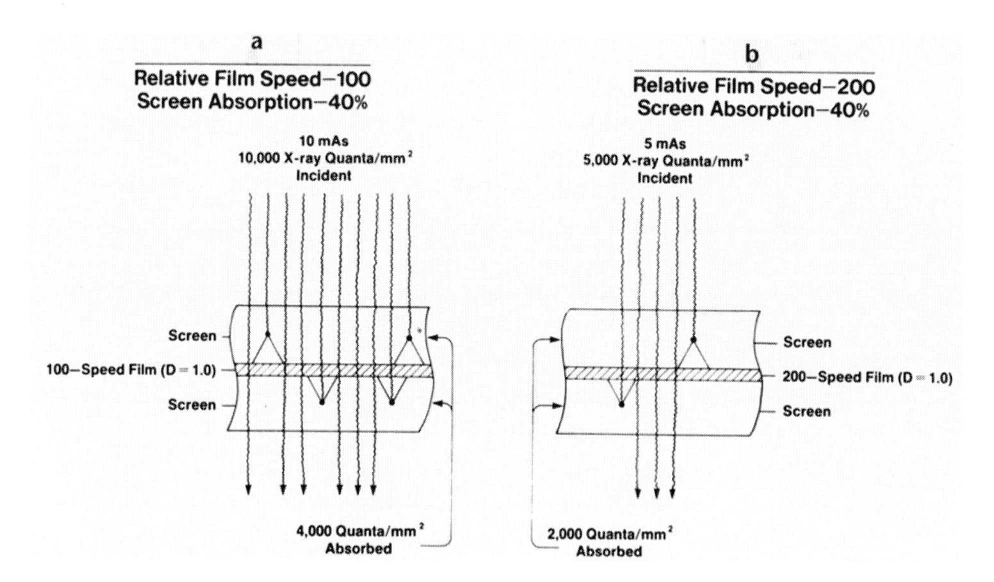


**Figure 4-20** Illustration of how increased film speed results in increased quantum mottle. In this hypothetical example, using the same screens in both cases, Film B requires only half as much screen light exposure as Film A, because its speed is twice that of Film A. Thus, the screen-Film B combination needs only 2,000 absorbed x-ray quanta per square millimeter ($mm^2$) to produce a density of 1.0, whereas the screen-Film A combination requires 4,000 quanta per square millimeter. The result is greater quantum mottle in radiographs that used fewer x-ray quanta.

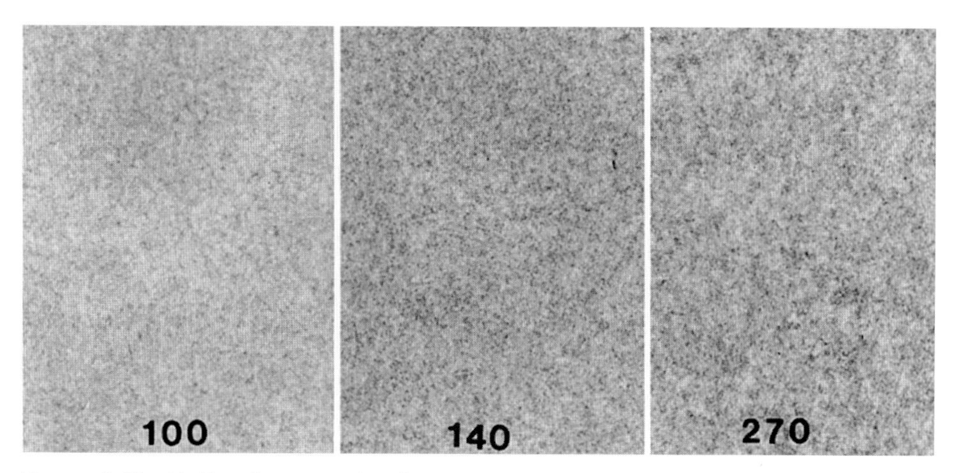


**Figure 4-21** Uniformly exposed radiographs using a 28-kVp setting demonstrate radiographic noise differences for mammographic screen-film combinations with relative speeds of 100, 140, and 270. The average gradients of these films are approximately 2.9, 3.2, 3.2, respectively. These images were photographically magnified 10x. Note the appearance of increased noise as system speed increases. This noise increase is primarily due to quantum mottle; as the speed of the screen-film combination increases (and radiation dose decreases), fewer x-ray photons are involved in producing the mammographic image.

5. *Film Contrast.* As stated earlier, when film contrast increases, quantum mottle increases, other factors remaining constant. For a given difference in brightness between two areas of an intensifying screen, a high-contrast film produces a larger density difference in the radiograph than does a low-contrast film. This is true whether the brightness differences are the result of differences in absorption by the subject or whether they are the result of small random fluctuations in energy conversion of the x-ray quanta absorbed in the screen. Accordingly, a high-contrast film will produce more quantum mottle than a low-contrast film because of its greater enhancement of brightness differences in the emission patterns of the screens. The effect is illustrated in Figure 4-22.

   Quantum mottle can affect the visibility of structures in a radiograph. For example, compare the images of the beads in Figure 4-23. Many more x-ray quanta (more x-ray absorption events) were required to produce the radiograph in Figure 4-23a than in Figure 4-23b. Consequently, the increased quantum mottle in Figure 4-23b has noticeably degraded the image quality. Quantum mottle has a relatively greater effect on the perceptibility of low-contrast objects, such as cysts or soft-tissue lesions.

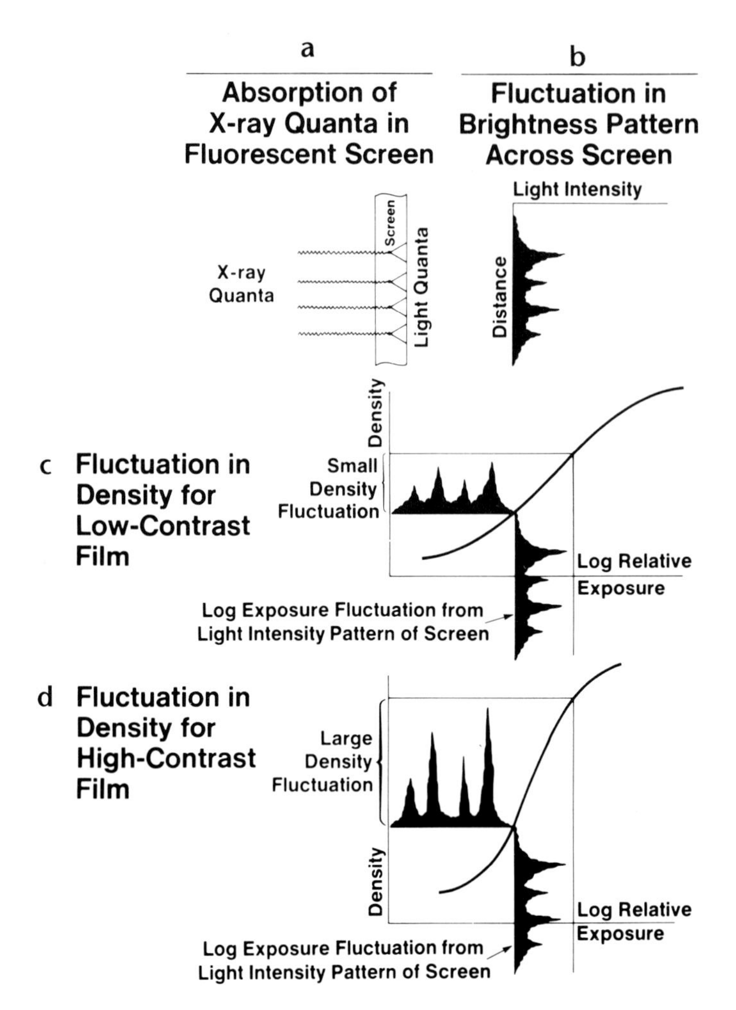


**Figure 4-22** Illustration of how increased film contrast results in increased quantum mottle. (a) The absorption of x-ray quanta and conversion of their energy to light in a fluorescent intensifying screen. (b) A hypothetical graph showing microscopic changes in light intensity produced by each of these absorption events as measured by a photometer scanning across the screen. (c) The characteristic curve for a low-contrast film. (d) The characteristic curve for a high-contrast film.

If a film having these characteristics is exposed to light having the intensity distribution or pattern shown in (b), which is represented by the log relative exposure pattern entering from the bottom of (c), the film-process combination will produce a corresponding density pattern—as shown emerging to the left of the characteristic curve in (c). When the intensity pattern of (c) is registered on a high-contrast film having a characteristic curve like that shown in (d), the resulting density differences are greater than for the low-contrast film in (c). These increased density differences increase the relative deviation in density in the radiograph made with the high-contrast film. Therefore, quantum mottle is greater for the high-contrast radiograph than for the low-contrast radiograph, regardless of whether the contrast difference is the result of choice of film type or of differences in processing conditions for the same film.

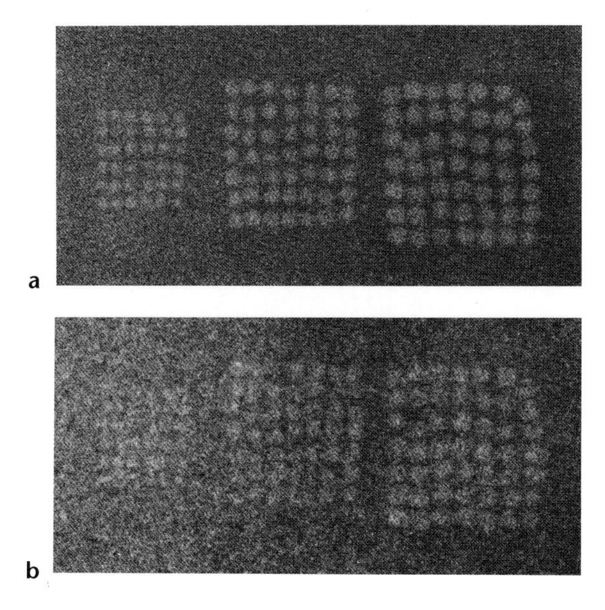


**Figure 4-23** Effect of quantum mottle on image quality. (a) A radiograph of a phantom containing moderate absorbing beads made with a slow screen-film combination using many x-ray quanta (low quantum mottle) is compared with (b) a radiograph made with a fast screen-film combination using relatively few x-ray quanta (high quantum mottle).

## Artifacts

Artifacts are unwanted density variations in the form of blemishes in the radiograph that arise from improper handling, exposure, processing, or housekeeping. Chapter six and Appendix C discuss a number of artifact causes. Some artifacts, such as those produced from dirty film processor rollers, produce images with a mottled appearance (i.e., wet pressure). These artifacts are not, however, due to random statistical fluctuations and, therefore, are not representative of quantum mottle.

## Viewing Conditions

Much of the information contained in the radiographic image cannot be seen if proper consideration is not given to viewing conditions. Radiographs should be interpreted under conditions that provide good visibility, comfort, and minimal fatigue.

The illuminator surface should provide diffused light of uniform brightness, and variations in surface luminance should be gradual. The luminance level must be sufficient to illuminate areas of interest in the radiograph. Ideally, all viewboxes should have lamps of the same color.

The contrast sensitivity of the eye (the ability to distinguish small luminance differences) is greatest when surroundings are of about the same brightness as the area of interest. Therefore, to see detail in a radiograph, it is important to reduce glare to a minimum, to avoid surface reflections, and to reduce the ambient light level to approximately that reaching the eye through the radiograph. Glare and reflections can be reduced by locating illuminators away from bright surroundings such as windows, by turning off surrounding viewboxes when not in use, and by using masks to cover unused portions of a viewbox or to cover low-density areas in the radiograph being examined. Subdued lighting is preferred in the viewing room. It is also important to have a variable high-output light source (with appropriate masks) to view

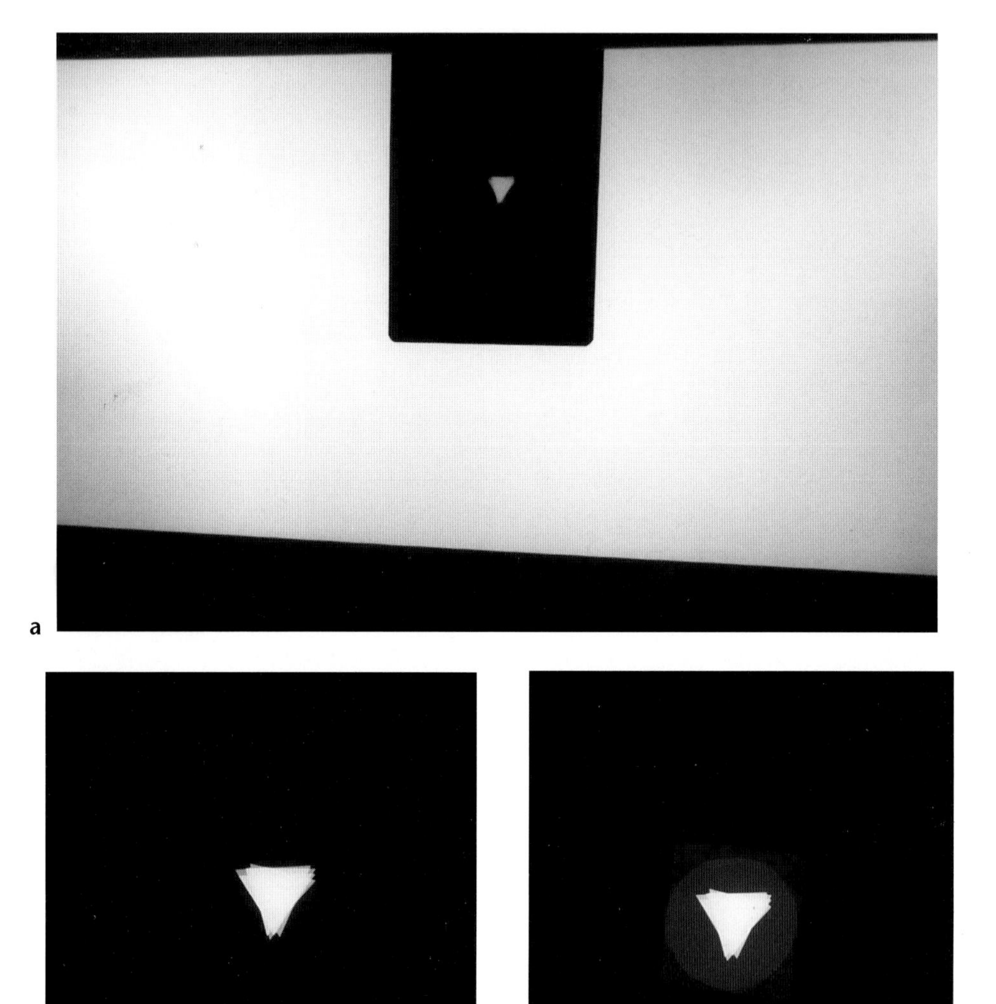


**Figure 4-24** Using the viewing conditions tool. (a) Place the film on a viewbox with all adjacent viewboxes illuminated and overhead lights turned on (no control of ambient lighting). (b) The four triangles typically seen when lighting is as described in (a). (c) After eliminating all extraneous light (room darkened) and placing a mask over the film, multiple objects on the film may be perceived, including the triangles, a circle, a square, and a star.

high-density areas on radiographs, and to make sure radiographs are properly exposed. A tool, consisting of a special film and mask, can be used to demonstrate the potential visual loss of diagnostic information when viewing radiographs with less than optimum viewing conditions (Figure 4-24).

## References

1. Arnold BA, Eisenberg H, Bjarngard BE. Measurement of reciprocity law failure in green-sensitive x-ray films. Radiology 126: 493–498, 1978.
2. Bollen R. Influence of ambient light on the visual sensitometric properties of and detail perception on a radiograph. SPIE Proceedings 273: 57–62, 1981.
3. Bunch PC, Huff KE, Van Metter RL. Analysis of the detective quantum efficiency of a radiographic screen-film combination. Journal of the Optical Society of America 4: 902–909, 1987.
4. Haus AG. Film processing systems and quality control. In: Gould RG, Boone JM, eds. A categorical course in physics: Technology update and quality improvement of diagnostic x-ray equipment. Oak Brook, IL: Radiological Society of North America, 1996: 49–66.
5. Haus AG. Historical developments in film processing in medical imaging. In: Haus AG, ed. Film processing in medical imaging. Madison, WI: Medical Physics Publishing, 1993: 1–16.
6. Haus AG. Measures of screen-film performance. RadioGraphics 16: 1165–1181, 1996.
7. Haus AG. Screen-film image receptors. In: Haus AG, Yaffe MJ, eds. Syllabus: A categorical course in physics: Technical aspects of breast imaging. Oak Brook, IL: Radiological Society of North America, 1992: 69–84.
8. Haus AG, Cullinan JE. Screen-film processing systems: A historical review. RadioGraphics 9: 1203–1224, 1989.
9. Haus AG, Dickerson RE. Characteristics of screen-film combinations for conventional medical radiography. Eastman Kodak Company publication no. N-319, 1995.
10. Haus AG, Gray JE, Daly TR. Evaluation of mammographic viewbox luminance, illuminance and color. Medical Physics 20 (3): 819–821, 1993.
11. Haus AG, Rossmann K, Vyborny C, Hoffer P, Doi K. Sensitometry in diagnostic radiology, radiation therapy, and nuclear medicine. Journal of Applied Photographic Engineering 3: 114–124, 1977.
12. Hendee WR. Variables that affect image clarity. In: Haus AG, ed. The physics of medical imaging: Recording system measurements and techniques. New York: American Institute of Physics, 1979: 427–441.
13. Huff KE, Wagner PW. Crossover and MTF characteristics of tabular-grain x-ray film. SPIE Proceedings 486: 92–98, 1984.
14. Introduction to medical radiographic imaging, Eastman Kodak Company publication no. M1-18, 1993.
15. Kimme-Smith C, Haus AG, DeBruhl N, Bassett LW. Effects of ambient light and viewbox luminance on detection of calcifications in mammography. American Journal of Roentgenology 168: 775–778, 1997.

16. Kodera Y, Doi K, Chan HP. Absolute speeds of screen-film systems and their absorbed energy constants. Radiology 151: 229–236, 1984.
17. Rao GU, Fatouros P. The relationship between resolution and speed of x-ray intensifying screens. Medical Physics 5: 205–208, 1978.
18. Seibert JA, Barnes GT, Gould RG. Specification, acceptance testing, and quality control of diagnostic x-ray imaging equipment. New York: American Institute of Physics, 1994.
19. Sprawls P Jr. Physical principles of medical imaging. Rockville, MD: Aspen, 1987.
20. Understanding graininess and granularity. Eastman Kodak Company publication no. F-20, 1973.
21. Van Metter RL. Describing the signal-transfer characteristics of asymmetrical screen-film systems. Medical Physics 19: 53–58, 1992.
22. Wagner RF. Toward a unified view of radiological imaging systems. II: Noisy images. Medical Physics 4: 279–296, 1977.

# 5

# Quality Control

- Background
- Processor QC
- Fixer Retention Test

## Background

Personnel in medical imaging facilities have been involved in quality assurance (QA) and quality control (QC) activities for many years.

Quality assurance is a plan that requires daily monitoring to ensure consistent film processing, regular testing to detect equipment malfunction, regularly scheduled equipment maintenance, and an ongoing assessment of variables that could affect image quality and diagnosis. An important facet of quality assurance is quality control.

Quality control refers to the tests and procedures used in a quality assurance plan to perform routine assessments of darkroom facilities, processors, equipment and accessories, viewboxes, and viewing conditions.

Early QA/QC activities were voluntary and primarily aimed at film processing.

Surveys conducted by the Food and Drug Administration, beginning in the late 1970s, highlighted under-processing, resulting in higher radiation exposure and lower contrast, as a serious problem associated with using automatic film processors.

In the late 1980s, as the emphasis on mammography was increasing because of rising breast cancer rates, the American Cancer Society (ACS) became concerned about recommending screening mammography to women without some assurance of the quality of mammography and the safety of the procedure. The ACS approached the American College of Radiology (ACR) about developing standards of quality for mammography. The ACR formally established its voluntary Mammography Accreditation Program (MAP) for mammography facilities in 1987 to address issues of image quality in mammography.

In 1990, the ACR developed QC manuals in mammography for the radiologist, radiologic technologist, and medical physicist. Revisions have been made to the manuals in 1992, 1994, and 1996. The tests that are performed encompass the darkroom, mammographic equipment, film, intensifying screens, cassettes, and processing equipment.

In 1992, the Mammography Quality Standards Act (MQSA) was passed to establish national quality standards for mammography. Under MQSA, all facilities performing mammography in the United States must be accredited, certified, and inspected annually. QA and QC are an integral part of achieving and maintaining accreditation and certification.

Today, personnel employed in medical imaging are aware that QC activities related to the processor are an important part of obtaining high-quality radiographs while minimizing radiation exposure to patients.

This chapter will focus on two of the QC tests that pertain specifically to processing: processor QC and the fixer retention test. A third test, darkroom fog, is discussed in Appendix D.

Note two additional points:

1. Correct, consistent methodology in the performance of QC tests must be employed every day by everyone in the facility involved with QC.
2. All QC procedures, including how the tests are done and evaluated and what corrective actions should be taken, should be documented in writing in the policy and procedures manual.

# Processor QC

In order to maintain consistent processor performance and to detect changes in image quality before they become significant enough to affect diagnosis and radiation dose to the patient, it is important to implement a processor quality control program.

## Tools Required for Processor QC

The tools necessary to perform processor QC (Figure 5-1) include the following:

- A thermometer.
- A simulated light sensitometer.
- A box of quality control film.
- A densitometer.
- Processor control charts.

### *Using the Thermometer*

After the processor has reached operating temperature first thing in the morning (usually signaled by the processor going into standby mode), a thermometer (preferably digital) with a known accuracy of at least ±0.5°F (±0.3°C) should be used to check the developer temperature.

The probe of the thermometer should be placed in the same location of the developer tank each time, according to the recommendations of the processor manufacturer. This is usually on the nondrive side of the processor between the side plate of the developer rack and the rack support.

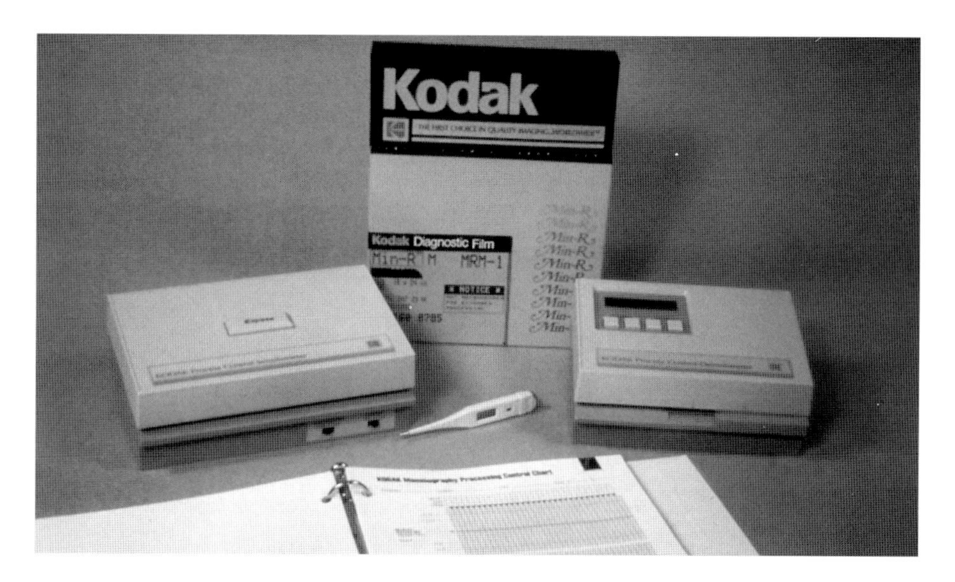


**Figure 5-1** The tools required for processor quality control include (clockwise from top) a box of QC film, a densitometer, processor QC charts, a sensitometer, and (center) a thermometer.

Thermometers that contain mercury should never be used. Refer to Chapter three for additional information on thermometers.

## *Using a Simulated Light Sensitometer*

A simulated light sensitometer is a device that is used in the darkroom to expose x-ray film to different light intensities in a "step" fashion. The incremental log exposure difference between settings on sensitometers is typically about 0.15. The color and range of brightnesses produced by the sensitometer simulate the range of light produced by the screens that expose the film inside a cassette. The processed film exposed by the sensitometer is called the sensitometric strip, the film strip, or the processor QC strip.

Note that the light emitted from simulated light sensitometers is different from the light emitted from intensifying screens. Therefore, the response of film to an exposure from a simulated light sensitometer differs significantly from the response of film to exposure from intensifying screens. Because of this, simulated light sensitometers should not be used to make comparisons of film speeds and contrast (Figure 5-2).

A simulated light sensitometer that exposes 21 density steps and allows single- or double-emulsion film to be exposed by green or blue light should be used. The required settings (single or dual, green or blue) should be verified prior to each use, especially if the sensitometer is carried between several darkrooms and used for general radiology as well as mammography film (Figure 5-3). It is important to use single-sided exposure for single-emulsion films and double-sided exposure for double-emulsion films. Incorrect or misleading sensitometric data may be obtained

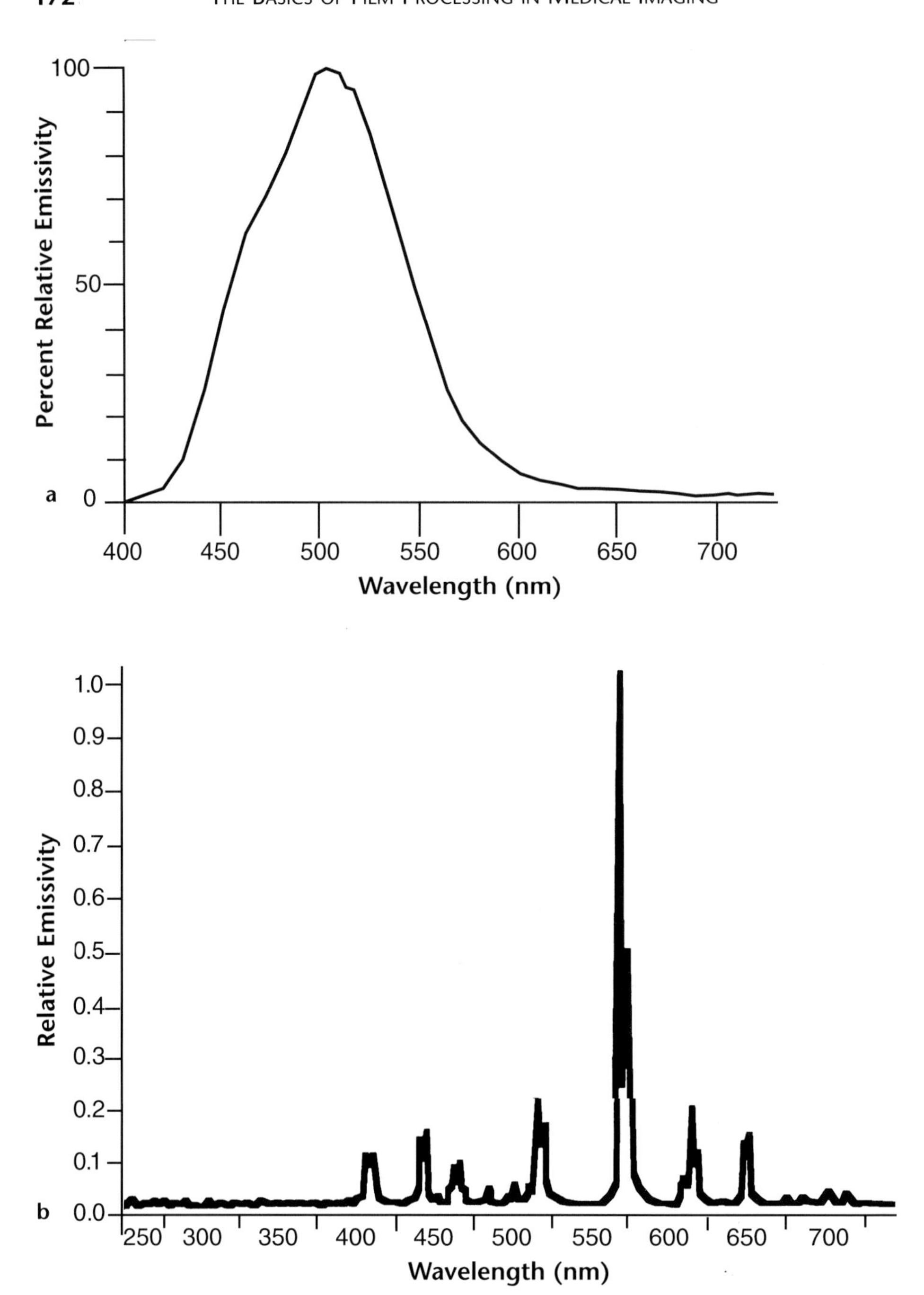


**Figure 5-2** (a) Typical broad band emission spectrum of a green electroluminescent panel with the peak emission at approximately 510 nanometers. (b) Typical nearly monochromatic spectral emission of a green radiographic intensifying screen.

**Figure 5-3** The settings on a sensitometer (single or dual and green or blue) should be verified prior to each use for processor QC.

for film contrast evaluation and processor quality control if a simulated light sensitometer with a single-sided, light-exposing device is used for double-emulsion films with very low light crossover (Figure 5-4).

The use of asymmetric near-zero crossover films requires special consideration when they are used with simulated light sensitometers for processor quality control. Because these types of films have different front and back emulsions, even simultaneous double-sided simulated light exposure does not generate the correct characteristic curve shape if the proper degree of asymmetry in light output is not achieved. The use of a neutral density filter placed between at least one of the light sources and the film emulsion can achieve a more similar front-to-back ratio of the light intensity produced by an asymmetric screen pair. Therefore, by using the neutral density filter, the proper characteristic curve shape can be obtained (Figure 5-5).

For single-emulsion mammography film, the sensitometer will usually be set on single and green. In addition, film must be inserted into the sensitometer so the emulsion side of the film is toward the light source.

Many sensitometers are exposure adjustable. Dip switches recessed into the back of these sensitometers allow the densities to be shifted, based on the speed of the particular type of film being used. A dip switch setting that positions an optical density of approximately 1.10 to 1.50 on step 11 is recommended because automatic scanning densitometers are often preset by the manufacturer to read the speed step on step 11. Consult the film or sensitometer manufacturer for dip switch recommendations for specific film types.

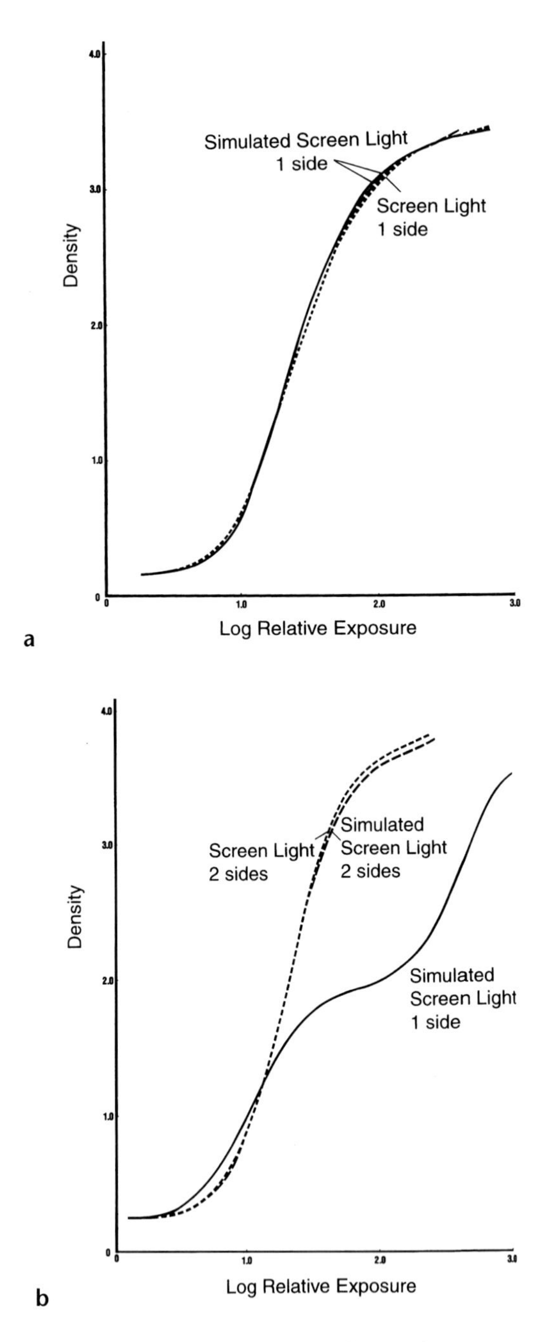


**Figure 5-4** (a) Characteristic curve of a single-emulsion mammography film. (b) Characteristic curves of a double-emulsion film with a crossover-control layer. A significant distortion (bump) occurs in the curve when the film is exposed with a single-sided simulated light sensitometer. This distortion in the characteristic curve is absent when a double-sided, simulated green-screen-light exposure is given. (Reprinted with permission from Medical Physics 1990; 17: 693–694.)

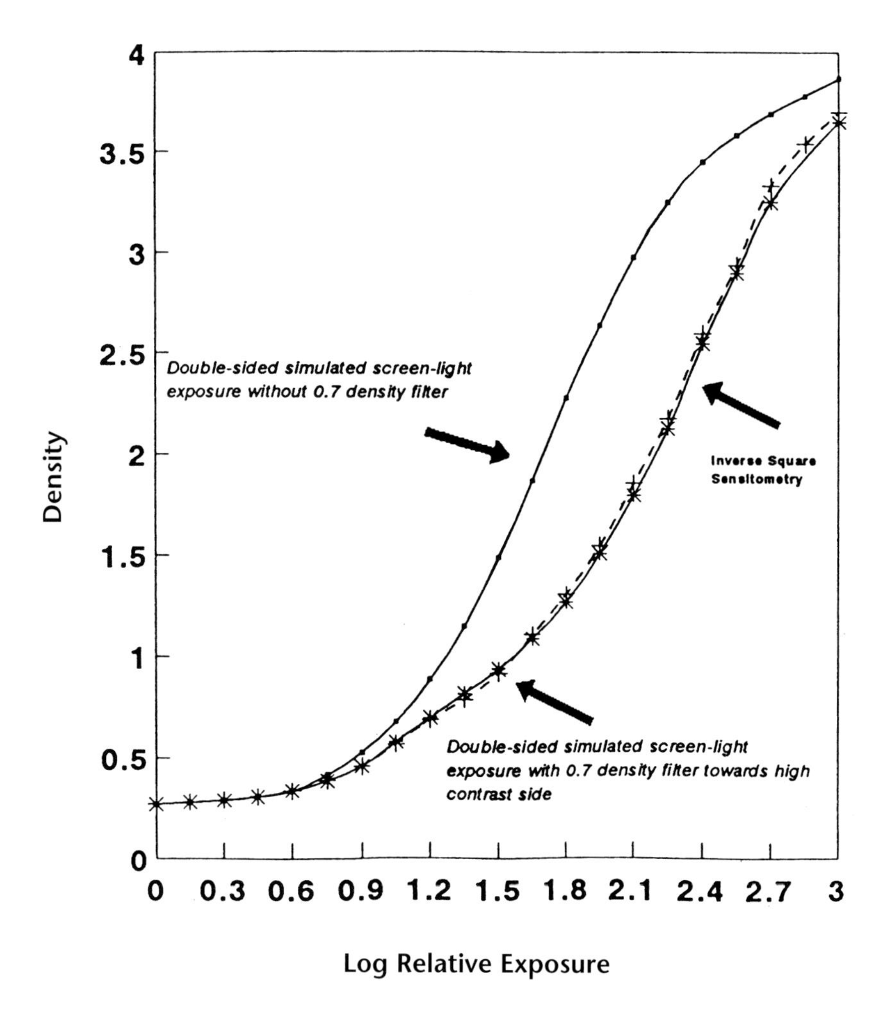


**Figure 5-5** Characteristic curves of an asymmetric film receiving a double-sided simulated screen-light exposure (with and without a 0.7 density filter) compared to inverse square sensitometry. (Reprinted with permission from Medical Physics 1994; 21: 527.)

The key requirement for simulated light sensitometers is that they must provide a reproducible exposure, which enables the daily processor QC test to be performed. Simulated light sensitometers generally meet this requirement.

Sensitometers should be used and stored in clean areas. Dust and dirt could affect film density and should be periodically removed from the exposing area of the sensitometer. Canned air, where the air is clean and free of moisture, may be used for this purpose. An extension tube for the canned air may be helpful for those instruments that feature a slot into which the film is inserted. Note that most canned air should be used in an upright position and not shaken prior to use.

Environmental temperature should also be controlled, typically between 59 and 86°F (15–30°C) in any areas where sensitometers are used.

Many sensitometers are powered by a standard battery that should be replaced periodically. If a problem is suspected with a sensitometer (i.e., no exposure), replace the battery before referring the unit to the factory for repairs.

In addition, it may be desirable to recalibrate simulated light sensitometers periodically because light output can diminish as the tungsten light source or the electroluminescent panel in the sensitometer ages. The sensitometer manufacturer can provide additional information on this service, including the recommended frequency. Minimally, perform periodic QC on the sensitometer by a direct comparison between two sensitometers by exposing film from the same box, processing through the same processor, and measuring optical density values using the same densitometer.

It is important that a single sensitometer be used for processor QC of a particular processor. If it becomes necessary to use a different sensitometer for whatever reason, adjustments to or reestablishment of the operating levels on the processor QC chart will be necessary, even if the sensitometer is the same make and model. It is recommended that crossover sensitometry be performed using both sensitometers for several days. This will verify that the processor is operating properly while reestablishing the processor QC chart.

Sensitometric strips from the last month, or longer if regulations require it, should be retained.

### *Quality Control Film*

When establishing a processor quality control program, a box of fresh film should be selected and set aside for processor QC as follows:

- For a processor dedicated to a particular type of film, such as mammography, the box of film should be the same type of mammography film (identical catalog number). Select a box of 18 x 24 centimeter film.
- For processors with a mix of different film types, such as mammography and general radiography, mammography film should be used. (Ideally, processor QC should be performed using all film types in a mixed processing environment.)
- For processors with a mix of different film types other than mammography, the most process-sensitive film of the types being processed should be selected for processor QC. Note that because mammography film is more process-sensitive than films used for conventional medical imaging, it is not recommended in this case because it may result in unnecessary control actions for the less process-sensitive films.

It is important that the emulsion number of the box of film be noted on the processor control chart. If an emulsion number log is maintained, as recommended for mammography film in Chapter seven, also record the number there. The emulsion number can usually be found on the front and side of each box of film (Figure 5-6).

The box selected for processor QC should be clearly labeled "QC" (Figure 5-7). As with all film, it should be protected from sources of radiation, light, and chemical

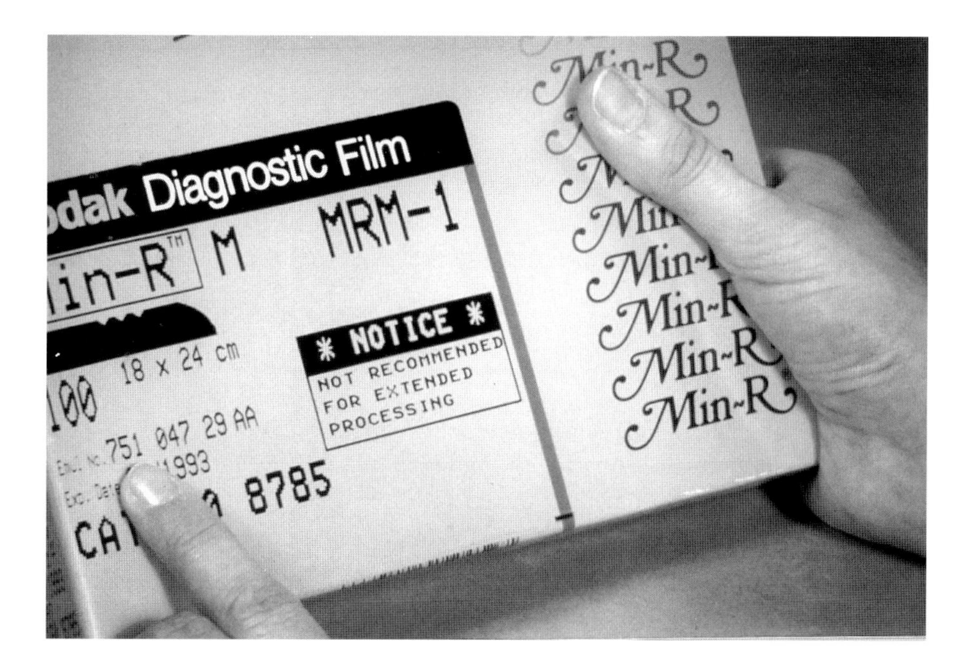


**Figure 5-6** The emulsion number can usually be found on the front and side of each box of film.

fumes and stored within the recommended temperature and relative humidity ranges necessary for film (50 to 70°F [10 to 21°C]/30 to 50 percent relative humidity).

If the box of film (100 sheets) is used to monitor a single processor in a single darkroom and films are processed at least five days per week, the box will typically last approximately three months. If environmental temperature and humidity cannot be controlled 24 hours/day, 7 days/week, film aging and sensitometric changes may be accelerated, and it may be necessary to select a new box of film for processor QC at shorter intervals. In this case, the balance of the film in the box can be used for other purposes (e.g., other quality control tests).

It has been common practice in medical imaging facilities to reserve a large quantity of film with the same emulsion number (i.e., an entire case [500 sheets], or enough to last for six to nine months) for processor QC. Such facilities usually have a large number of processors that are controlled with only one type of film, and they primarily wish to avoid having to do crossovers frequently.

A large supply of film should be reserved only if the film is kept frozen at an approximate temperature range of 0 to 32°F (-18 to 0°C). Before opening a new box of film that has been kept in the freezer for QC, remove the box from the freezer and allow it to remain, unopened, at room temperature (68 to 70°F [approximately 20°C]) for a minimum of 24 hours. A crossover should be done when the next supply of film is selected.

An alternative for large facilities with many processors using the same type of diagnostic film would be to select a large quantity of film with a single emulsion number and provide a box of QC film for each individual processor. This would limit having

**Figure 5-7** The box selected for processor QC should be clearly labeled "QC."

to do a crossover to approximately four times per year. Note that all processors should be at the same operating levels.

For mammography facilities, a crossover procedure should be performed with each new box of film selected, regardless of whether or not the emulsion number is the same as the current box.

## *Using the Densitometer*

A densitometer is a device that is used to measure the optical density of selected steps after the film exposed by a simulated light sensitometer has been processed.

Two types of densitometers are generally available:

- Spot-reading densitometers (manual transmission densitometers) (Figure 5-8).
- Automatic scanning densitometers (as shown in Figure 5-1, page 171).

Spot-reading densitometers allow the operator to measure the optical density on a film by positioning the area to be read over the light source aperture and then bringing the reading head of the unit into direct contact with the film. Care must be taken in positioning the film as close as possible to the center of each step on the film, or inaccurate readings could result. (Optical density measurements taken across any single step will differ slightly due to the step tablets installed in the densitometer.) Automatic scanning densitometers with many advanced features are also available, but at a much higher price. The advantages of such units include the ability to measure all sensitometric steps on the processed film quickly and easily, to obtain reproducible density readings, to program the unit to read selected steps to monitor the processor, to print processor QC charts automatically (regulatory bodies require

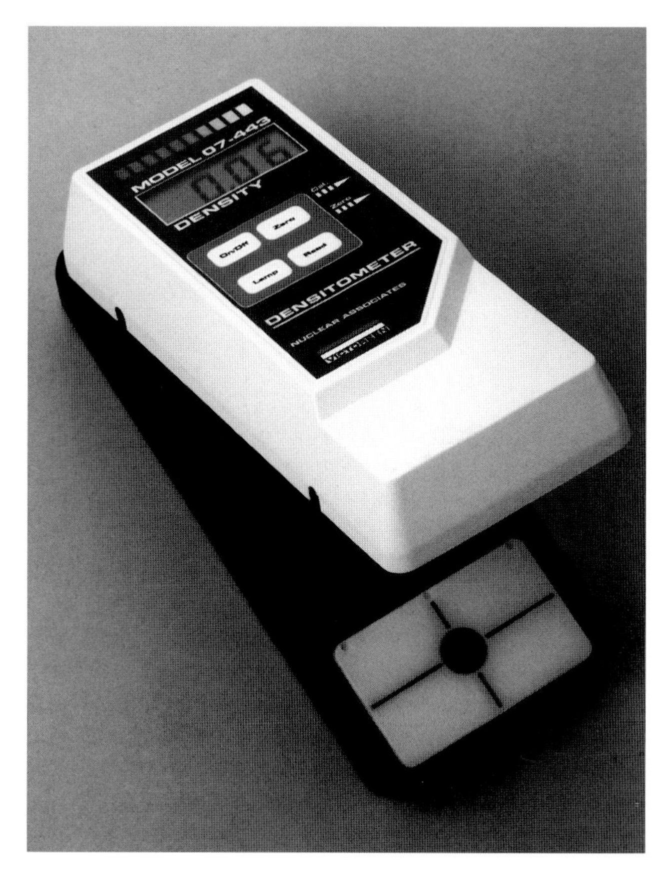


**Figure 5-8** Typical spot-reading densitometer. (Printed with permission from Nuclear Associates.)

this to facilitate trend analysis), and to download data into a computer so it can be stored and manipulated. A considerable disadvantage of automatic scanning densitometers appears to be the difficulty of understanding and using any but the most basic functions (i.e., how to change the programming, etc.) Familiarity with the operator's manual and additional training on the use of an automatic scanning densitometer are recommended.

Many spot-reading densitometers come with apertures in several diameters (i.e., 1, 2, and 3 millimeter). Any aperture may be used as desired; as the size of the aperture selected increases, the precision required in positioning the area on the sensitometric strip to be measured also increases. Generally speaking, although more care must be exercised in positioning the densitometer, the larger apertures produce a better signal-to-noise ratio and, therefore, provide more reliable and reproducible results.

It is important that a single densitometer be used for processor QC of a particular processor. If it becomes necessary to use a different densitometer for whatever reason, adjustments to or reestablishment of the operating levels on the processor QC chart will be necessary.

## *Checking the Calibration of the Densitometer*

The calibration of the densitometer should be checked at least annually. This is accomplished by using a calibration strip, which is a piece of clear-based film with several steps at different optical densities traceable to the National Institute of Standards and Technology (NIST). A calibration strip usually comes with the instrument when initially purchased and may also be purchased separately. The typical accuracy of the calibrated steps is ±0.02 optical density or 2.5 percent, whichever is greater. The optical density measurements of the densitometer should correlate to the optical densities of the calibrated steps.

A small calibrated step tablet, with five calibrated steps, is used with spot-reading densitometers.

A calibration strip, with 21 steps (5 of which are calibrated), is required for automatic scanning densitometers. (Note that automatic scanning densitometers are frequently self-calibrating.)

Calibration strips must be handled carefully by holding the strip near the edges of the film only. Fingerprints or other foreign substances on the measurement area will cause errors in optical density readings. A soft camel hair brush may be used to clean the surface of the strip. The use of any other cleaning agents may invalidate the strip.

Calibration strips are usually expiration dated for some period of time (i.e., 18 months from a stated date or whenever physical abuse of the strip is noted). The life of the strip may be prolonged by storing it in a dark, cool, dry place.

For more frequent QC checks of the calibration of the densitometer (i.e., daily or weekly), minimize the handling of the calibration strip by generating a sensitometric strip to be used for QC of the densitometer. If the densitometer accurately measures the calibration strip, the measurements taken on the sensitometric strip can also be considered accurate. Use the sensitometric strip until smudges or scratches on the film appear; then generate and calibrate another sensitometric strip to the calibration strip.

In addition, densitometers may conform to two different standards (ANSI PH2.19-1976 or ANSI PH2.19-1986). The optical density readings from densitometers set according to the 1986 standards cannot be directly compared with the optical density readings from densitometers set to the 1976 standards.

If it should be necessary to make a comparison, it is possible to convert from 1986 standards to 1976 standards by adding 0.03 to the optical density of step 11, 0.06 to the optical density of step 15, and 0.08 to the optical density of step 21 (no adjustments for steps 1 and 8 are needed). In addition, densitometers set according to the 1976 standards will not read the optical densities of high D-max films, such as mammography films, as accurately as densitometers set to the 1986 standards or, in the case of automatic scanning densitometers, may not accept the strip at all. Check the operator's manual or ask the densitometer manufacturer to determine which standards apply for a particular unit, and whether upgrades are available.

### *Using Processor Control Charts*

Processor QC involves the measurement of three parameters—speed, as indicated by the optical density of a mid-density (MD) step on a sensitometric strip; contrast, as indicated by the density difference (DD) between two selected steps on the sensitometric strip; and base plus fog, as indicated by the optical density of an unexposed region of the film. The MD, DD, and base plus fog values for each processor are plotted versus time on a special form known as a processor control, or QC, chart (see Figure 5-13, page 187). Many charts also contain space to either plot or merely record developer temperature. Control charts are a convenient way to maintain data and facilitate analysis of the data (i.e., looking for trends). Control charts may be filled in manually by hand, or they may be generated automatically by having the densitometer linked to either a printer or a computer and printer.

Other important areas of the processor control chart include the following:

- Dates and year the information on the chart reflects.
- Initials of the individual performing the charting.
- The type of processor and its location.
- The type of film used for QC and the complete emulsion number of the film.
- The control limits of speed, contrast, and base plus fog, and possibly also of developer temperature. Results that fall on, or outside, the control limits indicate that corrective action is required.

  Note that each line above or below the operating level for speed, contrast, and base plus fog usually indicates a change of 0.01 (1/100th). Each line above the operating level for developer temperature, on those charts that include it, usually indicates a change of 0.1 (1/10th) degree.
- The developer and fixer replenishment rates.
- The "remarks" section, where all documentation relating to what has happened on any particular day, such as service, a chemical change, a crossover, the solution to a problem, etc., has been noted.

Processor QC charts, whether filled in manually or created automatically, should contain all of the information required by any particular regulatory body under which the facility operates. Neatness and ease of interpretation are important.

Processor QC charts should be retained for at least one year or as required by local regulations.

Forms are commercially available or may be copied from quality control manuals, such as the ACR's QC manual, where permission is given to do so.

## Prior to Establishing a Processor QC Program

The processor QC charts of many medical imaging facilities indicate fairly consistent processing. If the processing environment has not been set up correctly initially, however, the consistency shown by the charts may actually reflect consistently poor processing.

Therefore, the first critical step prior to establishing the processor QC program is to set up the processor and darkroom environment correctly. This is best accomplished by checking that the processor, film, and chemical manufacturers' recommendations have been adhered to for the following:

- Processor installation.
- Processor and darkroom ventilation.
- Processor maintenance. The processor must be clean, functioning properly, and maintained at appropriate intervals.
- Safelighting. The darkroom environment should be evaluated, any deficiencies corrected, and the darkroom fog test performed satisfactorily before establishing the processor QC program (Appendix D).
- Processor cycle time.
- Fresh chemicals, recently mixed according to specifications, including the specified amount of the chemical manufacturer's developer starter. Note that the first day of establishing the processor QC program may be either the same day chemicals are changed (fresh start, as discussed in Chapter two), or may be delayed for several days while some amount of seasoning takes place. If the chemicals have been freshly mixed, it is best to wait at least one hour before processing a sensitometric strip.
- Developer and fixer replenishment rates, based on the volume and type of film.
- Solution and dryer temperatures.

If the processor will be used for single-emulsion mammography films, it should also be evaluated to determine if processing the films with the emulsion side of the film up or down provides the best overall image quality results (Appendix C).

## Establishment of a Processor QC Program

After the processor and darkroom are ready and at the time of day processor QC will normally take place, usually first thing in the morning, follow these steps to establish the processor QC program:

1. Check that the developer temperature has reached operating temperature with the thermometer.
2. Select, label, and open a fresh box of film for QC.
3. Set the simulated light sensitometer for the appropriate film spectral sensitivity.
4. Expose and immediately process the sensitometric strip.

   Under appropriate safelighting, remove a sheet of film from the QC box. Insert the film fully into the sensitometer (Figure 5-9). If using single-emulsion film, the emulsion side of the film must be toward the light source (down), and the location of the film notch as the film is inserted into the sensitometer should also be defined so that all personnel follow the same procedure. If an automatic scanning densitometer will be used, there must be

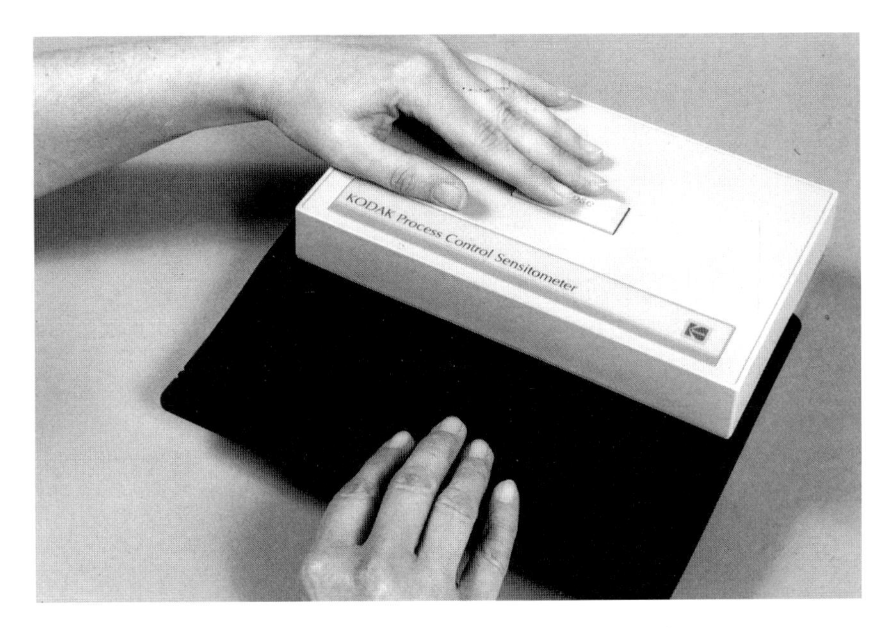


**Figure 5-9** A film must be inserted fully into the sensitometer for processor QC. If you are using single-emulsion film, as shown, the emulsion side of the film must be toward the light source (down).

enough unexposed area (leader) before the outside edge of the first step for the densitometer to accept the strip.

It is generally recommended that each sensitometric strip be exposed and processed immediately because variable time between exposing and processing may cause sensitometric changes due to latent image keeping (Appendix A). Using preexposed strips is not recommended.

For single-emulsion film, it is also important to define how the film will be placed on the film feed tray (if processing manually, i.e., a room light system is not being used). Include the location of the film notch, emulsion side up or down, and which side of the film feed tray (right or left) will be used. The film should be fed so the least exposed density step is fed into the processor first (Appendix C) (Figure 5-10).

Instructions for exposing and processing the sensitometric strip should be included in the policy and procedures manual and posted on the darkroom wall. This assists all personnel, including processor service personnel, in following the particular procedure used in the facility.

5. Follow the above procedure for five consecutive days. The data from the five days are required to determine an average and to establish a meaningful baseline. The five days must be representative of normal operations in the facility. Consider delaying the start of the processor QC program if events, such as a long holiday weekend, etc., will interrupt the normal five days. Also, avoid establishing the processor QC program using strips from fewer than five days.

6. Read and record the optical densities of the steps on each of the five strips. Minimally, reading the least exposed step and steps 8 through 14 will usually suffice. If using a spot-reading densitometer, make sure that optical density measurements are taken from the center of each step.
7. Determine the average of the optical density of each step. For example, the optical density values for step 11 from the five strips should be added together and divided by five to obtain the average value of step 11.
8. Determine which step will be used for speed or MD. Select the step with the average optical density closest to 1.20. This step ideally should also be located on the straight-line portion of the characteristic curve (Chapter four), the most speed-sensitive portion of the curve. Note that the optical densities of all 21 steps can be plotted, if needed, to show the straight line portion of the curve (Figure 5-11). Record the step number and the average value on the processor control chart on the line that represents the operating level for speed.

   This step number should be used consistently unless the processor QC program is reestablished and another step number is chosen.
9. Determine which two steps will be used for contrast or DD. Select the step with the average optical density closest to 2.20. Then select the step with the average optical density closest to, but not less than, 0.45. Subtract the lower density value from the higher one to determine the density difference. Record the step numbers and the density difference on the processor control chart on the line that represents the operating level for contrast.

   Note that the two points used to calculate density difference are located on either end of the straight line portion of the characteristic curve (Figure 5-11).

   These step numbers should be used consistently unless the processor QC program is reestablished and different step numbers are chosen.
10. The average of the least exposed density step (usually the first step) determines the base plus fog, or gross fog, of the film. Any clear area of the film may also be used to determine base plus fog. Measure the same area consistently.

    Base plus fog reflects the optical density of the film base and any inherent chemical fog that may be present in the emulsion. Because of the recent trend to use even bluer bases for film, typical base plus fog values are greater than seen previously. This is most notable in mammography films where typical values range between 0.18 and 0.20.

    Figure 5-12 shows a sensitometric strip with typical values for base plus fog, MD, and DD.
11. Establish the control limits for speed, contrast, and base plus fog. Calculate and record the values of the control limits on the processor control chart. This helps avoid making simple point plotting errors.

    Limits of ±0.15 for speed and contrast and +0.03 for base plus fog are typical. For mammography, it may be desirable to control the processing environment more tightly in general. If so, limits of ±0.10 for speed and

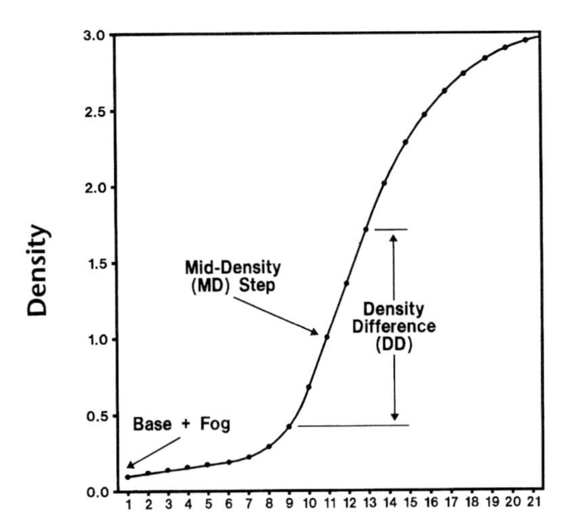


**Figure 5-11** A characteristic curve showing base plus fog at step 1, the mid-density (MD) step located approximately in the middle of the straight-line portion of the curve, and the two points used to calculate the density difference (DD) at either end of the straight-line portion of the curve.

contrast should also be included on the processor QC chart. The ±0.15 limits may then be used as the maximum acceptable limits, or the limits requiring immediate action. Some regulations, such as MQSA in the United States, stipulate that mammography films may not be processed if the speed or contrast reach or exceed the maximum control limits.

Developer temperature, which may be plotted or merely recorded occasionally on some charts, should be maintained within ±0.5°F (±0.3°C).

12. Record the complete emulsion number of the QC film and all other pertinent information on the control chart.
13. Plot the values from the last sensitometric strip as the first day on the processor control chart.

## Daily Processor QC

Processor QC should be performed daily after the parameters have been properly established.

Each morning, after the processor has reached normal operating temperature, use the simulated light sensitometer and film from the QC box to create a sensitometric strip. Safelight illumination in the darkroom is generally used when generating a sensitometric strip, but total darkness may provide even more consistent results. Process the film immediately according to the facility protocol.

Use the densitometer to measure the values for speed, contrast, and base plus fog, using the same steps recorded on the processor control chart. Plot these values on the control chart and evaluate the results immediately.

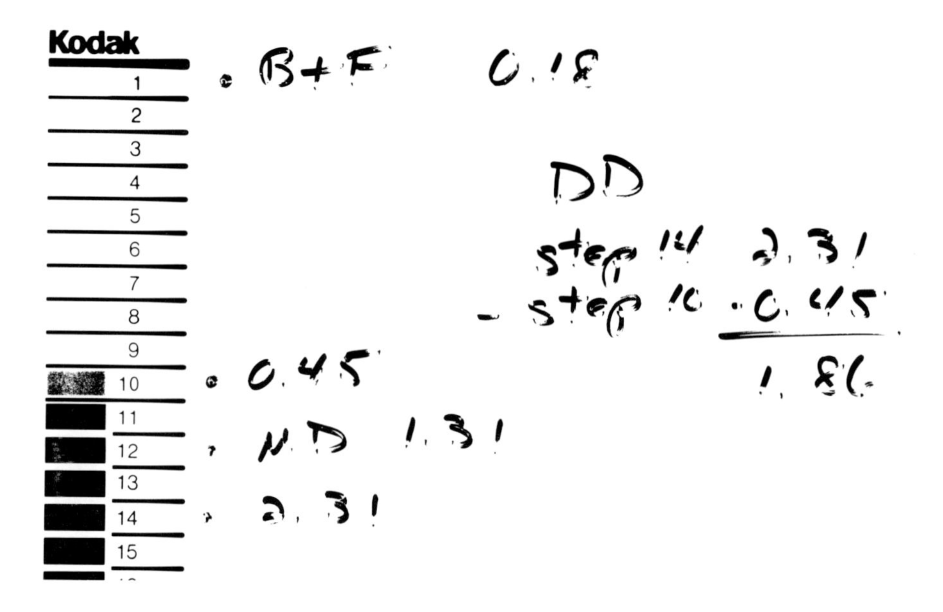


**Figure 5-12** A sensitometric strip with typical values for base plus fog, mid-density, and density difference.

## Evaluating the Processor QC Chart

Determine if any points plotted have reached or exceeded the control limits. If so, expose and process a second sensitometric strip, double-checking that correct procedure is followed. If the same results are obtained, the cause must be determined and corrected. Refer to Chapter seven for information on troubleshooting the processor.

The points that are out of control should be circled, notations as to the cause(s) and solution(s) recorded in the "remarks" section of the processor control chart, and the data from another sensitometric strip generated after the processor has been brought back into control plotted (Figure 5-13).

If the ±0.10 control limits for speed and contrast are used for mammography as well as the ±0.15 limits and any of the data points from the daily strip are between ± 0.10 and ±0.15, expose and process a second sensitometric strip for comparison. The processor should be closely monitored to make certain the ±0.15 limits are not exceeded while performing an initial assessment as to the cause of the problem.

Base plus fog values should generally remain within +0.03 of the operating level. Any points plotted for base plus fog that exceed +0.03 require immediate analysis.

The developer temperature should remain as close as possible to the temperature recommended by the film manufacturer for the film type and processing cycle. The developer temperature should not fluctuate more than ±0.5°F (±0.3°C). Note that situations necessitating deviation from the manufacturer's developer temperature recommendations may occur (i.e., different models of processors used for mammography within the same facility must be matched, etc.).

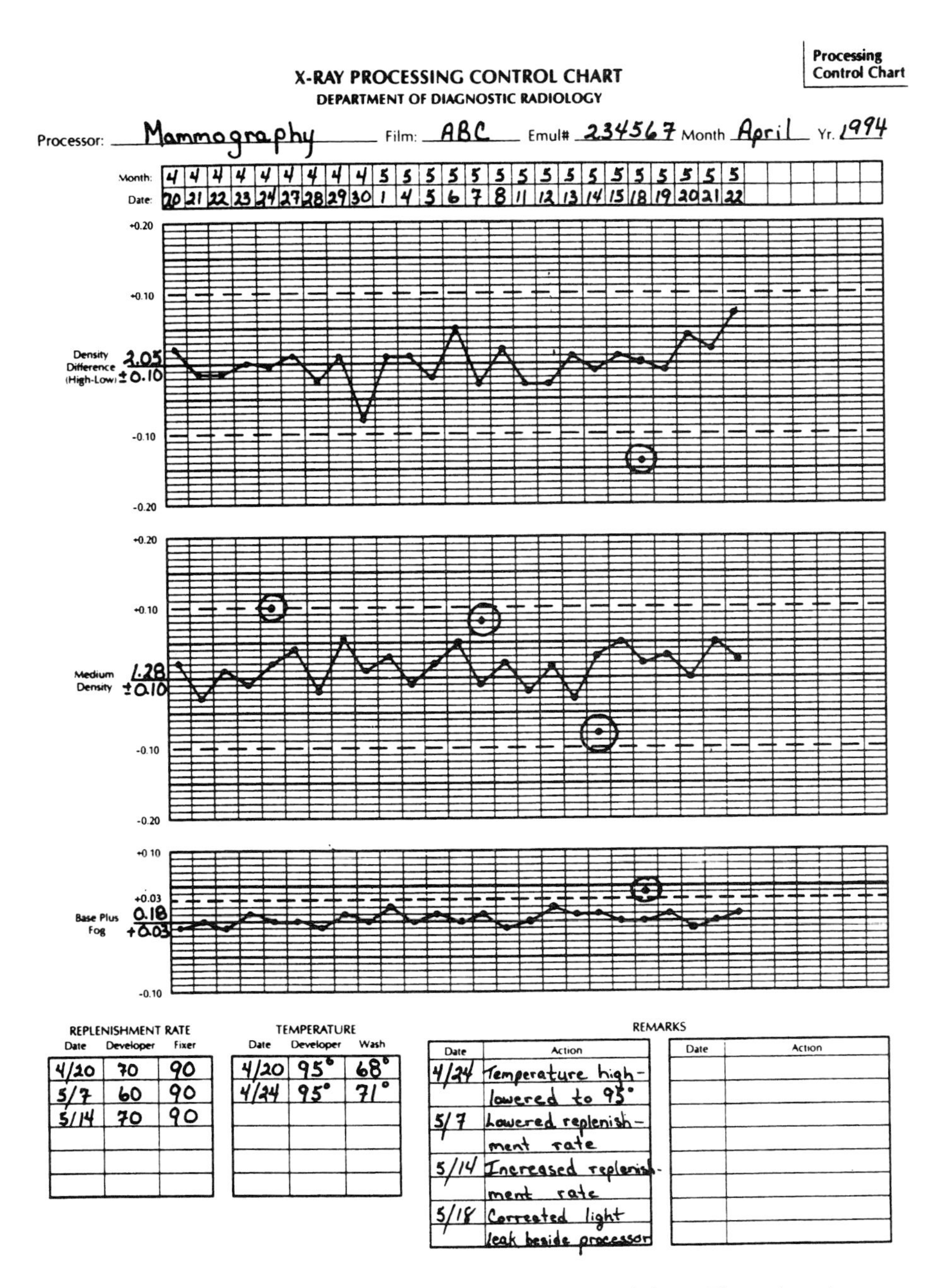


**Figure 5-13** A processor quality control chart with typical data. The points that are out of control should be circled, notations as to the cause(s) and solution(s) recorded in the "remarks" section of the processor control chart, and the data from another sensitometric strip generated after the processor has been brought back into control plotted. (Reprinted with permission from the American College of Radiology.)

Finally, evaluate the processor for any trends. A trend is a series of three or more consecutive points progressing steadily upward or downward. Trends may be watched for up to five days before taking action, unless the operating limits are reached or exceeded. It is important not to react too soon. However, if the processor *clearly* has a trend developing, immediate corrective action can be initiated, even before a control limit (e.g., ±0.10 for speed) is reached.

## The Crossover Procedure

Perform a crossover procedure when a new box of QC film is opened to adjust the originally established parameters for speed, contrast, and base plus fog on the processor QC chart for the characteristics of the new emulsion. An adjustment is required because film used in medical imaging is produced in batches, and there may be slight differences in the sensitometric characteristics from batch to batch. The conditions under which the film is stored (temperature, relative humidity, exposure to fumes from chemicals or to ionizing radiation, etc.) and the age of the film can also affect the sensitometric characteristics.

Performing a crossover procedure is necessary to maintain a controlled processing environment. As it has been a source of considerable confusion, note the following important points:

1. The crossover should be performed all on the same day rather than over five consecutive days as when the processor QC program was initially established.
2. Expose and process all the films required for the crossover:
   - At the same time of day that processor QC is normally performed because slight sensitometric changes due to fluctuations in film feeding patterns throughout the day are common.
   - All at the same time, without interruptions.
   - With the same delay (e.g., a few seconds) between exposure and processing that is normally used.
   - Alternating the films from the existing QC box and from the new box. Expose and process the first film from the existing box, followed by the first film from the new box, the second film from the existing box, the second film from the new box, etc. This distributes any sensitometric changes due to seasoning equally among the 10 films. The films from the existing box and new box can easily be distinguished from each other before processing by marking one set of films with a lead pencil or by cutting a corner of one set of films.

   The computations and adjustments to the control chart may be delayed to a more convenient time later in the day if preferred.
3. The chemicals (developer) in the processor should be seasoned when a crossover is performed on a single day. Generally, for this purpose, it is important to avoid performing a crossover immediately after a preventive maintenance procedure (PM). It is necessary to monitor both the number of films remaining in the QC box and when the next processor PM will be performed, so the crossover can be performed before the PM occurs and with

fully seasoned chemicals in the developer tank. A good rule of thumb is that a minimum of 10 to 15 films should remain in the existing box of QC film when beginning a crossover procedure. A piece of cardboard from the film box inserted between the last 15 films and the balance of the box makes a good divider and reminder that a crossover must be performed.

4. The processor should be in control (within the ±0.10 control limits for speed and contrast) for crossovers performed all on a single day.

   It is extremely unlikely, however, that the speed and contrast of the processor will be at the operating level originally established. It will be necessary, therefore, to adjust the new operating levels (taken from the characteristics of the new film) by the difference between the originally established operating levels and the existing box of film.
5. The developer in the processor must have reached its operating temperature before doing a crossover procedure.
6. The crossover procedure should be fully outlined in the policy and procedures manual, and all QC personnel should undergo training.

### *Performing a Crossover*

Follow these steps:

1. Alternately expose and immediately process five sensitometric strips each from the existing and new boxes of film.
2. Determine the average of the steps previously chosen for processor quality control (i.e., step 11 for speed, steps 13 and 9 for contrast, and step 1 for base plus fog from the five films from the existing box and from the five films from the new box).
3. Adjust the original operating levels on the control chart for speed, contrast, and base plus fog by the difference in the average values between the existing and new boxes of film, as shown in the following equations:

   Step 1: New box average value - existing box average value = difference.
   Step 2: Original operating level + difference = new operating level.

   *Example #1:* The original operating level for speed is 1.20; the average for speed of the five films from the existing QC box is 1.28; the average for speed of the five films from the new QC box is 1.32.

   $1.32 - 1.28 = 0.04$
   $1.20 + 0.04 = 1.24$

   *Example #2:* The original operating level for contrast is 1.80; the average for contrast of the five films from the existing QC box is 1.75; the average for contrast of the five films from the new QC box is 1.70.

   $1.70 - 1.75 = -0.05$
   $1.80 + (-0.05) = 1.75$

   A way to check that the correct adjustment has been made is that the offset, or distance, between the new operating level and the point that represents the average of the five films from the new box of QC film must be exactly the same as the offset, or distance, between the old operating level and the point that represents the average of the five films from the existing box of QC film.

In Example #1 above, the offset, or difference, between the existing QC box average value (1.28) and the original operating level (1.20) is 0.08, with the average being higher. The new QC box average value (1.32) is also 0.08 higher than the new operating level (1.24).

In Example #2 above, the offset, or difference, between the existing QC box average value (1.75) and the original operating level (1.80) is 0.05, with the average being lower. The new QC box average value (1.70) is also 0.05 lower than the new operating level (1.75).

4. Record the complete emulsion number of the new box of film on the control chart.
5. Make a notation in the remarks section of the date and that a crossover was performed.

Table 5-1 reviews the important points of the crossover procedure.

## Reestablishing the Processor QC Program

After a processor QC program has been initially established, it may be possible to proceed for a long period of time (i.e., a year or more) using the original operating levels if the processing environment is well controlled and all personnel operate in a consistent manner.

Most facilities, however, will benefit from a periodic reevaluation and reestablishment of the processor QC program. This might be done on an annual basis, at least. Some facilities are hesitant to reestablish a processor quality control program. If you do reestablish your program, be sure to make a notation in the remarks section of the processor QC chart that the program was reestablished and the reason why (e.g., annual reestablishment, etc.).

Table 5-1 Summary of the Crossover Procedure

| |
|---|
| A crossover procedure is necessary to make adjustments for slight differences in the sensitometric characteristics between different batches of film. |
| A crossover should be performed all on the same day and at the same time of day processor QC is normally performed. |
| The chemicals in the processor should be seasoned. |
| The processor should be in control (within the ±0.10 control limits for speed and contrast). |
| Adjust the operating levels on the control chart for speed, contrast, and base plus fog by the difference in the average values between the existing and new boxes of film. |
| The crossover procedure should be fully outlined in the policy and procedures manual, and all QC personnel should undergo training. |

There are a number of other events that dictate the reestablishment of the processor QC program. These include the following:

- A change of film type in a dedicated processor.
- A change of film mix in a nondedicated processor.
- A change made by a film manufacturer to an existing film (e.g., increase of the D-max) with the recommendation that the processor QC program be reestablished.
- A change in film volume.
- A change of brand(s) of chemicals used.
- A change of chemical form used (e.g., switching from premixed solutions to concentrates used in an automixer).
- A change in replenishment rates.
- A change in the settings on a specific gravity automixer.
- A change in the average optical density of the films being processed (e.g., the level of exposure on mammographic images is increased).
- A change in sensitometer and/or densitometer (use of a different instrument or recalibration of the same instrument).
- Running out of film or having less than five sheets of film in the QC box, thus preventing a crossover from being done correctly.
- A change in the way the crossover procedure is done. Note that it is particularly important to reestablish the processor QC program first (over five consecutive days) and then, upon opening the next new box of QC film, to implement a new way of doing the crossover.
- A restart of the processor (i.e., completely changing the chemicals) after contamination due to a film jam or a PM, and the processor is not in control.

In addition, over time the values for speed and contrast on particular steps may no longer be consistent with the guidelines for establishing processor QC (the speed step closest to 1.20, and the two steps used to calculate the density difference closest to 2.20 and closest to, but not less than, 0.45) due primarily to adjustments from the crossover procedure. If the step chosen to monitor speed, in particular, is now lower than 1.0 (down in the toe, not on the straight-line portion of the characteristic curve), the control chart will be relatively insensitive to processing changes that could affect clinical films. If the step chosen to monitor speed is now higher than 1.50, the control chart will be hypersensitive to processing changes, and it may be difficult to maintain a controlled environment.

The reestablishment of the processor QC program, including changing the step numbers used, if appropriate, is recommended.

## Average Gradient

The 21 points from the sensitometric strip may be plotted and used to calculate average gradient. Average gradient is the slope of a straight line between two specific points of the characteristic curve, the most diagnostically useful part of a practical image (approximately the minimum and maximum useful densities in the clinical setting). As discussed in Chapter four, for most medical x-ray films, average gradient is measured between the densities of 0.25 and 2.00 above base plus fog. It is calculated by the formula:

$$\text{Average Gradient} = \frac{(2.0 + \text{base plus fog}) - (0.25 + \text{base plus fog})}{\log E_2 - \log E_1}$$

where:

- Base plus fog is defined as the optical density of the film base plus any additional density to the emulsion coated on the film base.
- Log $E_2$ is defined as the log relative exposure value for the point on the curve represented by a density of 2.0 + base plus fog.
- Log $E_1$ is defined as the log relative exposure value for the point on the curve represented by a density of 0.25 + base plus fog.

Average gradient may also be described as the rise of a straight line between two points on the density axis over the run of the same straight line between two points on the log E axis.

Calculating the average gradient is not difficult, but it requires a precisely drawn characteristic curve and careful interpretation of the density and log E values if the results are to be meaningful and repeatable. Automatic scanning densitometers are programmed to make this calculation.

Small differences in the characteristic curve shape can result in significant differences in average gradient values. Two factors that contribute to these differences, aside from film-processing effects, are the following:

- Deviations from the nominal 0.15 optical density difference between the steps of a typical step tablet (installed in simulated light sensitometers).
- The assumption that the exposure difference between steps is 0.15 when characteristic curves are plotted, either manually or automatically.

To a limited extent, the plotting routines, or algorithms, in automatic scanning densitometers may also contribute to differences in curve shape.

The sensitometers and densitometers currently available do not address these variables. As a result, there are often significant discrepancies in average gradient values from one sensitometer/densitometer system to another. Therefore, it is not practical to compare average gradient values generated with different systems.

## *Using Average Gradient*

Average gradient is the most accurate method for determining contrast because it is derived from the relationship between two specific (optical) densities and the exposure required to produce them with any given film-developer combination.

The ways average gradient can be used include:

- Comparison of the published average gradients for films from a specific manufacturer to determine which film is higher or lower in contrast (i.e., to rank the films, all from the same manufacturer, in relation to one another).

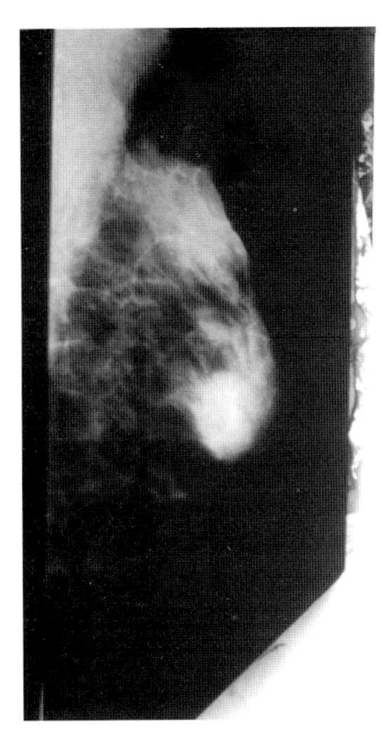

**Figure 5-14** A mammogram that was not properly washed.

- As a relative measure of contrast. After establishing the processor QC program (discussed earlier in this chapter), calculate the average gradient and use it as a benchmark for all future comparisons, provided the same film type, processor, chemical type, sensitometer, and densitometer are used.

Average gradient should not be used to compare the contrast of different screen-film combinations.

## Fixer Retention Test

The fixer retention test allows an estimate of the amount of retained ammonium thiosulfate (hypo) in film to be made. It is usually performed on a quarterly basis. Excessive amounts of retained hypo indicate insufficient washing and can significantly impact the long-term stability of film (Figure 5-14). Radiographic image stability is very important, particularly for mammography, where radiologists compare current mammograms with previously taken films to discern subtle changes in breast tissue, a possible sign of early breast disease.

### Tools Required for the Fixer Retention Test

The tools that are required include the following:

- A hypo estimator.
- A bottle of test chemical.

- A processed film (one processed immediately prior to performing the test, or one processed no longer than two weeks previously).

The hypo estimator and test chemicals are commercially available. The chemicals may also be mixed as follows:

1. Place 75 milliliters of distilled water in a container.
2. Add 12.5 milliliters of 28 percent acetic acid and stir. (To make approximately 28 percent acetic acid from glacial acetic acid, carefully add 3 parts of glacial acetic acid to 8 parts of distilled water. Caution: extreme care must be taken in handling glacial acetic acid. Special attention must be taken to ensure ventilation, eye protection, and skin protection. Pour the acid into the water; do not pour water into the acid.)
3. While continuing to stir, add 0.75 grams of silver nitrate.
4. Stir until all of the solid material has dissolved.
5. Add sufficient distilled water to make 100 milliliters of solution.
6. Place the solution in a dark bottle and store in a cool, dark area.

Acetic acid and silver nitrate are generally available from photographic supply stores. Both chemicals should be handled with care. If contact with skin occurs, wash thoroughly with cold water and seek medical assistance.

Appropriate measures for protection of eyes, skin, and clothing are recommended with use of either the homemade or commercially prepared solution.

## Performing the Fixer Retention Test

Follow this procedure:

1. Process one sheet of unexposed film in the processor.
2. Place the processed film on top of a white sheet of paper on the countertop. If using single-emulsion film, place the film so the notch(es) is (are) located toward the upper right-hand corner of the film. The emulsion side of a single-emulsion film will then be facing toward you.
3. Working under normal lighting conditions (avoid bright light or sunlight), place one drop of the test solution on the film (Figure 5-15).
4. Blot the area after two minutes.
5. Immediately place the estimator in its clear protective sleeve over the film (Figure 5-16). Assessment of the results of the test must take place while the test spot is still damp because it darkens as it dries. Position the test spot between the colored sections of the estimator. (Remove the paper backing sheet, if present, from the estimator before using.)
6. Determine which section on the estimator most closely matches the test spot. Note that the test spot may differ in color from the color patches printed on the estimator because the estimator is a tool designed to be used with many types of photographic papers and films. This can make it difficult to judge

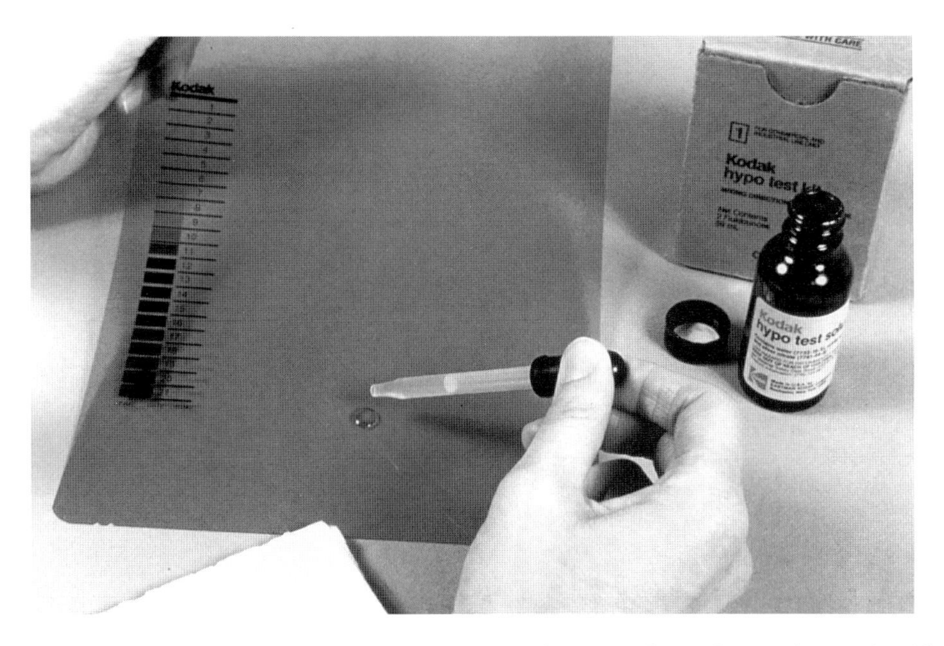


**Figure 5-15** Working under normal lighting conditions, a drop of test solution should be placed on the emulsion side of a film.

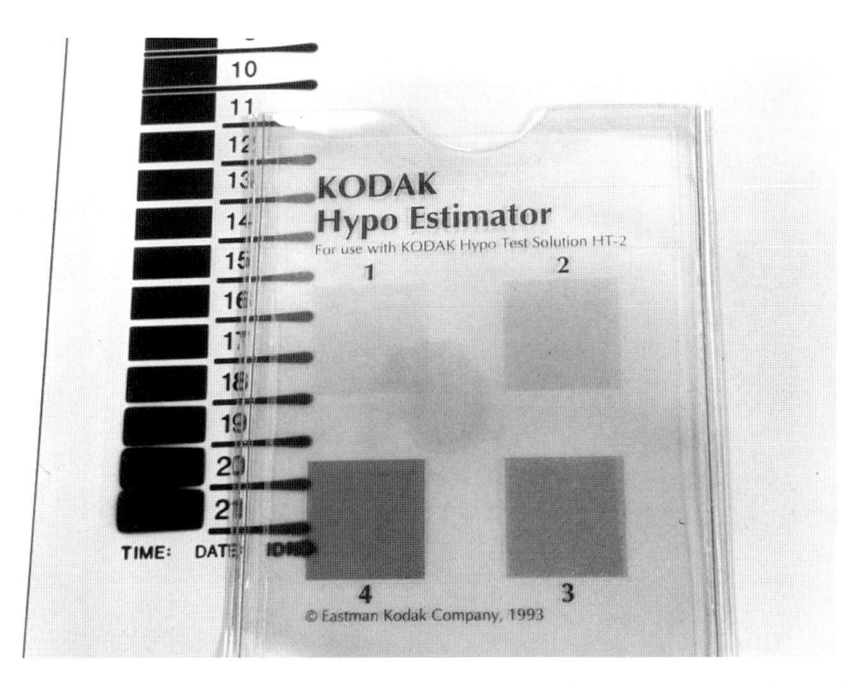


**Figure 5-16** Compare the damp test spot to the color patches on the hypo estimator by placing the estimator in its plastic sleeve over the spot on the film.

which color patch most closely matches the test spot. In this case, use a spot-reading densitometer to take several optical density readings as follows:

Measure the optical density of the test spot on the film. A clear area of the estimator and the plastic sleeve enclosing the estimator must be positioned on top of the test spot when taking the measurement.

Measure the optical density of several of the color patches. Include the film (any nonexposed area adjacent to the test spot) and the plastic sleeve that encloses the estimator when taking the measurements.

The optical density measurement that comes closest to the reading taken from the test spot indicates the best match.

7. If testing a double-emulsion film, turn the film over and repeat the test on the other side. Select a different area so that the test spots do not overlap before blotting.

   A spot equal in color intensity to the third darkest color printed on the estimator indicates an estimated 0.05 grams of retained ammonium thiosulfate per square meter of film. Images on film with this amount of retained fixer should last "long term," or approximately one hundred years.

   A test spot whose intensity is lighter or equal to the third darkest color on the estimator meets regulatory requirements. Be sure to check the instructions with a particular estimator to correlate specific color patches with the amount of retained fixer.

## References


1. American College of Radiology (ACR) Mammography quality control manual for radiologists, medical physicists and technologists. Reston, VA: American College of Radiology, 1994.
2. Andolina VE, Lillé SL, Willison KM. Mammographic imaging: A practical guide. Philadelphia, PA: J. B. Lippincott Company, 1992.
3. Dickerson RE, Haus AG, Baker CW. Method of simulated screen sensitometry for asymmetric, low-crossover medical x-ray film. Medical Physics 21: 525–528, 1994.
4. Haus AG, Dickerson RE. Problems associated with simulated light sensitometry for low-crossover medical x-ray films. Medical Physics 17: 691–695, 1990.
5. Gray JE. Photographic quality assurance in diagnostic radiology, nuclear medicine, and radiation therapy. Volume 1. In: Photographic processing, quality assurance and the evaluation of photographic materials; HEW publication no. (FDA)76-8043. Washington, DC: Bureau of Radiological Health, U.S. Department of Health, Education and Welfare, 1976.
6. Gray JE. Photographic quality assurance in diagnostic radiology, nuclear medicine, and radiation therapy. Volume 2. In: Photographic processing, quality assurance and the evaluation of photographic materials; HEW publication no. (FDA)77-8018. Washington, DC: Bureau of Radiological Health, U.S. Department of Health, Education and Welfare, 1977.
7. Gray JE, Winkler NT, Stears J, Frank ED. Quality control in diagnostic imaging. Baltimore, MD: University Park Press, 1983.
8. Introduction to medical radiographic imaging, Eastman Kodak Company publication no. M1-18, 1993.
9. Jaskulski SM. Procedure for processor quality control: A step-by-step approach. In: Haus AG, ed. Film processing in medical imaging. Madison, WI: Medical Physics Publishing, 1993: 225–233.
10. National Council on Radiation Protection and Measurements (NCRP) Report No. 99 on Quality Assurance. Bethesda, MD: National Council on Radiation Protection and Measurements, 1988.
11. Suleiman OH. Technical corner: The crossover procedure. In: Mammography matters 4: 9, 1997. Rockville, MD: Center for Devices and Radiological Health, Food and Drug Administration, Public Health Service, Department of Health and Human Services.
12. Suleiman OH, Conway BJ, Rueter FG, Slayton RJ. Automatic film processing: Analysis of 9 years of observations. Radiology 185: 25–28, 1992.
13. Wentz G. Mammography for radiologic technologists. 2nd Edition. New York: McGraw-Hill, 1997.

# 6

# Artifacts

- Artifacts on Single-Emulsion Film
- Tools for Artifact Analysis
- Minimizing Artifacts
- Examining Films for Artifacts
- Processing Artifacts
- Shadow Images
- Isolating Individual Processor Components

An artifact can be defined as any defect on processed film. Artifacts may distract radiologists from interpreting diagnostic information contained in radiographic images. All efforts should be made to keep them to a minimum. Artifacts can be caused by the film processing system, improper film handling, x-ray equipment, incorrect positioning of the patient, or even a dirty darkroom. Many of the artifact examples shown in this chapter occurred on single-emulsion film used for mammography. They may also occur on double-emulsion film. This chapter will focus primarily on processing artifacts.

## Artifacts on Single-Emulsion Film

Artifacts should be a concern for all radiographic specialties. In mammography, however, they are of special concern. This is due to the use of single-emulsion films for mammographic imaging—that is, films with emulsion coated on only one side of the film support instead of the double-emulsion films used in most other radiographic applications (Figure 6-1).

Artifacts are more readily apparent on mammography films for a number of reasons:

- Because there is emulsion on only one side of the film base, artifacts are less likely to be obscured while being reviewed on the viewbox than they are on double-emulsion film.

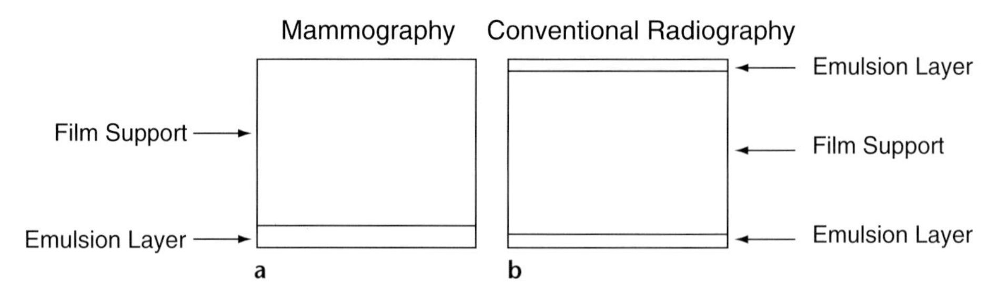


**Figure 6-1** Diagrams of the physical configurations of (a) single-emulsion film used for mammography and (b) double-emulsion film used for other radiologic procedures

- The emulsion coating on a single-emulsion film is thicker than the coating on either side of a double-emulsion film, causing it to swell more during processing and, therefore, making it more prone to processing and handling artifacts.
- Mammographic film images are more likely to be viewed with magnifiers than are conventional films.
- Mammography films today have higher contrast, higher maximum density (D-max), and increased optical density.
- The American College of Radiology (ACR) recommends exposing as much of the film as possible instead of collimating the x-ray beam to the breast in order to reduce extraneous light during viewing.
- The use of specialized viewboxes with significantly higher levels of luminance for viewing mammograms is increasing.
- Masking of radiographs and control of ambient room light to reduce extraneous light is also increasing (Figure 6-2).

Artifact evaluation is currently performed on an annual basis by a medical physicist in mammography facilities.

## Tools for Artifact Analysis

Knowing where to begin in analyzing artifacts is half the battle. This section provides a procedure for determining the origin of the artifact, general rules for processing, handling, and equipment-caused artifacts, information on artifact characteristics and categories, and other helpful information.

### Determining the Origin of Artifacts

The first step is to isolate the cause of a particular artifact to the processor or, at least, to eliminate the processor as the cause.

Note that the following procedure uses a mammography x-ray unit, mammography cassette, and single-emulsion film, which is more sensitive to all types of artifacts. If desired, the procedure may be modified to use a general purpose x-ray unit, cassette, and double-emulsion film.

a

b

**Figure 6-2** The American College of Radiology suggests that all mammography films should be completely masked and ambient room light controlled. (a) Mammography films being viewed with light from the viewboxes spilling around the edges of the films. (b) Mammography films being viewed in a darkened room and completely masked with no light spilling around the film edges.

First, you need to know:

- The orientation of the film when fed into the processor; that is, which edge was the leading edge, the trailing edge, the top surface, and the bottom surface. (Film should always be fed into the processor consistently.)
- For single-emulsion mammography films, whether the film was processed emulsion side up or emulsion side down. (It is particularly important that all single-emulsion film be fed consistently.)

To determine if an artifact originated during processing, use the mammography x-ray unit to expose two test films as follows:

1. Select a cassette that is known to have good screen-film contact.
2. In the dark, load a sheet of film into the cassette so that the film notch is positioned at the lower right corner or the upper left corner. The emulsion is upward facing if the notch or notches are located as described (Figure 6-3).
3. Using the x-ray unit (with the compression paddle removed from the x-ray beam path), expose the film to an optical density of 1.10 to 1.50 with a 1-inch (2.54-centimeter) thick uniform sheet of acrylic placed on top of the cassette. The exposure time used should be at least 0.5 seconds or longer to allow the grid time to move properly.

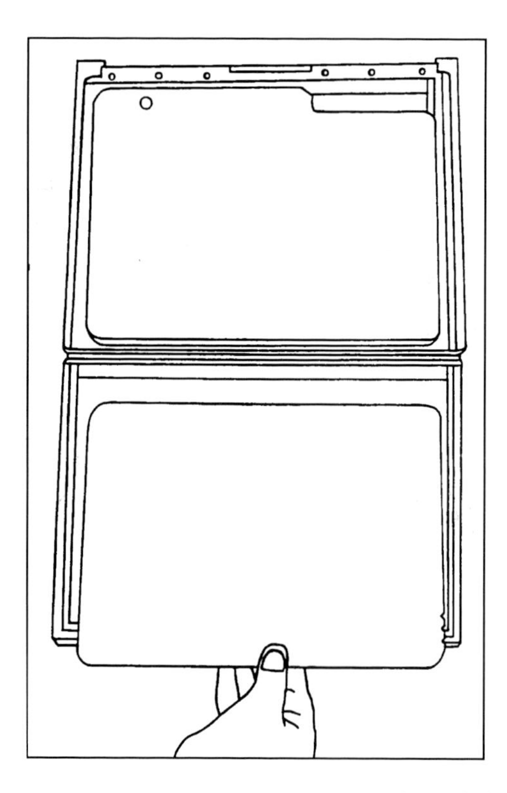

**Figure 6-3** Single-emulsion mammography film is correctly loaded into a cassette so the notches are located either at the lower right corner, as shown, or the upper left corner. (Reprinted with permission from Eastman Kodak Company. EKC publication no. N-326.)

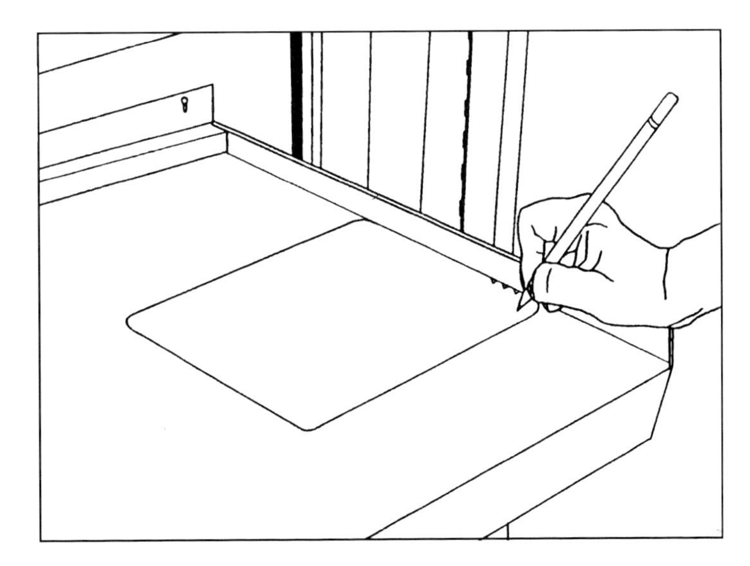

**Figure 6-4** One of the test films should be fed into the processor so that the widest dimension of the film is the leading edge. Use a pencil to mark an arrow and other important information on a corner of the film. Either the leading or trailing film edge corner nearest the film guide may be used. (Reprinted with permission from Eastman Kodak Company. EKC publication no. N-326.)

4. In the darkroom under safelight illumination, remove the exposed film from the cassette. Lay the film on the film feed tray in the darkroom so that the widest dimension of the film is the leading edge (Figure 6-4). Using a pencil, mark an arrow on a corner of the film nearest the film guide immediately before processing to indicate the direction of film travel. The leading or trailing film edge corner may be used, as desired, as long as consistency is employed.

   It is also helpful to mark the emulsion orientation (up or down), as well as which side of the processor feed tray (right or left) is being used, when feeding the film into the processor (e.g., "↑UR" indicates the direction of film travel and that the film was processed emulsion side up with the film edge abutted along the film guide on the right side of the processor).

   If sequential films are to be processed, they should be sequentially numbered.
5. Load a second sheet of film into the cassette. Be sure to orient the film notch in the same position.
6. Expose the second film in the same manner as the first.
7. In the darkroom under safelight illumination, remove the film from the cassette. Lay the film on the film feed tray so that the narrowest dimension of the film is the leading edge (Figure 6-5). Mark the corner of the film to indicate the direction of film travel, the emulsion orientation, and the side of the processor being used to process the film.
8. To evaluate the films, orient them on the viewbox so that the arrows point in the same direction (Figure 6-6). Artifacts that are parallel to each other on both films were created during processing. (Note that they may be either parallel or

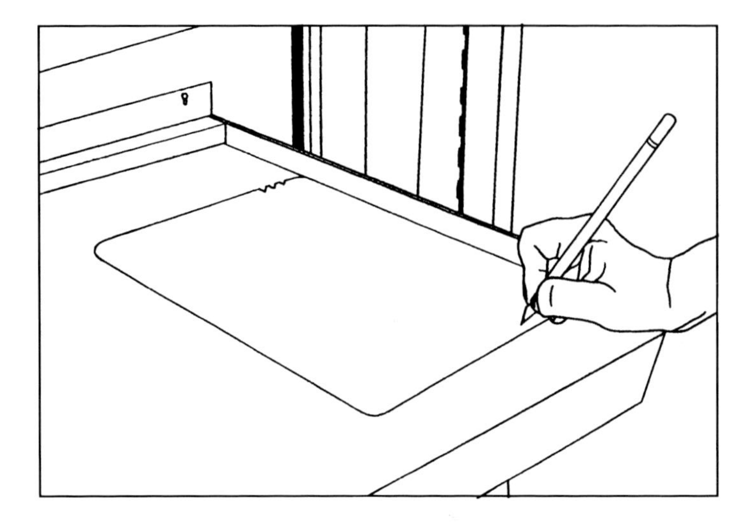

**Figure 6-5** The second test film should be fed into the processor so that the narrowest dimension of the film is the leading edge. (Reprinted with permission from Eastman Kodak Company. EKC publication no. N-326.)

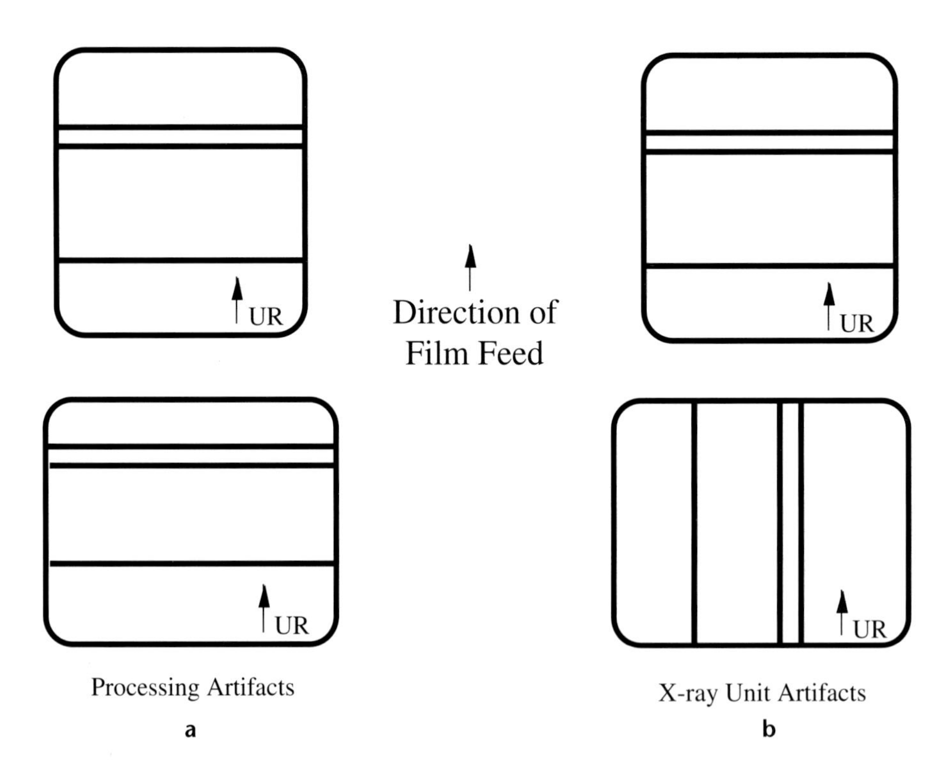


**Figure 6-6** To evaluate the films, orient them on the viewbox so that the arrows point in the same direction. (a) Artifacts that are parallel to each other on both films were created during processing. (b) Artifacts that are perpendicular to each other occurred during exposure from the x-ray unit and were caused by sources such as any part of the x-ray unit, the moving grid, or the cassette.

perpendicular to film travel.) Artifacts that are perpendicular to each other occurred during exposure from the x-ray unit and were caused by sources such as any part of the x-ray unit, the moving grid, or the cassette.

## General Rules for Processing Artifact Analysis

The following general rules can be used to analyze artifacts that occurred during processing:

1. Plus-density processing artifacts on film (those darker than the background) may be caused by problems with the developer.
2. Minus-density processing artifacts on film (those lighter than the background) may be caused by problems with the fixer.
3. Processing artifacts seen by reflected light may have been created by the wash rack or dryer assembly of the processor; severe dryer and wash rack artifacts may also be seen by transmitted light.
4. Artifacts that appear when single-emulsion film is processed emulsion side up or on the top surface of double-emulsion film may be caused by the inner path (inside) rollers of the racks (Figure 6-7) and by the crossover assemblies.
5. Artifacts that appear when single-emulsion film is processed emulsion side down or on the bottom surface of double-emulsion film may be caused by the outer path (outside) rollers of the racks (Figure 6-7) and by the turnaround assemblies.
6. To begin diagnosing the cause of a processing artifact, the leading edge, trailing edge, top surface, and bottom surface of the film must be identified.

### *Identifying the Leading Edge*

The leading edge of a sheet of film is the edge of the film that is fed first into the entrance rollers of the processor. As well as being the first edge fed into the

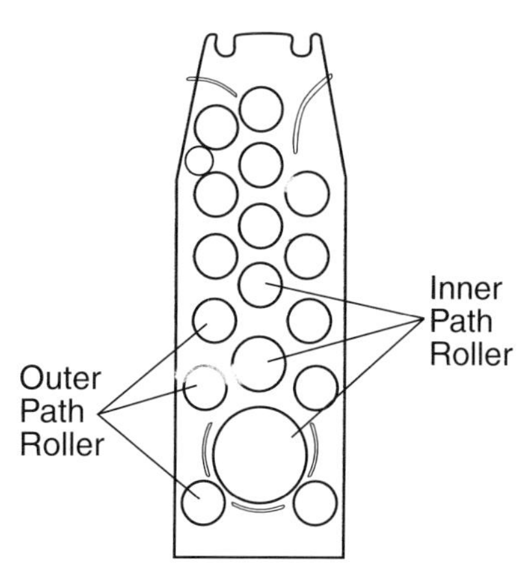


**Figure 6-7** Identifying inner and outer path rollers. (Reprinted with permission from Eastman Kodak Company. EKC publication no. N-1C0948.)

processor, the leading edge is also the first edge to exit from the exit slot in the dryer. Once you have removed a processed sheet of film from the receiving bin of the processor, it may be difficult to remember which end is the leading edge of the film. Without knowing which end is the leading edge, it also becomes difficult to identify which edge of the film was on the drive side of the processor and which edge was on the nondrive side of the processor. To help you find the leading edge of a film being used for artifact analysis, use a pencil to mark or scratch an arrow on the right corner of the top emulsion on the leading edge of the film before you feed it into the processor.

The trailing edge of the film is obvious if the leading edge has been identified as discussed above.

The identification system discussed earlier in this chapter may also be used to mark either the leading or trailing edge of the film.

### *Identifying the Top Surface*

The top surface of the film is the surface that faces up as you feed the film into the processor. Once you have removed a processed sheet of film from the receiving bin of the processor, it may be difficult to remember which side was facing up as you fed the film into processor. If you marked the right corner on the leading edge of the film, that mark can help you identify the top surface of the film.

The bottom surface of the film is obvious if the top surface has been identified as discussed above.

The identification system discussed earlier in this chapter may also be used to mark either the top (emulsion side up) or bottom (emulsion side down) surface of the film.

## General Rules for Handling Artifact Analysis

Mishandling film, especially single-emulsion mammography film, may cause handling artifacts. Occasionally, an entire box of film may be dropped or handled carelessly before the films are exposed or processed. Poor film handling can result in the appearance of consistent artifacts on multiple sheets of processed film.

Handling artifacts are generally caused by pressure on the film emulsion and may be either plus-density (dark) or minus-density (light) in appearance. The appearance of the artifact is dependent on whether the pressure or mishandling occurred before exposure or after exposure and on the type of grain used in the film emulsion. (Note that many different film technologies are used for medical imaging. Not all film emulsion grain types respond as indicated below.)

Pressure or mishandling before exposure can occur before the film box is opened, after an opened box has been placed into the film bin, or as film is removed from the box and placed into the cassette.

Pressure or mishandling after exposure can occur as exposed film is removed from the cassette and placed on the film feed tray for processing.

Minus-density, or light artifacts, caused by dirt, dust, and static are also considered handling artifacts. "Shadow images" are discussed later in this chapter. Static, which appears as plus-density marks that frequently resemble tree branches, is due to low relative humidity and personnel wearing synthetic fabrics. Static has other appearances, including looking like smudges, dots, dashes, or clusters.

The following general rules can be used to analyze artifacts that occurred during handling:

1. Pressure on three-dimensional-grain films before exposure may cause minus-density or light handling artifacts.
2. Pressure on cubic-grain and most tabular-grain films before exposure may cause plus-density or dark handling artifacts.
3. Pressure on all types of film after exposure may cause plus-density or dark handling artifacts.

Table 6-1 contains the above information on handling artifacts, and Appendix D contains additional information.

## General Rules for Equipment Artifact Analysis

If the films that were processed 90 degrees apart from one another as described earlier in this chapter showed artifacts that are perpendicular to each other, they occurred during exposure from the x-ray unit. Any part of the x-ray unit, the moving grid, the cassette, etc., could be the source.

Artifacts caused by the moving grid during mammographic imaging frequently resemble processor roller marks. Grid artifacts are usually subtle, plus-density shadings, usually seen in the uniform background of a mammographic phantom radiograph taken as part of the quality control activities.

Additional tests must be performed if the grid is suspected of causing the problem.

The following guidelines may be helpful when attempting to diagnose artifacts on film caused by the equipment:

1. If the equipment-caused artifact is minus-density, there is more mass to penetrate, or something is blocking the beam.
2. If the equipment-caused artifact is plus-density, there is less mass to penetrate, or there is an added source of radiation or light.
3. If the equipment-caused artifact is very unsharp and diffuse, it is magnified and farther from the film.
4. If the equipment-caused artifact is very sharp and distinct, it is very close to the film.

**Table 6-1** Handling Artifact Characteristics

| | |
|---|---|
| Pressure before exposure | Minus-density (light) artifacts on three-dimensional-grain films |
| | Plus-density (dark) artifacts on cubic-grain and most tabular-grain films |
| Pressure after exposure | Plus-density (dark) artifacts on all film grain types: three-dimensional, tabular, and cubic |

## Categories of Processing Artifacts

The organization of processing artifacts into categories according to their orientation on the film is extremely helpful in determining the cause of the artifact.

The three categories include:

1. Parallel to film travel artifacts.
2. Perpendicular to film travel artifacts.
3. Randomly occurring artifacts.

If artifacts occur randomly or sporadically, the method that employs two films processed 90 degrees apart, described earlier in this chapter, may not be useful. Additional testing will most likely be required.

Table 6-2 contains the most commonly occurring processing artifacts by classification.

## Artifact Spacing

The spacing between artifacts can also serve as a valuable clue. For example, if artifacts appear 3.14 or π inches (79.8 millimeters) apart and are parallel to the leading edge of the film, they are probably caused by a problem with a 1-inch (2.54-centimeter) diameter roller. If artifacts appear 6.28 inches (15.95 centimeters) apart and are parallel to the leading edge of the film, they are probably caused by a problem with a 2-inch (5-centimeter) diameter master roller, the large inner roller in the turnaround assemblies at the bottom of the developer, fixer, and wash tanks.

# Minimizing Artifacts

Roller transport processors have brought many benefits to modern diagnostic imaging, including the availability of dry, ready-to-read, consistently processed

**Table 6-2** Common Processing Artifacts

| | |
|---|---|
| Parallel to Film Travel | Delay streaks |
| | Entrance roller marks |
| | Guide shoe marks |
| Perpendicular to Film Travel | Chatter |
| | Slap lines |
| | Stub (hesitation) lines |
| Randomly Occurring | Bent corners |
| | Brown films |
| | Drying patterns and water spots |
| | Dye stain |
| | Flame patterns |
| | Pick-off |
| | Run back |
| | Skivings |
| | Wet pressure |

images in seconds and the removal of human variation from processing. No one would voluntarily return to the era of manual film processing. For all of its benefits, however, the current state of processing technology does not allow for the complete elimination of processing artifacts, even when using the latest, most advanced models.

Minimizing processing artifacts is dependent upon proper installation, ventilation, and maintenance of the processor, use of appropriate processing chemicals, and adherence to the processing recommendations of the film and processor manufacturers.

Minimizing handling artifacts is dependent upon the awareness and successful training of personnel regarding proper film handling techniques on the implementation of those techniques throughout the radiology department or mammography facility.

## Examining Films for Artifacts

It is important to check films regularly for artifacts. By checking films for artifacts as part of your daily routine, you are better able to identify and eliminate the source of artifacts as soon as a problem arises. The early detection of artifacts can help you reduce the number of films that are affected. This in turn provides two benefits:

1. Lowered operating costs.
2. Minimized patient exposure to x-rays.

You can incorporate the simple task of checking for film artifacts into your daily routine by following the two guidelines outlined below.

1. Check for film artifacts before and after you perform any of the procedures listed below:
   - Preventive maintenance checks.
   - Cleaning of the crossover assemblies and racks.
   - Inspection of components and assemblies.
2. Check for artifacts on films as they exit the processor. When examining films for artifacts, try to do the following:
   - Isolate the artifact.
   - Identify the cause of the artifact immediately.
   - Make any necessary corrections to eliminate the cause.

### Impact of Lighting Conditions

Another important factor to keep in mind when examining film for artifacts is the lighting conditions under which the films are being evaluated. Different lighting conditions can help identify different types of artifacts.

For example, by using reflected light, flaws in the surface quality of a film will be located more quickly and easily than if transmitted light is used. Therefore, when looking for unusual drying patterns or artifacts caused in the wash stage of the processing cycle, films should be examined with reflected light. Outdoor light or light from incandescent lightbulbs works best.

For artifacts caused in the development stage of the processing cycle, transmitted light should be used to examine films. Transmitted light will enable one to see through the film rather than just being able to see the surface. To examine films, use an appropriate viewbox in a darkened room. Mask the area around the film so that extraneous light from the viewbox does not affect viewing (Figure 6-2b, page 201).

## Magnification

Magnification should also be employed when examining film for artifacts. Standard magnifying lenses, such as are commonly used for viewing radiographs, typically have 1.5 to 2X magnification and are adequate. Higher power magnification, however, is recommended for analyzing artifacts. Loupes with at least 7 to 10X are particularly valuable and are available from precision optical equipment supply houses.

# Processing Artifacts

The basic design of the roller transport processor is shown in Figure 6-8. Film moves sinusoidally through each processing solution (developer, fixer, wash) and exits through the dryer section.

Figure 6-9 shows the individual components of the processor. Refer to Chapter three for additional information on processor components and function.

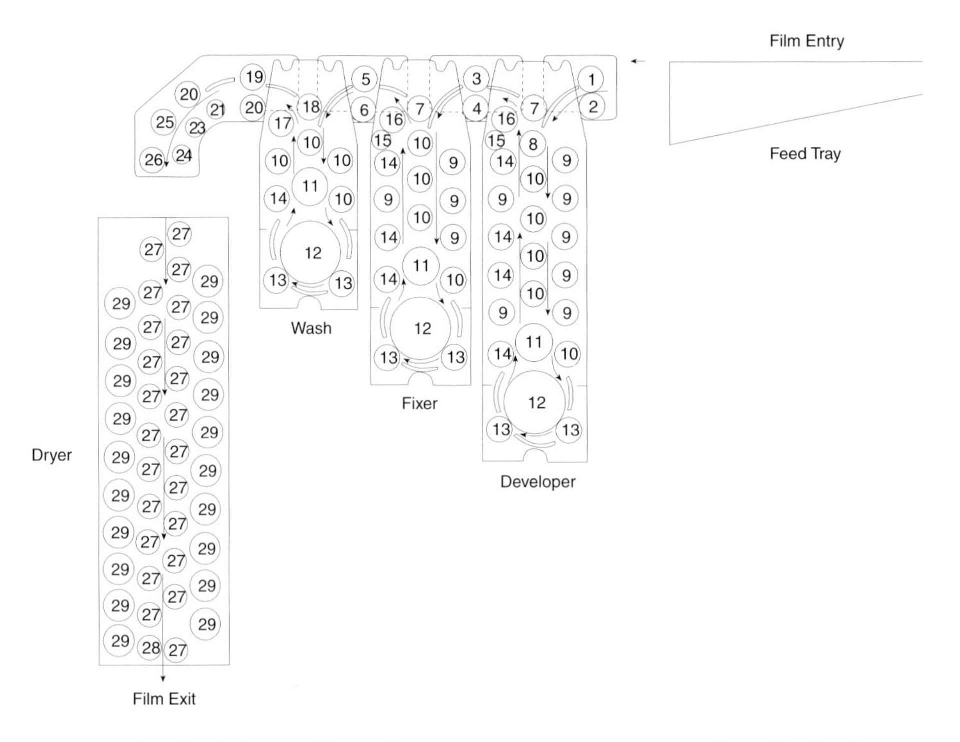


**Figure 6-8** The film is inserted into the processor transport system from the feed tray (film entry). The film is transported through the developer rack, the fixer rack, the wash rack, the dryer section, and exits dry and ready to read. (Reprinted with permission from Eastman Kodak Company. EKC publication no. N-326.)

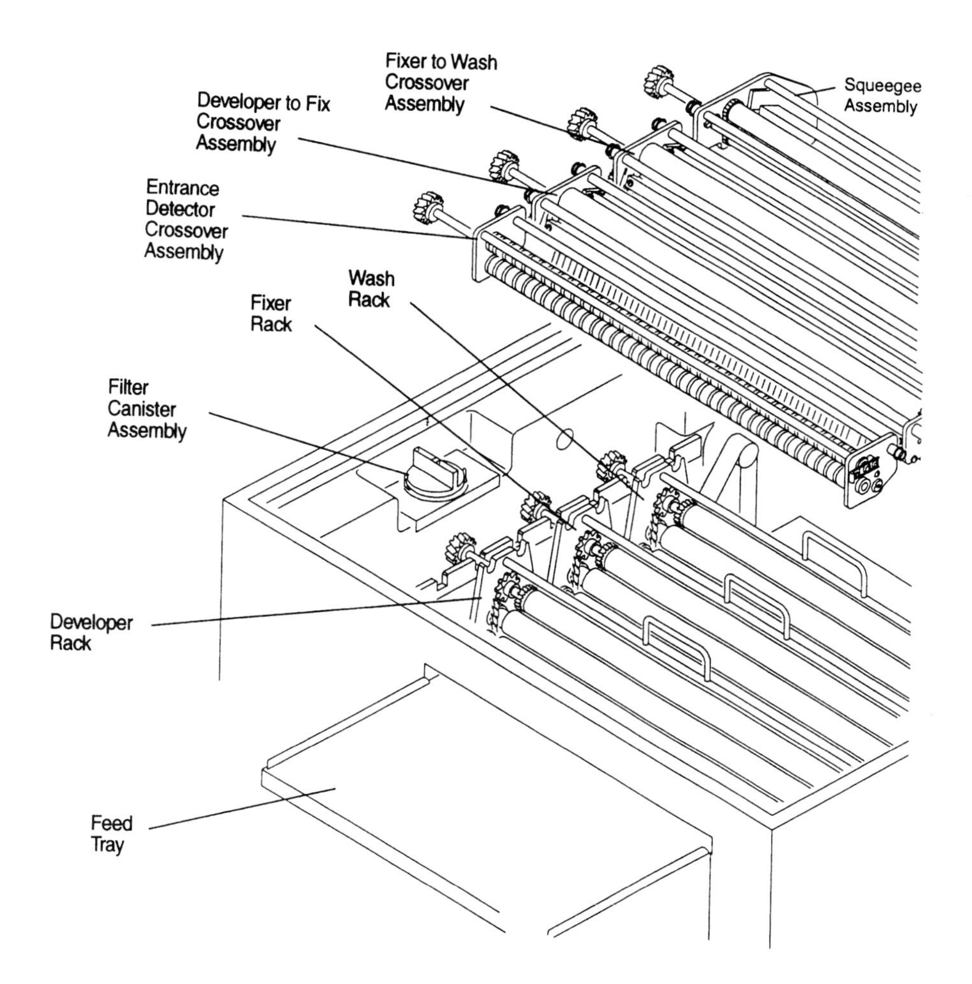


**Figure 6-9** Identifying the individual components of a typical processor. (Reprinted with permission from Eastman Kodak Company. EKC publication no. N-1C0948.)

## Parallel to Film Travel Artifacts

### *Delay Streaks*

Delay streaks appear as smooth wide lines or bands without sharp edges (Figure 6-10). They can appear as plus-density bands, minus-density bands, or a combination of both, and they can appear anywhere on the film.

Typically, once the processor has been in standby mode for an extended period of time, the streaks appear on only the first 3.14 or $\pi$ inches (7.98 centimeters) of film, which corresponds to one revolution of a roller. In severe cases, delay streaks may show for two or even three revolutions of a roller.

If small films are being processed, delay streaks may appear on more than one film. The occurrence of streaks on multiple films depends on the position of the film on the feed tray and the length of film travel.

Delay streaks are visible in transmitted and reflected light.

**Figure 6-10** Delay streaks. (Reprinted with permission from Eastman Kodak Company. EKC publication no. N-326.)

Delay streaks are usually caused by a buildup of oxidized developer on the developer-to-fix crossover assembly due to improper processor ventilation. (Chemical fumes accumulate under the lid of the processor, then dry and oxidize on the rollers above the solution level when a processor is not properly vented.) Oxidized developer can also build up on the developer rack re-wet roller. (The re-wet roller in the developer rack must remain in contact with the roller that is above the solution level; the re-wet roller transfers solution to the roller above the solution level in order to prevent chemical buildup.)

Specially designed roller transport cleanup film—14 x 17-inch (35 x 43 centimeter) film with a tacky, non-light-sensitive coating on both sides of the film base—is particularly useful to control delay streaks in all processors except those with area replenishment, which cannot sense this clear-based film. Roller transport cleanup film picks up lint, dirt, and other deposits and helps carry them out of the processor. For best results, one or two sheets should be processed.

Undeveloped fogged or expired single- or double-emulsion film may also be used as cleanup film. Note that mixing films from different manufacturers in the same environment should be avoided, as undesired sensitometric changes may occur. Any film used as cleanup film should be discarded after one use to avoid contaminating the developer solution with fixer and redepositing lint or dirt back onto the rollers in the processor.

**Table 6-3** Delay Streaks

*Appearance*

- Parallel to the film travel direction.
- Randomly spaced, narrow bands of varying widths on the leading edge of the first film processed after the processor has been on standby or inactive.
- Usually plus-density; may also be minus-density or combination.

*Causes*

- Buildup of oxidized developer on the developer-to-fix crossover assembly or on the re-wet roller.
- Improper processor ventilation.
- Low level of developer in the developer tank.
- Re-wet roller not functioning correctly.
- Evaporation covers not installed.

*Solutions**

- Process a sheet of roller transport cleanup film if the processor has been on standby or inactive for over 30 minutes.
- Clean any dirty rollers; replace damaged rollers.
- Clean, repair, or replace the developer re-wet roller.
- Check that the developer tank is full.
- Check that the processor is level.
- Check that the processor is vented correctly.
- Check that the evaporation covers are installed.

* All actions suggested under Solutions should be done by the most appropriate person, that is, one who has received the proper training to perform the necessary task.
(Reprinted with permission from Eastman Kodak Company. EKC publication no. N-326.)

Table 6-3 contains information on the primary appearance, causes, and solutions for delay streaks. Figure 6-11 shows the sections of the processor where delay streaks most frequently occur.

## *Entrance Roller Marks*

Plus-density bands approximately 1/8 inch (0.13 millimeter) wide may be caused by the rollers in the entrance detector crossover assembly as film is transported into the developer (Figure 6-12). The location of the marks on the film can vary with three factors: (1) pressure, (2) moisture, and (3) light.

Entrance roller marks may be caused by pressure from the rollers if one or both of the magnetic fields are not broken as film separates the two stepped rollers in the assembly. This may be due to a mechanical problem or spilled chemicals. Entrance roller marks from pressure are uniform over the length of the film. If small films are being processed, entrance roller marks may appear only on one of the two side edges of the film (drive side or nondrive side).

Entrance roller marks may be caused by moisture accumulating on the rollers due to improper processor venting, or from films that had already started to feed into the

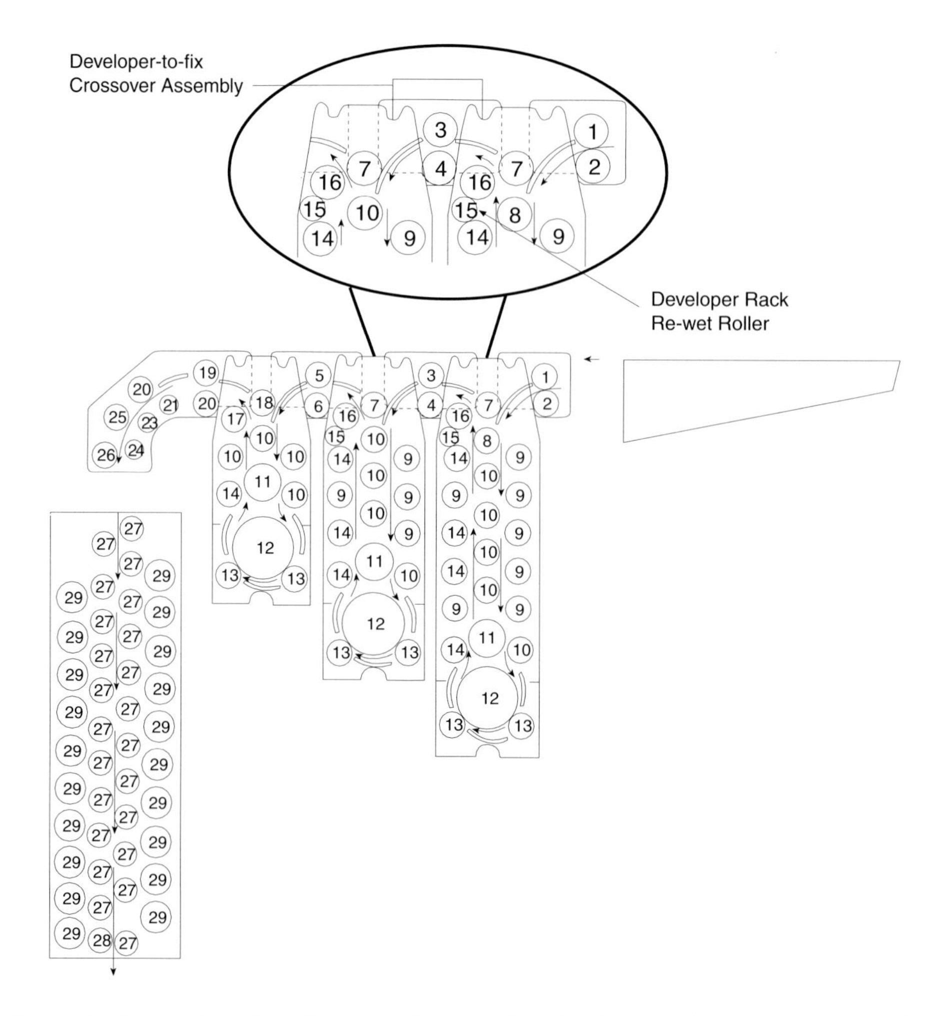


**Figure 6-11** Buildup of oxidized developer on the developer-to-fix crossover assembly or the developer rack re-wet roller may produce delay streaks. (Reprinted with permission from Eastman Kodak Company. EKC publication no. N-326.)

developer being pulled back. Entrance roller marks from moisture can be continuous or random. Generally, the marks will decrease in density across the film as the moisture is transferred from the rollers to the film.

They may result from turning on the light or opening the door before the entire film has been transported into the processor. Entrance roller marks caused by light appear near the trailing edge of the film.

Entrance roller marks are most visible in transmitted light, but in severe cases are also visible in reflected light.

Table 6-4 contains information on the primary appearance, causes, and solutions for entrance roller marks. Figure 6-13 shows the section of the processor where entrance roller marks occur.

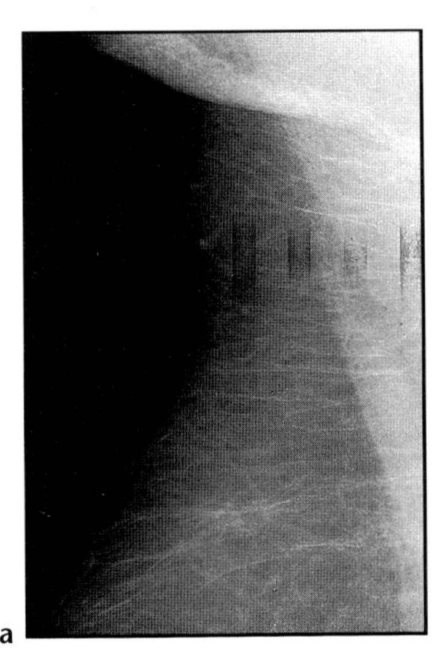

**Figure 6-12** Entrance roller marks, (a) on a mammogram and (b) on a uniform field radiograph. (Reprinted with permission from Eastman Kodak Company. EKC publication no. N-326.)

**Table 6-4** Entrance Roller Marks

*Appearance*

- Parallel to the film travel direction.
- Plus-density bands.
- Occur at evenly spaced intervals.

*Causes*

- Excessive pressure exerted on film by the entrance rollers.
- Moisture on entrance rollers.
- Trailing edge of film exposed to light.

*Solutions**

- Maintain the entrance detector crossover assembly according to the manufacturer's recommendations:
  - Check that the rollers are straight.
  - Check that the bearings have freedom of motion.
  - Make sure the assembly is square and properly seated in the processor.
- Check that the processor is vented correctly.
- Do not immerse the entrance detector assembly during cleaning; dry rollers thoroughly after cleaning with a damp, lint-free wipe.
- Do not process films after cleaning until entrance detector crossover rollers are thoroughly dry.
- Wait for signal that film has fed into processor before turning on light or opening darkroom door.

* All actions suggested under Solutions should be done by the most appropriate person, that is, one who has received the proper training to perform the necessary task.
(Reprinted with permission from Eastman Kodak Company. EKC publication no. N-326.)

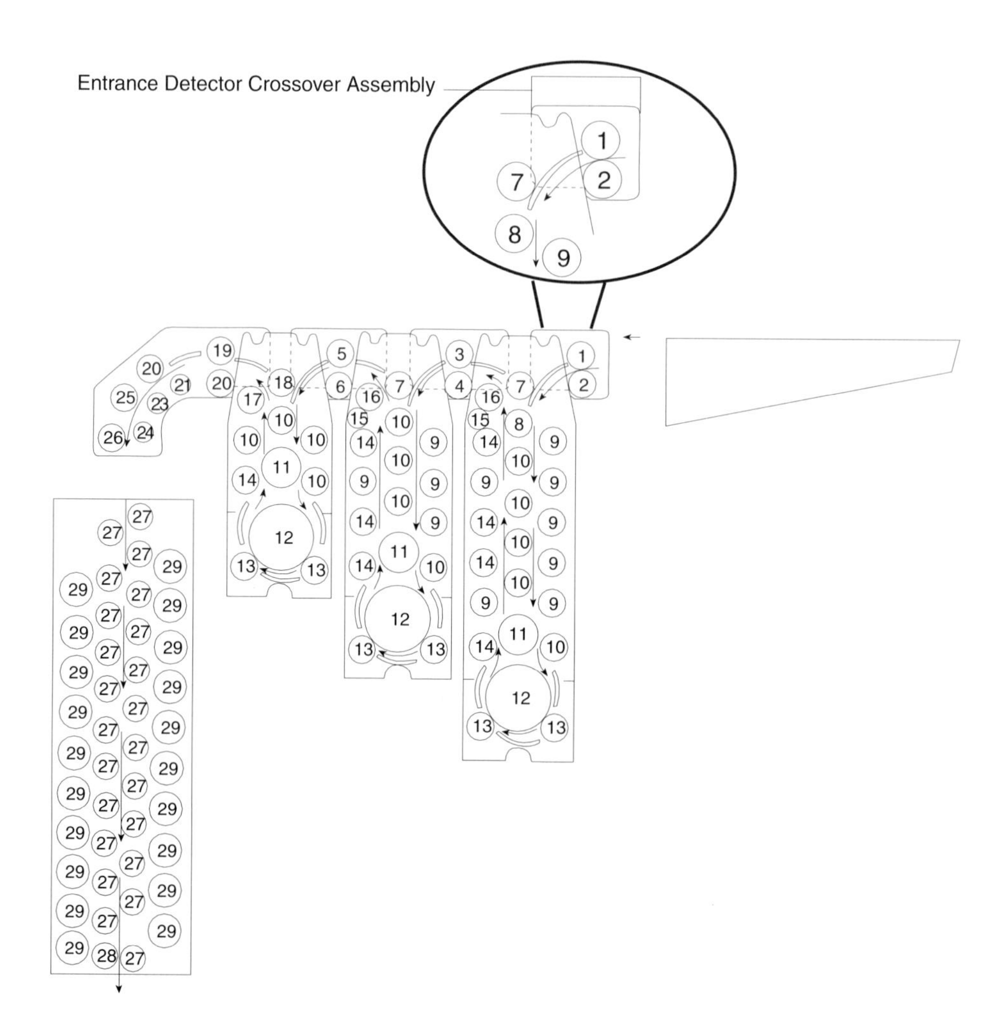


**Figure 6-13** Moisture on the entrance detector crossover assembly, or excessive pressure from it, may produce entrance roller marks. (Reprinted with permission from Eastman Kodak Company. EKC publication no. N-326.)

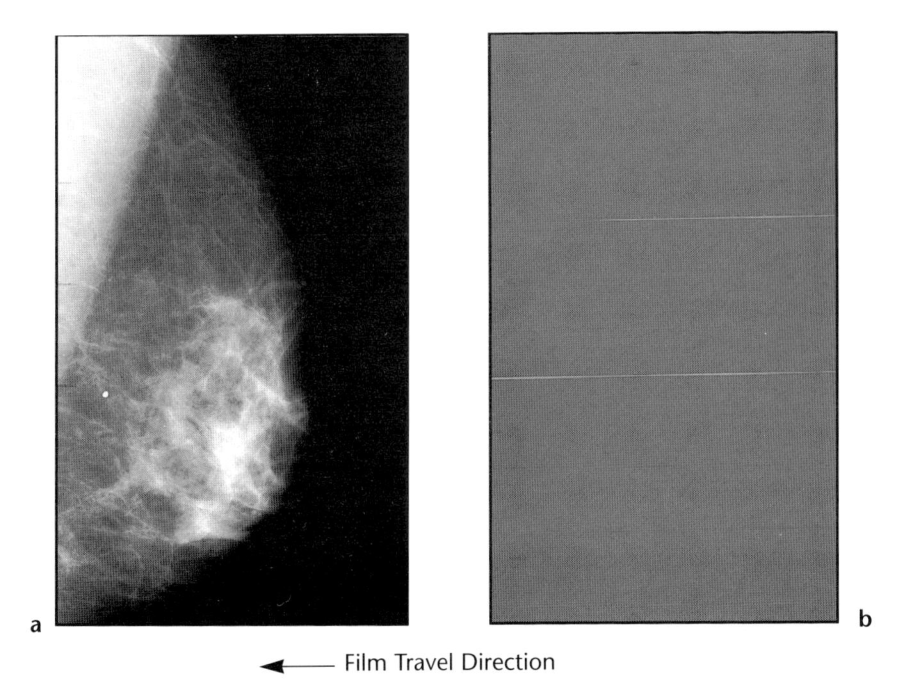


**Figure 6-14** Guide shoe marks: (a) plus-density along the chest wall of a mediolateral oblique mammogram and (b) minus-density. (Reprinted with permission from Eastman Kodak Company. EKC publication no. N-326.)

## *Guide Shoe Marks*

Guide shoe marks may appear as either plus-density, minus-density, or minus-density with film surface damage (Figure 6-14).

During normal operation of the processor, the film travels through several assemblies that contain guide shoes. As the film enters any one of the crossover, turnaround, or squeegee assemblies, the film responds in the same way: (1) the leading edge of the film contacts the guide shoes, (2) the remainder of the film conforms to the shape of the roller, and (3) the trailing edge of the film snaps against the guide shoe as it exits the assembly.

Any guide shoes that are bent, damaged, worn, not adjusted correctly, or not correctly installed may cause artifacts.

Guide shoes in the crossover assemblies can mark the top surface of the film; guide shoes in the turnaround assemblies can mark the bottom surface of the film.

As the developer-to-fix and fix-to-wash crossover assemblies should be removed daily for cleaning as discussed in Chapter three, it is extremely important that they be handled carefully to avoid damage to the guide shoes, replaced properly (fully seated) back in the processor, and replaced in their designated location. The washer mounted on the drive side of the crossover assembly identifies the developer-to-fix crossover assembly as "D/F" and the fix-to-wash crossover assembly as "F/W."

Table 6-5 contains information on the primary appearance, causes, and solutions for guide shoe marks. Figure 6-15 shows the sections of the processor where guide shoe marks may occur.

## Perpendicular to Film Travel Artifacts

### *Chatter*

Chatter appears as plus-density bands of variable density that occur at a uniform or consistent distance from each other (Figure 6-16). The spacing corresponds to the spacing of the sprocket or gear teeth in the developer rack. Chatter originating from the gears regularly occurs every 1/8 inches (3.2 millimeters). Chatter originating from the drive chain regularly occurs every 3/8 inches (9.5 millimeters).

**Table 6-5** Guide Shoe Marks

*Appearance*

- Parallel to the film travel direction.
- Can be plus-density, minus-density, or minus-density with film surface damage.
- Occur at evenly spaced intervals.
- Frequently found at the leading or the trailing edge of film.
- Length of normal guide shoe marks ≤1/4 inch (6.35 millimeters).

*Causes*

- Plus-density: incorrectly adjusted guide shoes in the developer section.
- Minus-density with no surface damage: incorrectly adjusted guide shoes in the fix-to-wash crossover assembly.
- Minus-density with surface damage: problem may be anywhere in the film transport path.
- Marks on top surface of film most likely caused by crossover assembly guide shoes.
- Marks on bottom surface of film most likely caused by turnaround assembly guide shoes.
- Chemical deposits on guide shoes.

*Solutions**

- Examine all guide shoes in the crossover and turnaround assemblies.
- Replace bent, damaged, or worn shoes.
- Remove any dried chemical deposits.
- Check to make sure that all processor rollers are straight and free-turning.
- Replace crossover assemblies and racks squarely following routine maintenance.
- Handle crossover assemblies carefully during daily and other maintenance.
- Always feed the short dimension of a mammography film as the leading edge when feeding film into the processor.

* All actions suggested under Solutions should be done by the most appropriate person, that is, one who has received the proper training to perform the necessary task.

(Reprinted with permission from Eastman Kodak Company. EKC publication no. N-326.)

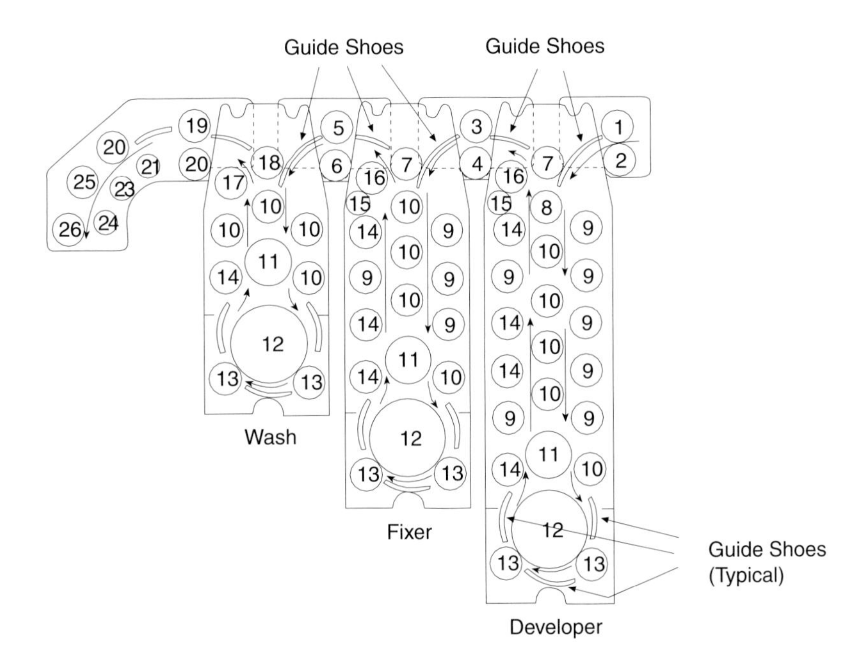


**Figure 6-15** Guide shoes anywhere in the processor may cause guide shoe marks. Incorrectly adjusted guide shoes in the developer section may cause plus-density guide shoe marks; incorrectly adjusted guide shoes in the fix-to-wash crossover assembly may cause minus-density guide shoe marks. (Reprinted with permission from Eastman Kodak Company. EKC publication no. N-326.)

Chatter is most visible in transmitted light, but it may be visible in reflected light in severe cases.

Chatter may be caused by any processor transport component that is not driving smoothly, improperly aligned, worn, or damaged. A drive chain that is too loose or too tight may also cause chatter. (Note that the tightness of drive chains on new processors generally should not be altered until after the break-in period.)

Table 6-6 contains information on the primary appearance, causes, and solutions for chatter. Figure 6-17 shows the sections of the processor where chatter may occur.

## *Slap Lines*

A slap line appears as a plus-density, broad band that is located approximately 3/4 to 2 1/4 inches (1.9 to 5.7 centimeters) from the trailing edge of the film (Figure 6-18). The band can be up to 1/4-inch (6.35-millimeters) wide and does not have sharp edges.

Slap lines are normally visible only in transmitted light.

They occur when the trailing edge of the film abruptly releases from the developer-to-fix crossover assembly and slaps the partially immersed top center roller of the fixer rack.

**Figure 6-16** Chatter. (Reprinted with permission from Eastman Kodak Company. EKC publication no. N-326.)

Slap lines can be eliminated by using properly mixed fixer and adequate fixer replenishment rates and by ensuring that the processor is maintained at regular intervals as recommended by the manufacturer.

Table 6-7 contains information on the primary appearance, causes, and solutions for chatter. Figure 6-19 shows the section of the processor where slap lines may occur.

## *Stub (Hesitation) Lines*

Stub (hesitation) lines appear as plus-density and minus-density lines or bands (Figure 6-20). The marks or lines can appear anywhere on the film at regular or irregular intervals.

Stub (hesitation) lines are usually visible in transmitted light. They may occur whenever there is a change in the velocity of film travel. This change is usually due to any interference within the film path. The most common causes of interference with film transport include the hesitation of film at the first guide shoe in the

**Table 6-6** Chatter

*Appearance*

- Perpendicular to the film travel direction.
- Plus-density marks of variable density.
- Spaced at a uniform or consistent distance from each other; spacing like sprocket or gear teeth separation.

*Causes*

- Improperly adjusted developer rack drive chain or gears.
- Improperly adjusted developer-to-fix crossover assembly drive.
- Rusty drive chain.
- Chemical buildup on the gears.

*Solutions**

- Perform preventive maintenance as recommended by the manufacturer.
- Adjust developer rack and/or developer-to-fix crossover assembly drive closely following the manufacturer's instructions.
- Check that all gears mesh correctly.
- Check for proper chain orientation and tension.
- Replace the chain if it is rusty or appears worn.

* All actions suggested under Solutions should be done by the most appropriate person, that is, one who has received the proper training to perform the necessary task.
(Reprinted with permission from Eastman Kodak Company. EKC publication no. N-326.)

developer turnaround assembly and the hesitation of film encountering rollers throughout the processor that are warped, rough, dirty, or encrusted with dried chemical buildup.

Guide shoes on some processors are adjustable. Check with the processor manufacturer for additional information on whether adjustments will reduce stub (hesitation) lines.

As in many types of processing artifacts, following the manufacturer's recommendations in terms of processor maintenance, chemicals, and replenishment rates is also required to reduce the incidence of stub (hesitation) lines.

Table 6-8 contains information on the primary appearance, causes, and solutions for stub (hesitation) lines. Figure 6-21 shows the section of the processor where stub (hesitation) lines may occur.

## Randomly Occurring Artifacts

### *Bent Corners*

Excessive recirculation within the processor may alter the forward inline path of the film as it is transported through the processor. This can result in a corner of the film being damaged or bent. Use of a smaller orifice or orifices in the recirculation system will eliminate this problem.

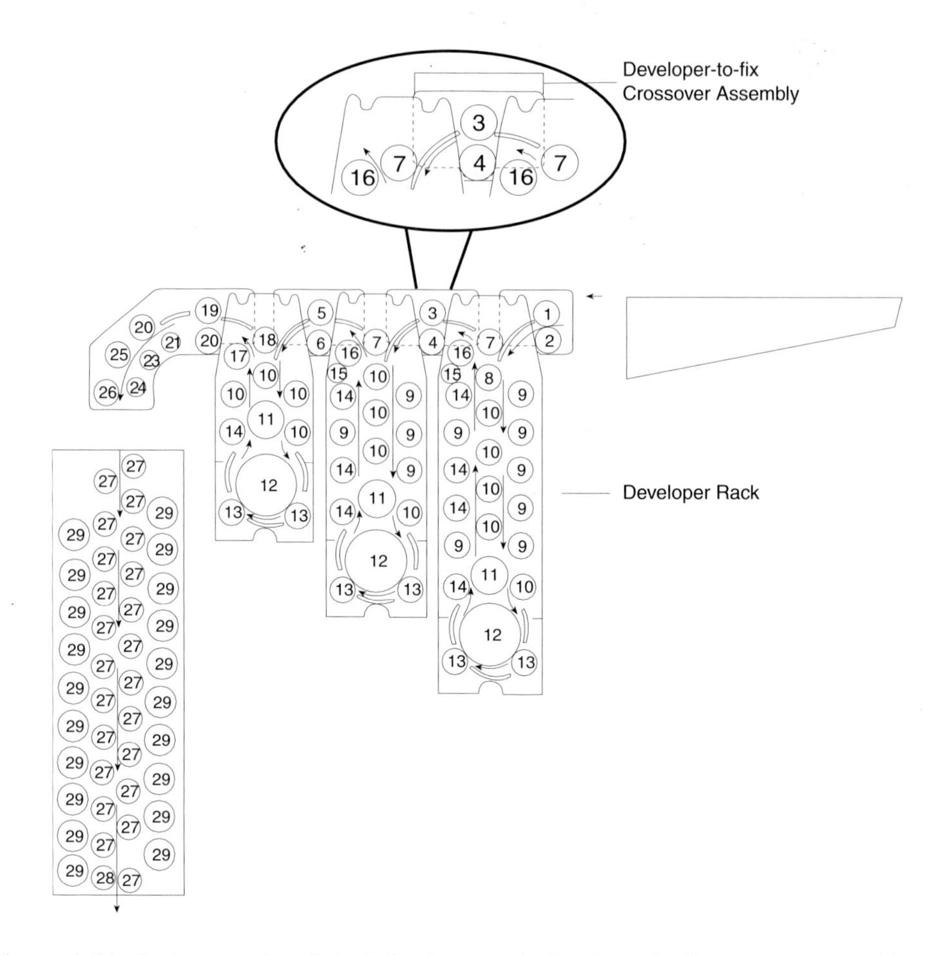


**Figure 6-17** An improperly adjusted developer rack, developer-to-fix crossover assembly drive chain or gears may produce chatter. (Reprinted with permission from Eastman Kodak Company. EKC publication no. N-326.)

Table 6-9 contains information on the primary appearance, causes, and solutions for bent corners.

## *Brown Films*

Brown films may result from inadequate washing or fixing of the film due to a number of problems.

Film will not be properly washed if the water is inadvertently left off or if the water filter is clogged. Film will not be properly fixed if poor quality or improperly mixed fixer is used. In addition, fixer replenisher not reaching the fixer tank or the fixer replenishment rate set too low may also contribute to brown films.

Solution temperatures, particularly water and fixer, must also be as specified by the manufacturer.

**Figure 6-18** Slap lines (trailing edge of film). (Reprinted with permission from Eastman Kodak Company. EKC publication no. N-326.)

**Table 6-7** Slap Lines

*Appearance*

- Perpendicular to the film travel direction.
- Plus-density.
- Broad line located 3/4 to 2 1/4 inches (1.9 to 5.7 centimeters) from trailing film edge.

*Causes*

- Trailing edge of film abruptly releases from the developer-to-fix crossover assembly and slaps the top immersed center roller of the fixer rack.

*Solutions**

- Use properly mixed fixer and adequate fixer replenishment rate.
- Check that the developer exit squeegee is operating uniformly.
- Verify film transport speed is uniform from the developer to fixer solutions.
- When replacing rollers, use rollers recommended by the manufacturer.
- Check the developer-to fix crossover assembly guide shoes for damage or improper positioning; replace as appropriate.

* All actions suggested under Solutions should be done by the most appropriate person, that is, one who has received the proper training to perform the necessary task. (Reprinted with permission from Eastman Kodak Company. EKC publication no. N-326.)

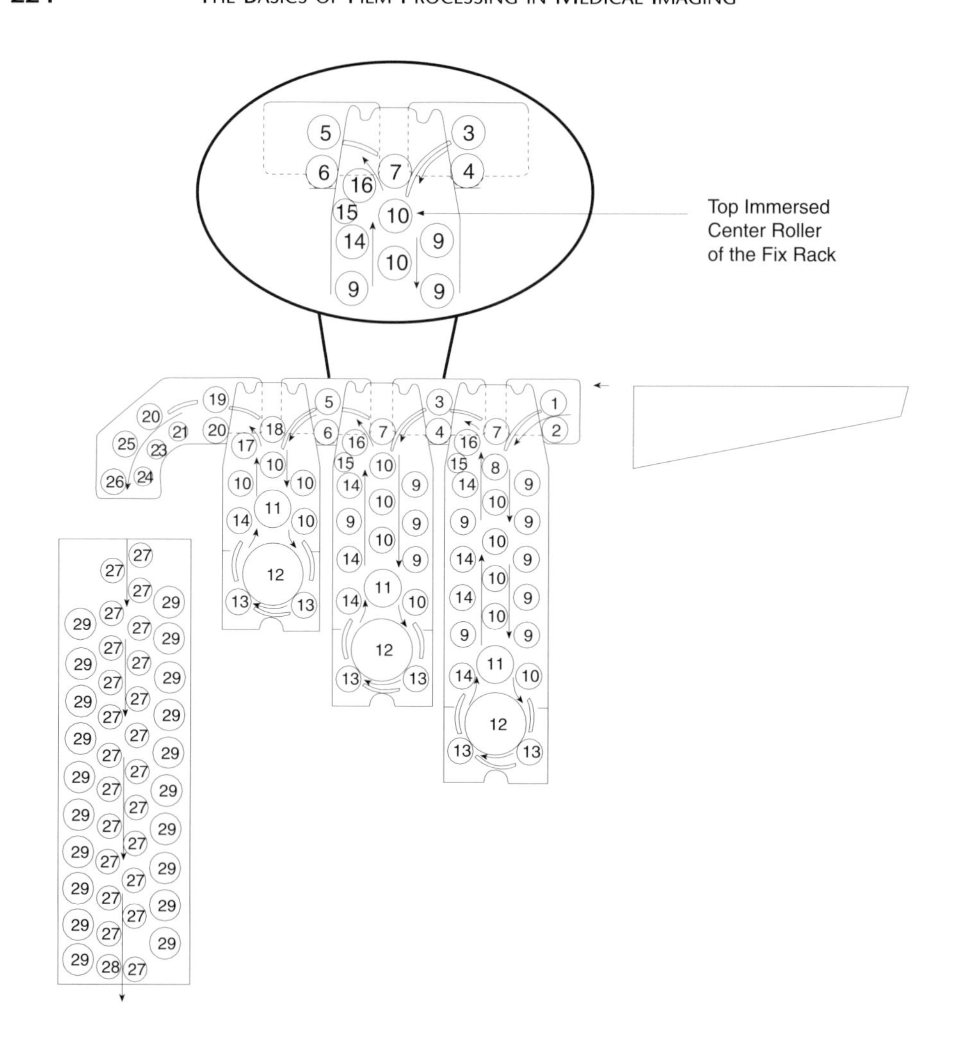


**Figure 6-19** The trailing edge of the film slapping the top immersed center roller of the fix rack may produce slap lines. (Reprinted with permission from Eastman Kodak Company. EKC publication no. N-326.)

Performing a fixer retention test (hypo test) will provide an estimate of the amount of retained ammonium thiosulfate or hypo in processed films. Refer to Chapter five for additional information on the fixer retention test.

Table 6-10 contains information on the primary appearance, causes, and solutions for brown films.

## *Drying Patterns and Water Spots*

Drying patterns appear as bands or spots of nonuniform density on the film surface (Figure 6-22). The streaks can contain areas that appear mottled, washed out, shiny, or glossy. Drying patterns may also be referred to as "shoreline patterns" due to their resemblance to the patterns left by water along a shoreline.

**Figure 6-20** Stub (hesitation) lines (leading edge of film; magnified). (Reprinted with permission from Eastman Kodak Company. EKC publication no. N-326.)

**Table 6-8** Stub (Hesitation) Lines

*Appearance*

- Perpendicular to the film travel direction.
- Plus-density and minus-density.
- May appear anywhere on the film at regular or irregular intervals.
- Frequently seen near the leading edge of the film.

*Causes*

- Poor-quality, exhausted, or contaminated processing chemicals.
- A malfunctioning film transport component, such as a roller, gear, or chain.
- Dirt or chemical buildup on rollers.
- Warped or rough rollers.
- Improper guide shoe positioning in the developer turnaround assembly.
- Improperly adjusted rack drive chain.

*Solutions**

- Clean processor.
- Replace processing chemicals with fresh solutions from a quality processing chemical supplier.
- Check replenishment rates and set to the film manufacturer's recommendations.
- Replace guide shoes.
- Clean, repair, or replace rollers.
- Adjust rack chains.

* All actions suggested under Solutions should be done by the most appropriate person, that is, one who has received the proper training to perform the necessary task.
(Reprinted with permission from Eastman Kodak Company. EKC publication no. N-326.)

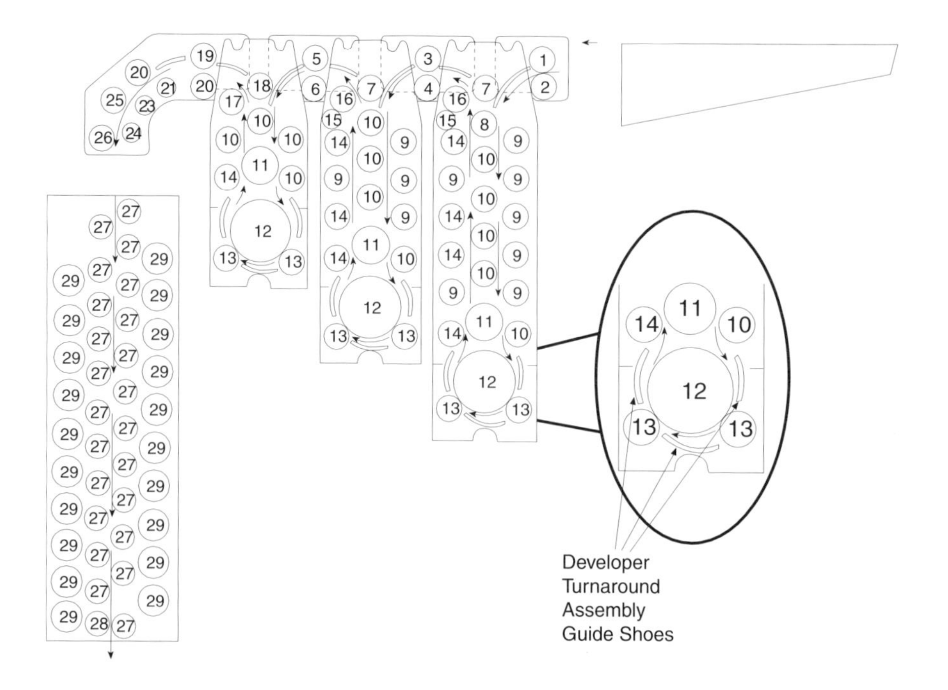


**Figure 6-21** Improper guide shoe positioning in the developer turnaround assembly may produce stub (hesitation) lines. (Reprinted with permission from Eastman Kodak Company. EKC publication no. N-326.)

Drying patterns and water spots are normally only visible in reflected light, but in severe cases they may also be seen in transmitted light.

Slight drying patterns along one or more film edges that are visible only in reflected light are considered normal and acceptable. Drying patterns that extend far into the image area and are visible in transmitted light as well as reflected light are considered abnormal and unacceptable.

Drying patterns are frequently caused by excessively high dryer temperatures, which cause film to dry too fast. They are also caused by the uneven removal of water from the film. This may be due to nonuniform airflow in the dryer section due to clogged, missing, or out-of-place dryer air tubes. Drying patterns can also result if the film is not dry as it exits the dryer section. Damp film may pick up patterns from the exit rollers, the dryer bin, or other films.

Drying patterns are also influenced by chemical quality in combination with high dryer temperatures. A difficulty in obtaining dry film, causing the dryer temperature to be raised and exacerbating drying patterns, may actually stem from the use of depleted, oxidized, or poor-quality processing chemicals. Use of chemicals as recommended helps eliminate film drying difficulties and drying patterns.

After following the manufacturer's recommendations for processing, chemicals, replenishment rates, etc., the dryer temperature should be lowered until film is coming out still damp or tacky. Temperature may then be raised in 5°F (3.5°C) increments until film is exiting the processor perfectly dried.

**Table 6-9** Bent Corners

*Appearance*

- Randomly occurring.
- Usually affects one corner of the film.

*Causes*

- Excessive recirculation.

*Solutions**

- Install a smaller orifice.

* All actions suggested under Solutions should be done by the most appropriate person, that is, one who has received the proper training to perform the necessary task.

**Table 6-10** Brown Films

*Appearance*

- Randomly occurring.
- Brown radiographs.
- Odor may also be present.

*Causes*

- Inadequate washing of the film.
  - Incorrect wash water level.
  - Inadequate water flow.
- Inadequate fixing of the film.
  - Depleted fixer solution.
  - Poor chemical quality of fixer.
  - Fixer improperly mixed.
  - Fixer replenishment rate set too low.
  - Fixer and/or water temperature too low.

*Solutions**

- Ensure that the water is turned on in the processor.
- Check if the water filter is clogged and replace if necessary.
- Check the quality of the fixer.
- Check the fixer replenishment rate.
- Check that the line from the fixer replenisher tank to the processor is not kinked or air-locked.
- Check that the recirculation pump is operational.
- Check that the water and fixer are within the temperature ranges required for proper processor function.

* All actions suggested under Solutions should be done by the most appropriate person, that is, one who has received the proper training to perform the necessary task.

**Figure 6-22** Drying patterns and water spots. (Reprinted with permission from Eastman Kodak Company. EKC publication no. N-326.)

**Table 6-11** Drying Patterns and Water Spots

*Appearance*

- Randomly occurring.
- Narrow wavering bands or spots that appear mottled, washed out, shiny, or glossy.
- Readily seen by reflected light.
- Severe artifacts may be apparent when viewed by transmitted light.

*Causes*

- Depleted, oxidized, or poor-quality processing chemicals.
- Poor squeegee action at the wash rack exit.
- Clogged, missing, or out-of-position dryer air tubes.
- Non-uniform or excessive airflow in dryer.
- Excessively high dryer temperature.
- Very cold wash water.
- Inadequate processor venting.

*Solutions**

- Reduce dryer temperature to minimum required to produce dry films.
- Replace processing chemicals with fresh solutions.
- Check replenishment rates and adjust according to the film manufacturer's recommendations.
- Use processing chemicals recommended by the film manufacturer.
- Provide wash water within the recommended temperature range.
- Adjust or replace the wash rack exit squeegee.
- Replace/clean air tube(s).

* All actions suggested under Solutions should be done by the most appropriate person, that is, one who has received the proper training to perform the necessary task.
(Reprinted with permission from Eastman Kodak Company. EKC publication no. N-326.)

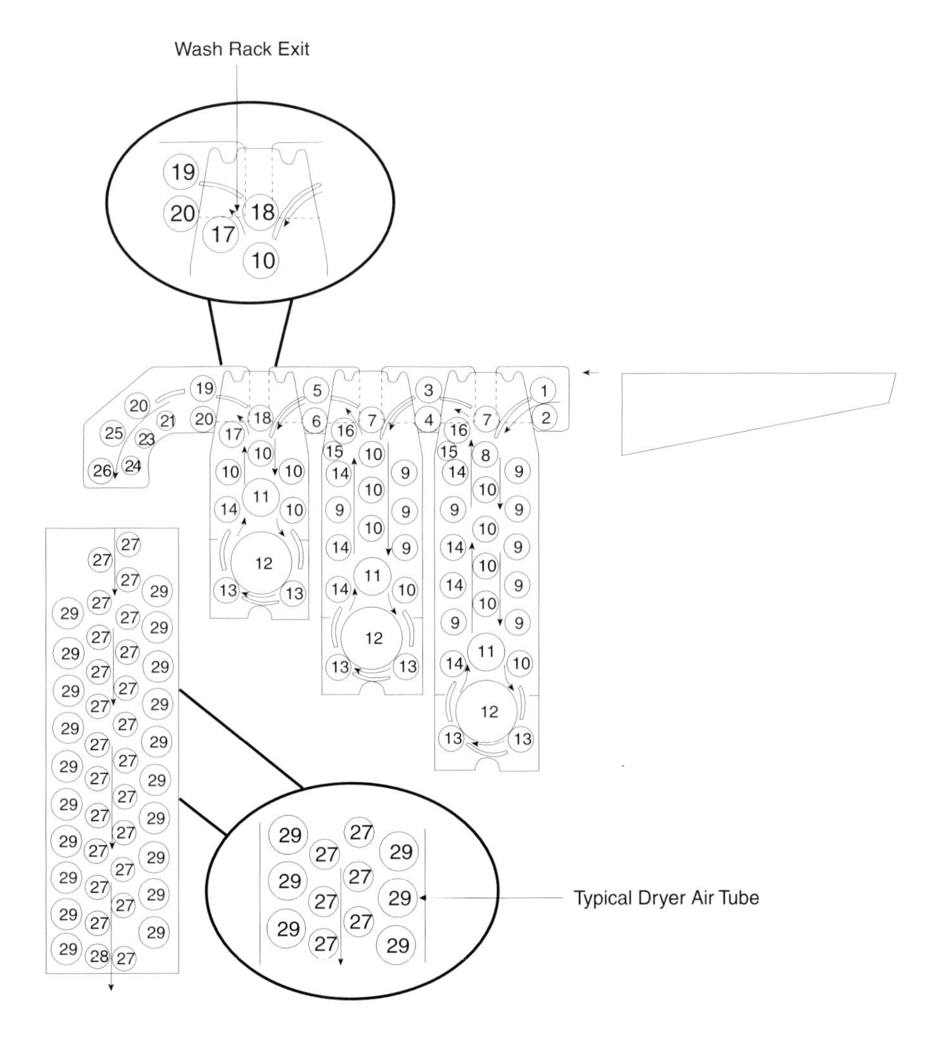


**Figure 6-23** Poor squeegee action at the wash rack exit and clogged or missing dryer air tubes may produce drying patterns and water spots. (Reprinted with permission from Eastman Kodak Company. EKC publication no. N-326.)

Table 6-11 (page 228) contains information on the primary appearance, causes, and solutions for drying patterns and water spots. Figure 6-23 shows the sections of the processor where drying patterns and water spots may occur.

## *Dye Stain*

Dye stain usually appears as pink or lavender areas or blotches in the D-min area of the film and is caused by the incomplete removal of sensitizing dye in the film. The color of the dye stain may vary with film type.

Dye stain is visible when you view the film using transmitted light.

The occurrence of dye stain is usually connected to fixer quality and replenishment rates that are set too low. Dye stain may also occur if fixer or water temperatures are set too low or if film is transported at an increased speed through the fixer section of the processor. The proper cycle for a specific film should be used to avoid dye stain.

The practice of reusing fixer may also increase the incidence of dye stain.

Table 6-12 contains information on the primary appearance, causes, and solutions for dye stain.

## *Flame Patterns*

Flame patterns are variations in the density of the film, which resemble the flame of a candle (Figure 6-24). Flame patterns are typically visible in transmitted light and are more difficult to see in reflected light.

Flame patterns are less likely to be seen when using seasoned chemicals. Seasoned developer may be reserved and reused indefinitely when the processor is thoroughly cleaned, typically on a monthly basis, provided the processor quality control chart indicates that the processor is "in control" both before and after the cleaning. Refer to Chapter seven for additional information.

Note that drying patterns, discussed earlier in this chapter, can also resemble flame patterns.

Flame patterns may be caused by low recirculation of the developer solution. This may be because of mechanical problems, such as a dirty developer filter, an incorrect

**Table 6-12** Dye Stain

*Appearance*

- Randomly occurring.
- Pink or lavender color in clear (D-min) area of the film.
- Color of stain may vary with film type.

*Causes*

- Incomplete removal of sensitizing dye.
- Inadequate fixing and washing.
  - Poor chemical quality of fixer.
  - Low fixer replenishment rate.
  - Fixer reused/recycled.
  - Low fixer or water temperature.
- Increased transport speed through fixer.
  - Film processed at cycle faster than recommended for specific film type.

*Solutions**

- Replace processing chemicals with fresh solutions.
- Check fixer replenishment rate and adjust/increase.
- Check method used for recycling fixer.
- Check fixer and water temperatures and adjust to recommended range.
- Process film at cycle required for film type.

* All actions suggested under Solutions should be done by the most appropriate person, that is, one who has received the proper training to perform the necessary task.

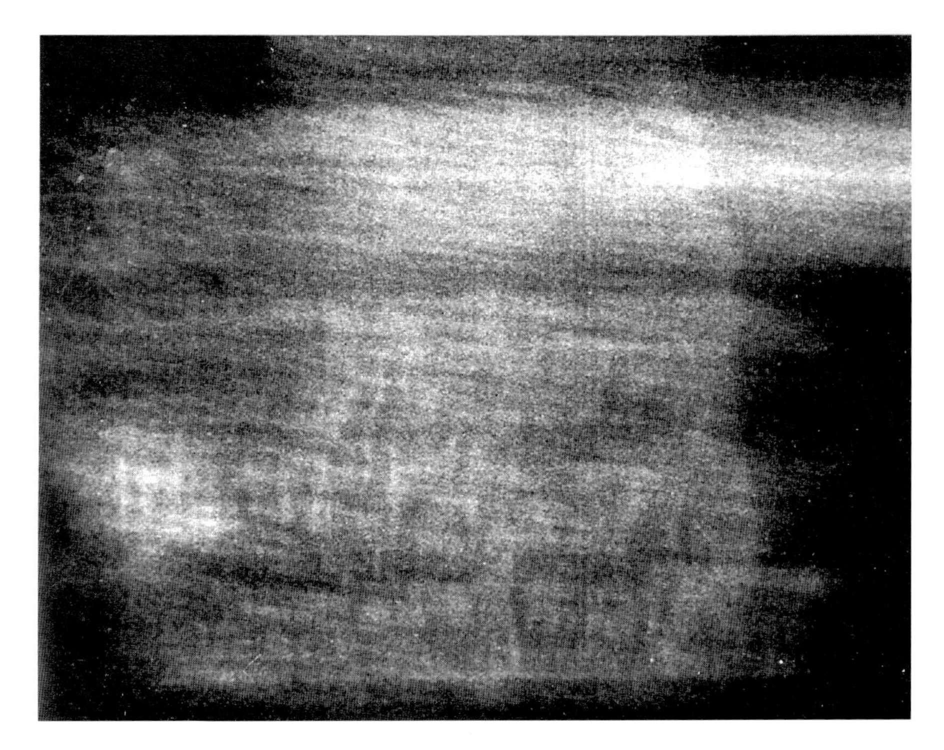

**Figure 6-24** Flame patterns. (Reprinted with permission from Eastman Kodak Company.)

or missing orifice in the recirculation line, a kinked recirculation line, etc., or the recirculation rate itself may be incorrect.

The developer filter should be changed regularly. The orifice recommended by the manufacturer should be installed in the developer recirculation system.

Table 6-13 contains information on the primary appearance, causes, and solutions for flame patterns.

## *Pick-Off*

Pick-off appears as very small minus-density (light) spots on the film where the emulsion has been removed down to the film base (Figure 6-25). Often a piece of the emulsion that is removed may be deposited on the film near the trailing edge of the minus-density mark or may be deposited randomly on the film. This deposit, or piling up, of the pick-off emulsion may easily be felt by passing the sensitive pad of a finger over it.

Pick-off is most apparent on single-emulsion films used for mammography. The term "pick-off" is, in fact, erroneously used to describe all minus-density spots on mammography films. Most of these minus-density spots are really handling artifacts caused by dirt and dust superimposed between the intensifying screen and film emulsion. The proper term for these artifacts is "shadow images," and they are discussed later in this chapter. Because most commonly used mammography films are high in contrast, it is difficult to distinguish between pick-off and shadow images.

**Table 6-13** Flame Patterns

*Appearance*

- Randomly occurring.
- Variations in density.
- Close resemblance to flame of candle.

*Causes*

- Low recirculation of developer.
  - Restricted, plugged, or incorrect configuration.
  - Dirty developer filter.
  - Recirculation hose kinked or flow impeded by air bubbles.
  - Missing, wrong, or misplaced recirculation hose orifice.
  - Inoperative recirculation pump.
- Low developer recirculation rate.

*Solutions**

- Change the developer filter at the recommended intervals.
- Use the filter type recommended by the manufacturer.
- Check the operation of the recirculation pump.
- Check that fresh developer replenisher is flowing into the processor developer tank.
- Check the developer recirculation rate, and adjust/increase as needed.
- Institute program of reserving seasoned developer with monthly processor cleaning provided processor is "in control" at time of cleaning.

* All actions suggested under Solutions should be done by the most appropriate person, that is, one who has received the proper training to perform the necessary task.

(Refer to Appendix D and Appendix E for information on troubleshooting the darkroom for dirt and dust and intensifying screen cleaning.)

Pick-off is most visible in transmitted light as are shadow images. Unlike shadow images, pick-off can also be seen in reflected light as a defect (hole) on the surface of the film. The pick-off hole can also be felt by passing a fingernail over the area.

Figure 6-26 shows artifact analysis using (a) transmitted light and (b) reflected light.

Pick-off is usually caused by poor processor maintenance. Rough or dirty rollers are capable of pulling emulsion from the film base. Pick-off can also occur if poor quality or exhausted processing chemicals are used.

Problems with the film emulsion may also cause pick-off.

Table 6-14 contains information on the primary appearance, causes, and solutions for pick-off.

## Run Back

Run back usually appears as dark, wavy areas, or a dribble or scallop on the trailing edge of the film (Figure 6-27). Occasionally, the marks may extend up to 3 inches

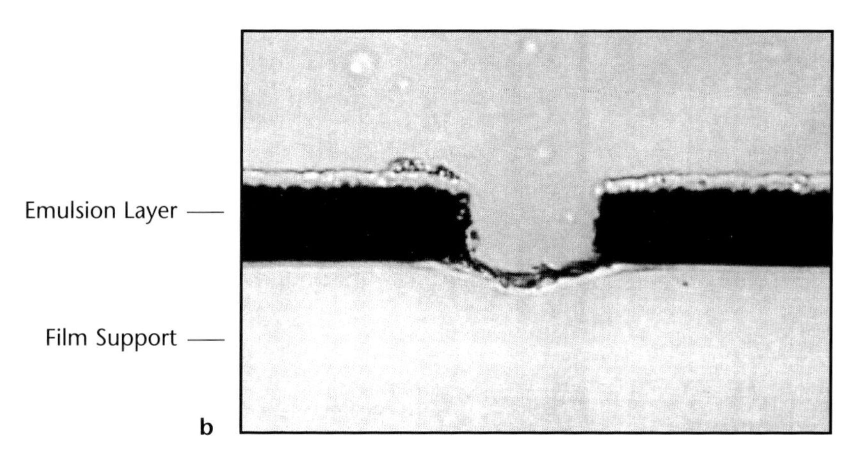


**Figure 6-25** (a) Pick-off. (b) Magnified cross-sectional view (850X) shows the surface defect that remains when film emulsion has been "picked-off" the film support. A portion of the emulsion has been deposited on the trailing side of the defect (visible immediately to the left of the hole). (Reprinted with permission from Eastman Kodak Company. EKC publication no. N-326.)

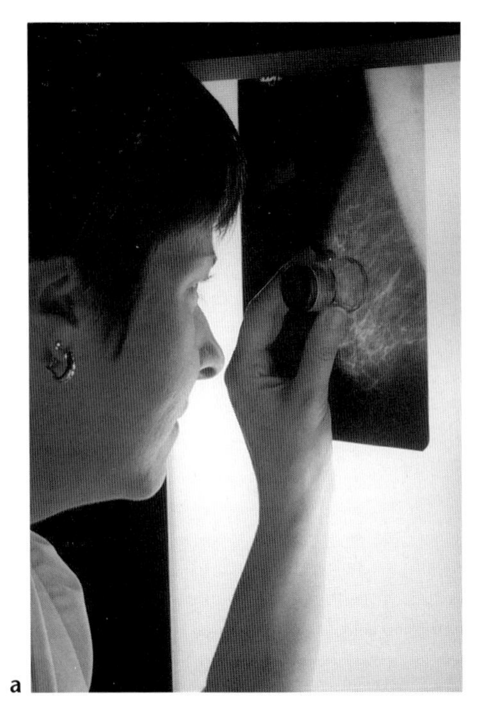

**Figure 6-26** (a) Viewing an artifact on a mammography film using transmitted light and a 7X magnifying loupe. (b) Examining the film surface using reflected light. (Reprinted with permission from Eastman Kodak Company.)

(7.62 centimeters) from the film edge. Their length is dependent upon the transport speed of the processor. They are more pronounced at slower speeds.

Run back is more common on single-emulsion films and is visible in transmitted light.

Run back is caused by the incomplete removal of developer solution from the film as the film exits the developer-to-fix crossover assembly. This causes developer solution to run down the trailing edge of the film.

Processors must be appropriately maintained to eliminate run back. Daily cleaning of the developer-to-fix crossover assembly and adequate spring tension on the squeegee rollers in the developer rack and crossover assembly are particularly helpful.

Table 6-15 contains information on the primary appearance, causes, and solutions for run back. Figure 6-28 shows the section of the processor where run back may occur.

## *Skivings*

Skivings are thin threads or slivers (whisker-like) of plus-density (dark) emulsion that have been removed from the film edge and redeposited back onto the film (Figure 6-29). Typically, skivings appear 3.14 inches (7.98 centimeters) from the leading edge of the film, but they may appear anywhere on it.

**Table 6-14** Pick-Off

*Appearance*

- Randomly occurring.
- Small, minus-density spots.
- Readily detected on single-emulsion film.
- Emulsion removed down to the film base.
- Characteristic "piling up" of emulsion on one side of the spot may be seen and felt.
- Visible on both transmitted and reflected light.

*Causes*

- Rough or dirty rollers.
- Inconsistent film transport speed.
- Poor quality or exhausted processing chemicals.

*Solutions**

- Use 7 to 10X magnification to confirm pick-off, not shadow images.
- Perform maintenance according to the manufacturer's recommendations.
- Establish program of regular preventive maintenance.
- Replace processing chemicals with fresh solutions.
- Use processing chemicals and replenishment rates recommended by the film manufacturer.

* All actions suggested under Solutions should be done by the most appropriate person, that is, one who has received the proper training to perform the necessary task.
(Reprinted with permission from Eastman Kodak Company. EKC publication no. N-326.)

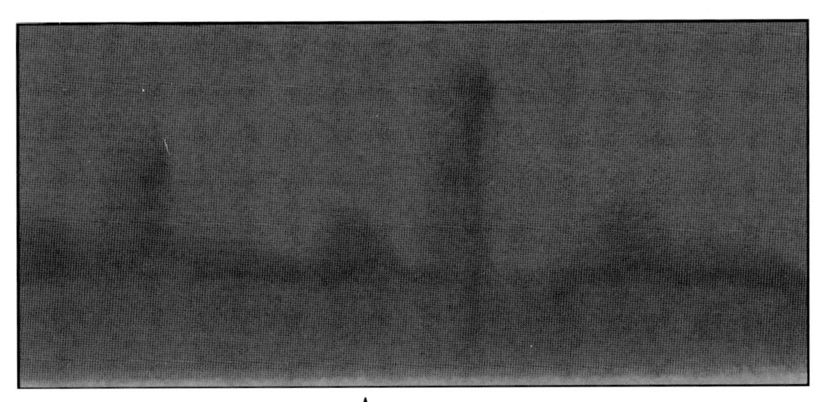


**Figure 6-27** Run back (trailing edge of film; magnified). (Reprinted with permission from Eastman Kodak Company. EKC publication no. N-326.)

**Table 6-15** Run Back

| |
|---|
| *Appearance*<br>• Randomly occurring.<br>• Plus-density "dribble" or "scallop."<br>• Appears on trailing edge of film. |
| *Causes*<br>• Insufficient squeegee action at exit of the developer rack or in the developer-to-fix crossover assembly.<br>• Developer solution runs down the trailing edge of the film as the film enters the fixer, causing increased and uncontrolled development. |
| *Solutions**<br>• Clean developer-to-fix crossover assembly daily.<br>• Maintain the crossover assembly squeegee as recommended by the manufacturer to reduce developer carryout.<br>• Replace worn or damaged parts.<br>• Check with the manufacturer to see if roller modifications are possible.<br>• Process single-emulsion films emulsion side up. |

* All actions suggested under Solutions should be done by the most appropriate person, that is, one who has received the proper training to perform the necessary task.
(Reprinted with permission from Eastman Kodak Company. EKC publication no. N-326.)

Skivings are visible in transmitted light and may also be seen in reflected light. Skivings can also frequently be felt on the surface of the film, and it may even be possible to remove them from the film by using a knife.

Skivings may be caused when the smooth transport of film through the processor is interrupted. The first guide shoe of the turnaround assembly in the developer tank is a frequent location.

Skivings may be minimized by maintaining the processor and using the processing chemicals and replenishment rates suggested by the processor and film manufacturers.

The periodic use of cleanup film is also helpful in removing skivings and other dirt from processor rollers.

In addition, the film manufacturer may be contacted for suggestions on minimizing skivings.

Table 6-16 contains information on the primary appearance, causes, and solutions for skivings.

## Wet Pressure

Wet pressure appears as plus-density fluctuations that may resemble noise or quantum mottle (Figure 6-30). These fluctuations are created from pressure applied

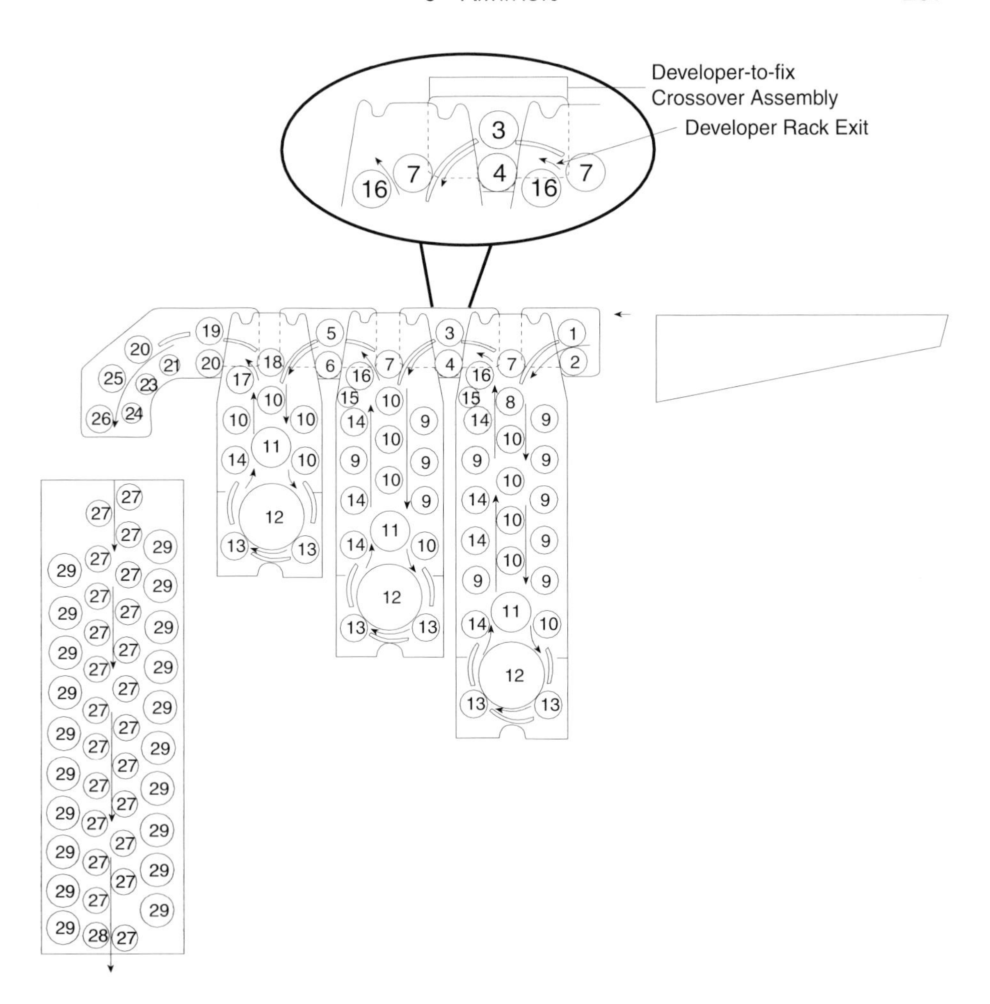


**Figure 6-28** Insufficient squeegee action at the exit of the developer rack or in the developer-to-fix crossover assembly may produce run back. (Reprinted with permission from Eastman Kodak Company. EKC publication no. N-326.)

to the film emulsion while the film is in the developer rack or the developer-to-fix crossover assembly.

Wet pressure may be caused by rough, blistered, warped, or dirty rollers in the developer rack or the developer-to-fix crossover assembly.

Wet pressure may be increased by using depleted or contaminated chemicals and by using a developer temperature higher than recommended for a specific film type and processing cycle.

Proper processor maintenance and adequate replenishment rates and chemical quality are required to minimize wet pressure. All processor rollers must be clean and free from defects.

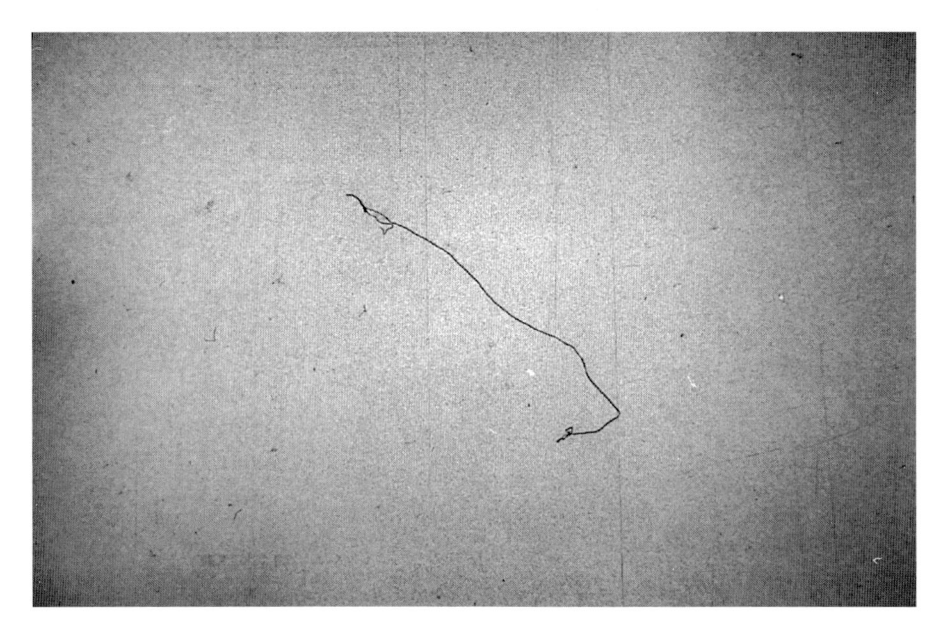

**Figure 6-29** Skivings (magnified). (Reprinted with permission from Eastman Kodak Company.)

**Table 6-16** Skivings

*Appearance*

- Randomly occurring.
- Thin, plus-density threads or slivers, (whisker-like).
- May be felt as well as seen.

*Causes*

- Unsmooth transport of film through processor causes roller to pick up thread of emulsion from film edge.

*Solutions**

- Use processing chemicals and replenishment rates recommended by film manufacturer.
- Perform maintenance recommended by processor manufacturer.
- Use cleanup film periodically to remove dirt and other matter from processor rollers.

* All actions suggested under Solutions should be done by the most appropriate person, that is, one who has received the proper training to perform the necessary task.

**Figure 6-30** Wet pressure. (Reprinted with permission from Eastman Kodak Company. EKC publication no. N-326.)

**Table 6-17** Wet Pressure Marks

*Appearance*

- Randomly occurring.
- Plus-density fluctuations that may resemble noise or quantum mottle.

*Causes*

- Depleted or contaminated processing chemicals.
- Pressure applied to the film emulsion by rough, blistered, warped, or dirty rollers in the developer rack or the developer-to-fix crossover assembly.

*Solutions**

- Replace processing chemicals with fresh solutions.
- Check developer replenisher quality.
- Follow the film manufacturer's recommendations for the developer replenishment rate and developer temperature setpoint.
- Check developer replenishment line for kinks.
- Clean, repair, or replace rollers.
- Use the rollers specified by the manufacturer.

* All actions suggested under Solutions should be done by the most appropriate person, that is, one who has received the proper training to perform the necessary task.
(Reprinted with permission from Eastman Kodak Company. EKC publication no. N-326.)

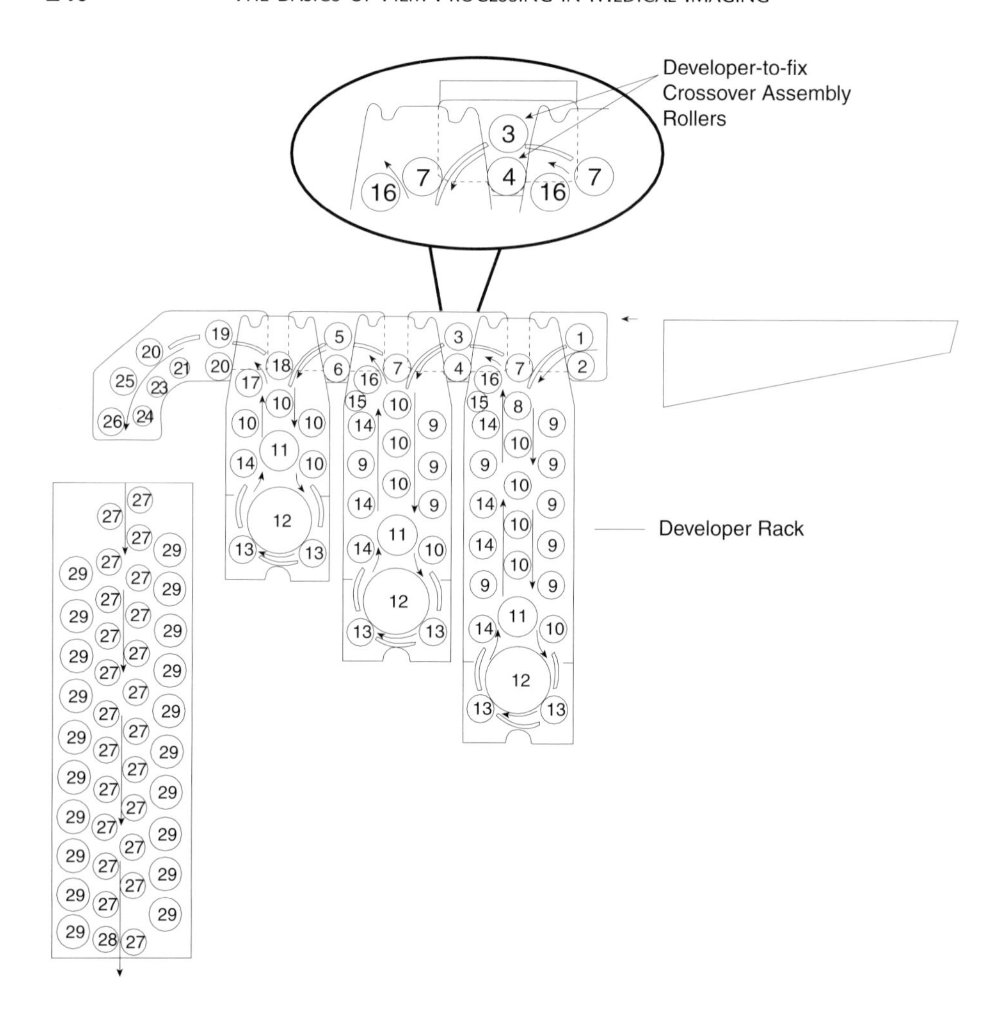


**Figure 6-31** Rough, blistered, warped, or dirty rollers in the developer rack or in the developer-to-fix crossover assembly may produce wet pressure marks. (Reprinted with permission from Eastman Kodak Company. EKC publication no. N-326.)

Table 6-17 (page 239) contains information on the primary appearance, causes, and solutions for wet pressure. Figure 6-31 shows the sections of the processor where wet pressure may occur.

## Shadow Images

Shadow images are the only handling artifact that will be covered in this chapter because they are frequently confused with pick-off.

Shadow images appear as small, random, minus-density (light) spots on the film (Figure 6-32). They are most apparent on single-emulsion films such as mammography films. It is estimated that shadow images may account for as much as 95 percent of minus-density artifacts seen on mammographic images.

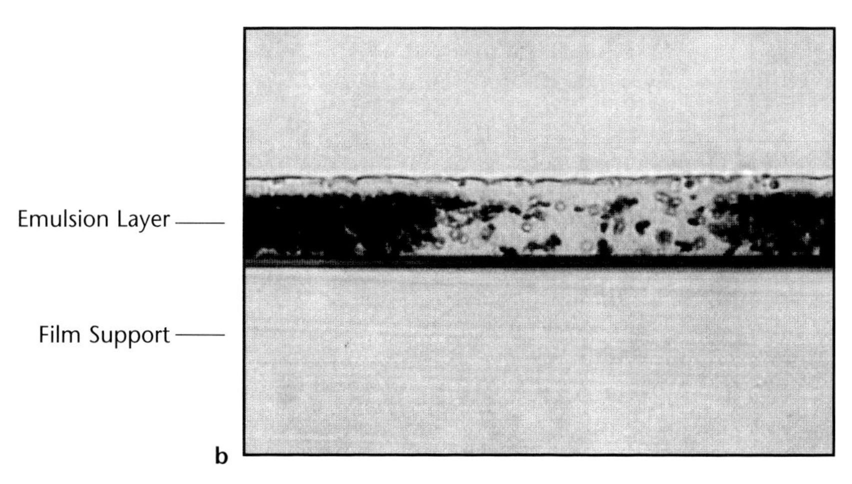


**Figure 6-32** (a) Shadow images. (b) Magnified cross-sectional view (850X) shows intact emulsion layer. (Reprinted with permission from Eastman Kodak Company. EKC publication no. N-326.)

Shadow images are visible only in transmitted light. When shadow images are viewed using 7 to 10X magnification and transmitted light, there is no piling-up of the emulsion on the trailing side of the spot nor is there a surface defect, such as are seen with pick-off. (Pick-off, a processing artifact, was discussed earlier in this chapter.)

Shadow images are caused by dirt or dust on the intensifying screens or in the cassette.

Cassettes and screens, especially those used for mammography, should be thoroughly cleaned and kept free of dirt and dust to eliminate shadow images. The darkroom should be kept as clean as possible. Incoming air should be clean and filtered.

**Table 6-18** Shadow Images

*Appearance*

- Randomly occurring handling artifact.
- Small, minus-density spots.
- Most visible on single-emulsion films.
- Visible with transmitted light.
- No surface defect when examined with reflected light.

*Causes*

- Dirt or dust on intensifying screens or in cassette.
- Dirt or dust on darkroom surfaces.
- Dust suspended in darkroom air.

*Solutions**

- Clean screens and cassettes thoroughly.
- Clean darkroom surfaces.
- Filter darkroom air.
- Use an ultraviolet light device to detect areas where dust or dirt accumulate; observe all safety precautions during use of light.

* All actions suggested under Solutions should be done by the most appropriate person, that is, one who has received the proper training to perform the necessary task.
(Reprinted with permission from Eastman Kodak Company. EKC publication no. N-326.)

The filtration system should be serviced regularly as recommended by the manufacturer. (Refer to Appendix D for a discussion on darkroom features that help eliminate dirt and dust from films. Refer to Appendix E for a discussion on screen cleaning.)

An ultraviolet light device is helpful to evaluate the darkroom for features that may contribute to ongoing dirt and dust problems. Observe appropriate safety precautions for eyes and skin.

Table 6-18 contains information on the primary appearance, causes, and solutions for shadow images.

## Isolating Individual Processor Components

When analyzing artifacts, additional testing of the processor may be desired. If so, refer to Figure 6-9 (page 211) to identify individual components of the processor and follow the procedure(s) outlined below for each section. These procedures employ 14 x 17-inch (35 x 43-centimeter) sized double-emulsion film.

### Isolating the Entrance Detector Crossover Assembly

1. Remove the entrance detector crossover assembly.
2. Manually feed a pre-exposed, unprocessed sheet of 14 x 17-inch (35 x 43-centimeter) film so that the 17-inch (43-centimeter) edge of the film enters the developer rack first.
3. Allow the film to complete the normal processing cycle.
4. Install the entrance detector crossover assembly.
5. Check the film for artifacts.

   If the artifact you are diagnosing does not appear, then the entrance detector crossover assembly is most likely the cause of the artifact.

   If the artifact does appear, continue with the next procedure.

### Isolating the Developer-to-Fix Crossover Assembly

1. Remove the developer-to-fix crossover assembly.
2. Feed a pre-exposed, unprocessed sheet of 14 x 17-inch (35 x 43-centimeter) film so that the 17-inch (43-centimeter) edge of the film enters the entrance detector crossover assembly first.
3. Remove the film as it exits from the developer rack.
4. Rotate the film 90 degrees.
5. Manually feed the film into the fixer rack so that the 14-inch (35-centimeter) edge of the film enters the fixer rack first.
6. Allow the film to complete the normal processing cycle.
7. Check the film for artifacts.

   If the artifact you are diagnosing does not appear, then the developer-to-fix crossover assembly is most likely the cause of the artifact.
8. Compare the sheet of film that just exited the processor with the original film displaying the artifact.

   If the position of the artifact on the two sheets of film is the same, the artifact most likely occurred before you rotated the film and therefore was caused in the developer section of the processor.

   If the position of the artifact on the two sheets of film is different by 90 degrees, the artifact most likely occurred after you rotated the film. The artifact may then have been caused by any of the components located after the developer-to-fix crossover assembly—fixer rack, fix-to-wash crossover assembly, wash rack, squeegee assembly, or dryer rack.
9. Install the developer-to-fix crossover assembly.
10. To determine which of the components listed above may have caused the artifact, continue with the next procedure.

## Isolating the Fix-to-Wash Crossover Assembly

1. Remove the fix-to-wash crossover assembly.
2. Feed a sheet of 14 x 17-inch (35 x 43-centimeter) film so that the 17-inch (43-centimeter) edge enters the entrance detector crossover assembly first.
3. Remove the film as it exits from the fixer rack.
4. Rotate the film 90 degrees.
5. Manually feed the film so that the 14-inch (35-centimeter) edge enters the wash rack first.
6. Allow the film to complete the normal processing cycle.
7. Check the film for artifacts.

   If the artifact you are diagnosing does not appear, the fix-to-wash crossover assembly is most likely the cause of the artifact.
8. Compare the sheet of film that just exited the processor with the original film displaying the artifact.

   If the position of the artifact on the two sheets of film is the same, the artifact most likely occurred before you rotated the film and therefore was caused in the fixer section of the processor.

   If the position of the artifact on the two sheets of film is different by 90 degrees, the artifact most likely occurred after you rotated the film. The artifact may then have been caused by any of the components located after the fix-to-wash crossover assembly—wash rack, squeegee assembly, or dryer rack.
9. Install the fix-to-wash crossover assembly.
10. To determine which of the components listed above may have caused the artifact, continue with the next procedure.

## Isolating the Wash Rack, the Squeegee Roller Assembly, and the Dryer Assembly

1. Remove the squeegee roller assembly. Note that not all processors contain a squeegee assembly.
2. Feed a sheet of 14 x 17-inch (35 x 43-centimeter) film so that the 17-inch (43-centimeter) edge enters the entrance detector crossover assembly first.
3. Remove the film as it exits the wash rack.
4. Allow the film to air dry.
5. Check the film for artifacts.

   If the artifact you are diagnosing does not appear, then the squeegee roller assembly or the dryer assembly is most likely the cause of the artifact.

   If the artifact does appear, then the wash rack is most likely the cause of the artifact.
6. Inspect the wash rack, the squeegee roller, and any components in the dryer section that may be causing the artifact.

## References

1. Film artifact diagnostics guide for Kodak X-Omat automatic film processors, Eastman Kodak Company publication no. 1C0948, 1996.
2. Widmer JH, Lillie RF, Jaskulski SM, Haus AG. Identifying and correcting processing artifacts. Eastman Kodak Company publication no. N-326, 1994.
3. Widmer JH, Lillie RF. Roller transport processing artifact diagnosis. In: Haus AG, ed. Film processing in medical imaging. Madison, WI: Medical Physics Publishing, 1993: 115–129.

# 7

# Troubleshooting

- Basic Principles for Troubleshooting the Processor
- General Troubleshooting Guide
- Fixer Retention
- Processing Artifacts
- Tools for Troubleshooting
- Information Needed by Manufacturers for Troubleshooting

Medical imaging involves the use of many different radiographic films, chemicals, and models of processors. Often products from many different manufacturers are used together with varied results. The knowledge of facility personnel assigned responsibility for maintaining a controlled processing environment ranges from very little to advanced.

With increasing regulations concerning film, processing, image quality, and quality control (QC) for medical imaging, especially for mammography, it is important to have a basic understanding of troubleshooting the film processor.

This chapter will provide concepts, guidelines, and tools to approach troubleshooting logically, to minimize processing errors, to maintain a controlled processing environment, and to achieve consistently high-quality radiographic images while minimizing radiation exposure to patients.

## Basic Principles for Troubleshooting the Processor

For simplification and ease of discussion, troubleshooting concepts have been organized as basic principles.

### #1 Be Knowledgeable about Processor Components and Functions

Chapter three covers processor components and function. It is important to review component parts and systems in the processor in order to perform effective troubleshooting related to the processor QC chart, image quality, fixer retention, artifacts, etc.

The key sections and components of the processor include the following:

- Developer.
- Fixer.
- Wash.
- Dryer.
- Crossover Assemblies.
- Turnaround Assemblies.
- Transport System.
- Replenishment System.
- Recirculation System.
- Temperature Control System.

## #2 Follow the Processing Recommendations of Film, Chemical, and Processor Manufacturers

The amount of time needed to troubleshoot the processor will be significantly minimized if the installation, ventilation, preventive maintenance, and processing recommendations of the film, chemical, and processor manufacturers are followed.

For best results, the film, chemicals, and processor should be treated as a system, with as many components as possible from the same manufacturer, since they are designed to work together and are tested accordingly. This provides time and cost reduction benefits.

### *The Ideal Processing Environment*

An assessment of how closely the processing environment adheres to manufacturers' recommendations is a good place to start. In other words, can it be classified as ideal or less than ideal?

An ideal processing environment can be described by the following:

- One that is dedicated to a particular film type, such as mammography film.
- The film, chemicals, and processor are all from the same manufacturer or as recommended by the film manufacturer.
- A consistent high volume of film, such as 100 sheets of film, is processed per eight-hour day.
- Processing occurs at least five days a week.

An acceptable situation usually results when the following apply:

- The processor is not dedicated to a particular type of film.
- Only films from the same manufacturer are processed.
- The film manufacturer's chemicals are used.
- A consistent high volume of film is maintained at least five days a week.

Potentially unacceptable situations may occur when the following apply:

- The volume of film processed is extremely low.

- Different film types and films from different manufacturers are processed together.
- The chemicals used are purchased from one or more different manufacturers.
- Films are not processed every day.
- The day-to-day film volumes per eight-hour day are inconsistent.

Most facilities are not aware of the possible detrimental effects of the above situations. If at all possible, patient schedules should be arranged to provide a consistent volume of film spread throughout an eight-hour day. If the processor will not be used overnight, it should be shut down; leaving the processor on without processing films (and replenishment not occurring) will negatively impact sensitometry due to chemical oxidation. If a consistent volume of films from different manufacturers is processed together and the environment is stable, care should also be taken not to introduce other new film types or to deviate from the volume without first testing to ensure there are no detrimental effects.

Flooded replenishment (Chapter two) may offset most of the problems encountered in potentially unacceptable situations.

Changing situations from potentially unacceptable to acceptable and ideal should be the goals to achieve the desired image quality consistently and minimize troubleshooting time.

## #3 Understand Processor QC Concepts

Processor quality control is discussed in Chapter five. Its purpose is to identify changes or trends in processing so corrections can be made before the clinical images are affected.

To ensure that the processor QC chart accurately reflects the actual state of the processing environment, the processor quality control program must have a meaningful start. This can only be accomplished by ensuring that the processing environment is set up correctly and by closely following standard procedures for the establishment of a processor QC program. These procedures are outlined in detail in this book and are available from many other sources. In addition, the QC technologist, medical physicist, and dealer service personnel can assist with the proper establishment of the processor QC program.

Some key features of a properly established QC program are the following:

- The processor must initially be cleaned and filled with fresh, properly mixed chemicals.
- Solution temperatures and replenishment rates must be correct.
- The numbers for the QC charts must be established based on the average of five sensitometric strips done over five consecutive days (one each day).
- The right tools and reference materials must be available and used correctly.
- Correct, consistent methodology is employed every day by everyone in the facility involved with QC.

It is also important not to play a numbers game; correlate the numbers from processor QC sensitometric testing with the clinical images. If the processor QC chart indicates an out-of-control situation, but there is no effect on the clinical images, it is likely that mistakes in procedure may have been made or the establishment of the processor QC chart operating levels (or aims) are not valid.

## #4 Different Films React in Different, and Sometimes Opposite, Ways

This is especially true today as film and chemical technology continues to evolve. It is important to avoid drawing conclusions based on practical knowledge accumulated over years of working in radiology with involvement in processor QC. The formulation of films used today is vastly different, and films may not always respond as expected.

Table 7-1 illustrates the above point. Four typical mammography films, Film A, Film B, Film C, and Film D, generally react differently, as shown, to the following situations:

- Developer temperature higher than recommended (within 5°F [3°C]).
- Developer temperature lower than recommended (within 5°F [3°C]).
- Water temperature higher than recommended (if developer temperature above set point).
- Water temperature lower than recommended (if developer temperature below set point).
- Developing time shorter than recommended.
- Developing time longer than recommended.
- Over-replenishment.
- No starter (added at the time of a fresh start).
- An insufficient amount (<50 percent) of starter (added at the time of a fresh start).
- Too much starter (>50 percent) (added at the time of a fresh start).

Note that throughout this book, speed and contrast are used interchangeably with mid-density (MD) and density difference (DD), respectively.

Another situation in which different films used for medical imaging may react differently is typically seen on the processor QC chart following a properly carried out preventive maintenance procedure in which the chemicals are completely changed (fresh start) and the correct amount of starter is added. Assuming that the replenishment rates are properly set for the volume of film being processed, the first several points plotted for speed will be lower than the aim point with some films. As films are processed and seasoning occurs, the level will rise. With other films, the first several points plotted for speed will be higher than the aim point. As films are processed and seasoning occurs, the level will fall.

**Table 7-1** Sensitometric Variability for Four Typical Mammography Films: Film A, Film B, Film C, and Film D

These four typical mammography films generally react *differently* to the following problems.

| | *Expected Change in MD, DD, and B + F* | | | |
|---|---|---|---|---|
| *Problem* | *Film A* | *Film B* | *Film C* | *Film D* |
| Developer Temperature Higher Than Recommended (within 5°F [3°C]) | MD ↑<br>DD ↑<br>B + F → | MD ↑<br>DD ↑<br>B + F → | MD ↑<br>DD ↑<br>B + F → | MD ↑<br>DD ↓<br>B + F ↑ |
| Developer Temperature Lower Than Recommended (within 5°F [3°C]) | MD ↓<br>DD ↓<br>B + F → | MD ↓<br>DD ↓<br>B + F → | MD ↓<br>DD ↓<br>B + F → | MD ↓<br>DD →<br>B + F → |
| Water Temperature Higher Than Recommended (if developer temperature above set point) | MD ↑<br>DD ↑<br>B + F → | MD ↑<br>DD ↑<br>B + F → | MD ↑<br>DD ↑<br>B + F → | MD ↑<br>DD ↓<br>B + F ↑ |
| Water Temperature Lower Than Recommended (if developer temperature below set point) | MD ↓<br>DD ↓<br>B + F → | MD ↓<br>DD ↓<br>B + F → | MD ↓<br>DD ↓<br>B + F → | MD ↓<br>DD →<br>B + F → |
| Developing Time Shorter Than Recommended | MD ↓<br>DD ↓<br>B + F → | MD ↓<br>DD ↓<br>B + F → | MD ↓<br>DD ↓<br>B + F → | MD ↓<br>DD →<br>B + F → |
| Developing Time Longer Than Recommended | MD ↑<br>DD ↑<br>B + F → | MD ↑<br>DD ↑<br>B + F → | MD ↑<br>DD ↑<br>B + F → | MD ↑<br>DD ↓<br>B + F ↑ |
| Over-Replenishment | MD ↓slight<br>DD ↓slight<br>B + F → | MD ↓slight<br>DD ↑<br>B + F → | MD ↑<br>DD ↓slight<br>B + F → | MD ↑slight<br>DD ↓slight<br>B + F ↑slight |
| No Starter | MD ↓slight<br>DD ↓<br>B + F → | MD ↓slight<br>DD ↑<br>B + F → | MD ↑slight<br>DD ↓<br>B + F → | MD ↑slight<br>DD ↓<br>B + F ↑slight |
| Insufficient Amount of Starter (<50 percent) | MD →<br>DD →<br>B + F → | MD ↓slight<br>DD ↑slight<br>B + F → | MD →<br>DD →<br>B + F → | MD ↑slight<br>DD ↓slight<br>B + F → |
| Too Much Starter (>50 percent) | MD →<br>DD →<br>B + F → | MD ↑<br>DD ↓<br>B + F → | MD →<br>DD →<br>B + F → | MD →<br>DD ↑<br>B + F → |

Key: MD: Mid-Density (Speed) DD: Density Difference (Contrast) B + F: Base Plus Fog
↑: Increasing ↓: Decreasing →: No Change

Table 7-2 shows that the same four typical mammography films generally react identically as shown to the following problems:

- Severe under-replenishment.
- Fixer temperature higher than recommended.
- Fixer temperature lower than recommended (if film clear).
- Fixer temperature lower than recommended (if film cloudy).
- Water temperature higher than recommended (if developer temperature unaffected).
- Water temperature lower than recommended (if developer temperature unaffected).
- Slight contamination of developer with fixer.
- Major contamination of developer with fixer.
- Severe oxidation of chemicals.
- Exhausted developer.
- Exhausted fixer (if film clear).
- Exhausted fixer (if film cloudy).
- Expired film.
- Film stored above the recommended temperature and relative humidity.

Rapid advancements in technology dictate that tomorrow's films will use new and different technologies, necessitating even more new ways of doing quality control testing.

## #5 Expect Fluctuations in Sensitometry Values

Speed, contrast, and base plus fog are used to monitor the processor on a daily basis to determine if the processing environment is in-control. Day-to-day fluctuations are to be expected.

The points plotted on the control chart should normally range above and below the value or line that represents aim. A processor that is *tightly* controlled (ideal situation) will generally stay between ±0.10 control limits for speed and contrast, and within +0.03 of the aim for base plus fog. For processing of most radiographic films, control limits of ±0.15 for speed and contrast are adequate, especially for acceptable or potentially unacceptable situations as defined earlier in this chapter.

A processor QC chart with little fluctuation is not possible without deliberate human intervention.

Note that for the processing of mammographic films, regulatory bodies may require that films not be processed if speed and/or contrast reach or exceed the ±0.15 control limits.

## #6 Repeat QC Testing When Questionable Results Are Obtained

When the fluctuations for speed and/or contrast reach or exceed the operating limits (i.e., ±0.15), first make sure they are real. Repeat the test, checking that all aspects of the test are properly performed.

**Table 7-2** Sensitometric Variability for Four Typical Mammography Films: Film A, Film B, Film C, and Film D

These four typical mammography films generally react *identically* to the following problems.

| *Problem* | *Expected Change in MD, DD, and B + F* | | |
|---|---|---|---|
| | MD | DD | B + F |
| Severe Under-replenishment | ↓severe | ↓severe | → |
| Fixer Temperature Higher Than Recommended | → | → | → |
| Fixer Temperature Lower Than Recommended (if film clear) | → | → | → |
| Fixer Temperature Lower Than Recommended (if film cloudy) | → | ↓ | ↑ |
| Water Temperature Higher Than Recommended (if developer temperature unaffected) | → | → | → |
| Water Temperature Lower Than Recommended (if developer temperature unaffected) | → | → | → |
| Slight Contamination of Developer with Fixer | ↑ | ↑slight | → |
| Major Contamination of Developer with Fixer | ↓ | ↓ | → |
| Severe Oxidation of Chemicals | ↓ | ↓ | → |
| Exhausted Developer | ↓ | ↓ | → |
| Exhausted Fixer (if film clear) | → | → | → |
| Exhausted Fixer (if film cloudy) | → | ↓ | ↑ |
| Expired Film | ↓ | ↓slight | ↑slight |
| Film Stored Above the Recommended Temperature and Relative Humidity | ↓slight | ↓ | ↑ |

Key: MD: Mid-Density (Speed) DD: Density Difference (Contrast) B + F: Base Plus Fog
↑: Increasing ↓: Decreasing →: No Change

If the same result is obtained and not caused by a procedural error or quirk, the decision to handle it or request outside help from the processor service provider or medical physicist can then be made.

## #7 Take Corrective Action for an Out-of-Control Processor

If the ±0.15 control limits (the maximum operating limits for speed and contrast) are reached or exceeded, corrective action is required.

For mammography, immediate corrective action is required before processing any mammography films.

In this situation, having a backup processor suitable for processing mammograms is advantageous. If a backup processor is not readily available, it may be possible to avoid canceling any appointments by exposing films and holding them protected from light until the processor can be used again. The medical physicist can help develop the procedure to be used, based on the expected delay and the latent image keeping characteristics of the mammography film used (Appendix A). This procedure should be included in the policies and procedures manual for the facility.

## #8 Perform a Few Quick and Easy Check Procedures

There are some obvious things that can be easily and quickly checked before calling in outside help (i.e., the processor service provider, medical physicist, or film representative).

The following checks should be standard procedure and included in the policy and procedures manual:

- Check the developer temperature of the processor, using an accurate digital thermometer. The temperature gauge on the processor should not be relied upon in this situation.
- Look at the processor and around the darkroom to determine if anything looks unusual.
- Listen to the processor to determine if anything sounds unusual.
- Familiarity with the processor, the darkroom, and the correct procedure for measuring developer temperature greatly assists with this exercise.

## #9 Check the Processor QC Chart for Trends

A trend is a series of three or more consecutive points progressing steadily upward or downward. Trends may be watched for up to five days before taking action, unless the control limits are reached or exceeded. It is also important not to react too soon.

If the processor *clearly* has a trend developing, immediate corrective action can be initiated, even before a control limit (e.g., ±0.10 for speed) is reached.

The most frequently occurring trend is a gradual downward drift throughout the month. This usually indicates that replenishment rates are not correct and that the chemicals are oxidizing. It may be necessary to reevaluate the film volume and replenishment rates. Whenever changes are made to the processing environment, such as a change in replenishment rates, it is always acceptable to reestablish the processor QC program.

## #10 Many Problems on the QC Chart Manifest Themselves in Relatively Few Symptoms

Three values—speed, contrast, and base plus fog—are used to monitor the processor. In addition, developer temperature may be measured and charted daily.

There are many problems that can cause any particular combination of speed, contrast, and base plus fog to increase, decrease, or remain unchanged.

Table 7-3 gives the most common causes of speed and contrast increasing and decreasing. Table 7-4 gives the most common causes of increases in base plus fog. Note that base plus fog changes may be more pronounced for double-emulsion tabular-grain films used for conventional radiography than for many single-emulsion films used for mammography (Figure 7-1).

Table 7-5 shows the most common reasons for variable trends, including speed and contrast increasing or decreasing, speed increasing while contrast decreases, and speed decreasing while contrast increases.

Additional troubleshooting information related to developer temperature, chemicals, and replenishment, are included in Tables 7-6 through 7-9.

### *Looking for Clues*

Additional clues and knowledge about what could be detrimental to processor stability are necessary in order to proceed with troubleshooting.

First, inspect the processed films.

- Do they appear to be properly processed or not?
- Are they too light or too dark?
- Are they dry, damp, or overdried?
- Are they properly cleared or milky in appearance?

Now look for other important clues to determine what could be causing the problem. The remarks section of the processor QC chart is particularly valuable; complete information should be routinely noted.

Check the processor maintenance logs.

- When was service last performed, what kind of service was performed, and did the appearance of the problem coincide with the service?
- Did the same service person perform the work, or was it someone new to the facility?
- Was a film jam recently cleared? Contamination of the developer with fixer may have occurred.
- Were fresh chemicals recently added to the replenisher holding tanks or automixer?
- Was a complete preventive maintenance procedure (PM) performed in which the chemicals in the processor were discarded and fresh chemicals added (fresh start)?
- Are splash guards routinely used?

**Table 7-3** Causes of Increases (↑) and Decreases (↓) of Speed and Contrast

*Speed ↑, Contrast ↑*

Increased developer temperature.

Increased developer immersion time.

Combination of both:

- Increased developer temperature, increased developer immersion time and/or combination of both may be used for extended-cycle processing.

Increased water temperature if the developer temperature increases as a result.

Slight contamination of the developer with fixer.

*Speed ↓, Contrast ↓*

Decreased developer temperature.

Decreased developer immersion time.

Combination of both.

Decreased water temperature if the developer temperature decreases as a result.

Under-replenishment of the developer.

Major contamination of the developer with fixer:

- Film jam; all films manually fed should be processed with a film edge alongside the guide on the film feed tray.
- Splash guards should always be used.

Severe oxidation of chemicals:

- Use of chemicals mixed more than 14 days previously.
- Floating lid not used in developer replenisher tank.
- Chemicals exposed to extreme heat.

Exhausted developer:

- No developer replenisher reaching processor.

Use of expired film:

- The expiration date on the film box should be routinely checked before the box is opened or before a shipment of film is accepted from the dealer.
- The film stock should be rotated on a first-in, first-out basis.
- Expired film should not be used clinically for medical/legal reasons.

Storage of film and chemicals above the temperature and relative humidity recommendations of the manufacturer:

- The temperature and relative humidity where film and chemicals are stored and used must be controlled 24 hours a day, 7 days a week.

**Table 7-4** Causes of Increases in Base Plus Fog

Major contamination of the developer with fixer.

Use of expired film.

Storage of film and chemicals above the temperature and relative humidity recommendations of the manufacturer.

Films incompletely cleared or "fixed":
- Check that the fixer replenishment rate and quality of the fixer is adequate.

Faulty safelighting:
- Check that the safelighting in the darkroom has been evaluated and is correct.
- Check with the manufacturer(s) for the safelighting requirements of the film types used in a particular darkroom.

Prolonged exposure to safelights:
- Handle films, both unexposed and exposed, as quickly and carefully as possible to minimize safelight exposure and to avoid causing film handling artifacts.

White light leaks in the darkroom:
- Check that the darkroom is light tight.
- Check the results of the darkroom fog test.
- Consider if the darkroom can be further improved.

Over-replenishment of general radiology films and some mammography films:
- Are replenishment rates correct for all film types?

Increased developer temperature or increased developing time for some films.

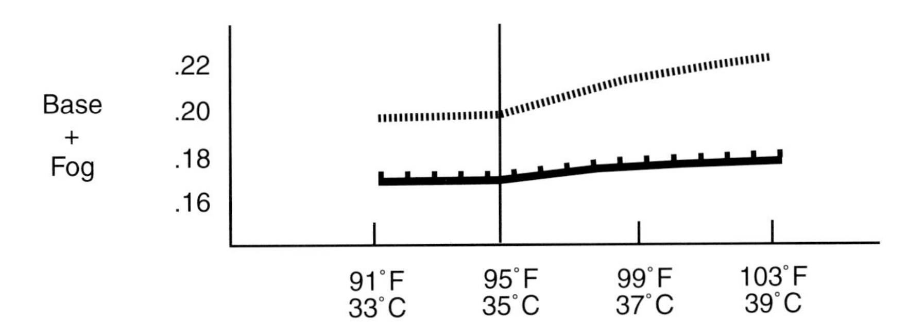


**Figure 7-1** Effect of developer temperature on base plus fog for a double-emulsion tabular-grain film (broken line) and two single-emulsion mammography films (3D grains: solid line; cubic grains: dotted line).

**Table 7-5** Causes of Variable Trends in Speed and Contrast

Fresh chemicals in the processor:

- Inconsistencies in how the chemicals were mixed either by the dealer or by various people within the facility.
- Chemicals not stored properly before or after being mixed (temperature of environment).

Slight contamination of developer with fixer:

- Film jam; all films fed manually should be processed with a film edge alongside the guide on the film feed tray.
- Splash guards should always be used.

Mixing different film types:

- Most likely if the particular mix of different films varies from week to week or a new type of film not previously used is introduced.

Incorrectly set replenishment rates:

- Especially in conjunction with fluctuating film volumes with no corresponding adjustment in replenishment rates.

Fluctuating developer temperature:

- Developer temperature should be checked with a hand-held calibrated thermometer or one that is known to be accurate.

Calibration or use error of the sensitometer and/or densitometer:

- Incorrect settings on sensitometer.
- Incorrectly exposing the QC film with the emulsion side of the film away from the light source.
- Densitometer out of calibration; a calibration check strip, available from the manufacturer of the densitometer, should be used at least annually to check the calibration.
- Instruments not used consistently; skill levels on the use of automatic scanning-type densitometers too low.
- Different sensitometer or densitometer used.

Inconsistent methods:

- QC protocols for everyone in the department have either not been established, are not clear, or everyone has not been trained.
- Wrong film used for processor QC.
- Wrong box of the correct type of film used for processor QC.
- Sensitometric strip not processed the same way each time.
- Variable times between exposing and processing the sensitometric strip from day to day.
- QC charts not established correctly, e.g., based on the films from fewer than five consecutive days.
- Crossover procedure not done or not done properly.

Variable Trends Include: speed and contrast ↑; speed and contrast ↓; speed ↑ and contrast ↓; and speed ↓ and contrast ↑.

**Table 7-6** Causes of Higher or Lower Developer Temperatures

Developer temperature set higher or lower than recommended for processor and cycle:
- To achieve a different "look" or speed from the film, especially film used for mammography; increased noise and increased propensity for artifacts may occur.

Thermostat malfunction.

Imprecise or malfunctioning thermometer.

Developer recirculation pump malfunction.

Incoming water temperature outside the recommended range:
- Higher than 85°F (29.5°C), for many processors.
- Lower than 40°F (4.5°C), for many processors.
- Inadequate water flow: low water pressure, clogged filter, or other blockage in the line.

Probe of thermometer not inserted in same place each time developer temperature is measured.

**Table 7-7** Factors That Affect Chemical Quality

Developer and fixer not properly mixed:
- Incorrect mixing sequence of chemicals by dealer or facility personnel.
- Incorrect amount of water added.
- Incorrect water temperature.
- Inadequate agitation.

Key components missing in developer and fixer (chemical quality):
- Increased replenishment rates may be required to compensate.

Chemicals stored outside the recommended temperature range.

Chemicals out of date or oxidized:
- Mixed chemicals, provided they are properly stored, will last 10 to 14 days.
- Oxidation of the developer can be minimized by the proper use of a floating lid on top of the developer in the replenisher holding tank; check that the floating lid is always used; purchase a new floating lid, if missing.

- Was the seasoned developer reserved and returned to the developer tank? Reusing seasoned developer, discussed later in this chapter, is an excellent way to reduce fluctuations in the processor control chart after a PM.
- Was the correct amount of the chemical manufacturer's starter added (no more, no less)? In addition, a graduated cylinder with milliliter markings should be used to measure the starter.

Check the patient schedule log to see if the total average daily volume of films has recently changed. Replenishment rates must be altered when the total volume of film processed changes. Rates must be increased as volume decreases; rates must be decreased as volume increases.

**Table 7-8** Assurance and Testing of Chemical Quality

*Assurance of Chemical Quality*

Buy chemical concentrates and mix them yourself:
- Generally results in better quality, if one person is consistently mixing the chemicals in the amount that will be used within 10 to 14 days.
- Could result in increased variability if more than one person mixes the chemicals.

Buy premixed chemicals from a reputable solution service firm:
- Most facilities buy premixed chemicals.

Use a reliable automixer and quality concentrates designed for the specific mixer.

Control environmental temperature between 40 and 85°F (5 and 30°C).

Mixed chemicals must be used within the recommended time frame of 10 to 14 days.

Eliminate practices that accelerate oxidation.

*Testing of Chemical Quality*

pH measurement:
- Indicates the acidity or alkalinity of a solution at a specified temperature.
- pH values outside the acceptable range indicate the chemicals were not mixed properly.
- pH values within the acceptable range do not guarantee properly mixed chemicals.

Specific gravity measurement:
- Indicates the ratio of the mass of a body to the mass of an equal volume of water at a specified temperature.
- Gives a good indication of over- or underdilution only.
- Values within the acceptable range do not guarantee the presence of all required components.

Laboratory analysis of components:
- A service available from some companies.
- Verifies chemical components and proper mixing.
- May be costly and time-consuming.

Processor QC:
- May be used to monitor your chemical quality besides monitoring the processor.
- Can quickly alert you to check chemical quality if changes occur on the chart in correlation to chemicals being added, etc.
- Easiest and most cost-effective way to monitor the chemicals.

Note that if a significant change is made to the processing environment (different film, altered version of the same film, different chemicals, different processing conditions such as altered replenishment rates), the processor QC program should be reestablished (Chapter five).

Check the technologist schedule:

- Was the QC technologist absent from work?
- Did other technologists assume processor QC responsibilities while he or she was gone?

**Table 7-9** Causes of Over- and Under-Replenishment

*Under-replenishment*

Replenishment pumps malfunctioning.

Developer overdiluted:
- Too much water added.

No developer replenisher reaching the processor:
- Developer strainer clogged.
- Replenisher tank empty (level of replenisher below level needed to flow through line).
- Kink or air lock in line.

Replenishment rate set too low for the average daily volume of film.

Decrease in number of films processed (decreased patient volume) without corresponding adjustment of rate (increased replenishment rate).

Faulty detection system (microswitches at the entrance detector crossover assembly not correctly adjusted).

*Over-replenishment*

Developer replenishment rate set too high.

Developer underdiluted:
- Not enough water added.
- Excessive evaporation.

Increase in number of films processed (increased patient volume) without corresponding adjustment of rate (decreased replenishment rate).

Faulty detection system (microswitches at the entrance detector crossover assembly not correctly adjusted).

Check into circumstances unique to the facility:

- Do all the technologists process films the same way? All films should be placed on the film feed tray consistently.
- Are outside films, such as from a mobile van, processed occasionally in the processor?
- Is the processor left on 24 hours a day even though films are only processed during the day?
- Did you just start using outdated film from a different manufacturer as cleanup film? Using only film from a single manufacturer is preferable, or use film designed for cleanup.
- Did you make any recent changes in the processing environment?
- Did you change chemical and/or service providers?
- Did you just start exposing more of the film (collimation opened to the film edge rather than adjusted by breast size)? If so, replenishment rates must be increased.
- Did you just increase the optical density on the phantom and all patient exposures? If so, replenishment rates must be increased.

- Did you advise your equipment, processor, and chemical service personnel, as appropriate, of any changes? Full and open communication is important in order to maintain a controlled processing environment.
- Did you just start using a different type of film or an improved version of your previous film? Replenishment rates may need to be changed, and the processor QC program should be reestablished over five days.
- Do you experience extremes in temperature and humidity throughout the year within your facility, including the area where patient films are archived?

  Processed and unprocessed film should be stored and used at a temperature between 50 and 70°F (10 and 21°C) and at a relative humidity between 30 and 50 percent. Chemicals should be stored and used at a temperature between 40 and 85°F (5 and 30°C). Temperature and humidity values outside these ranges may be detrimental to film and chemical quality.

### *Single or Multiple Causes*

There is a strong tendency to consider one variable as the sole cause of the problem. This is seldom the case. Most problems occur by the interaction of several factors working together, which makes a strong case for following the manufacturer's processing recommendations.

Keep in mind that when making changes in the processing environment in response to a problem, however, it is important to change only one thing at a time.

## #11 Restart the Processor If the Problem Cannot Be Immediately Diagnosed

If you cannot diagnose the problem in a short period of time (i.e., 20 minutes), discard the chemicals and restart the processor. This is especially appropriate when the cause (or causes) of the problem is not obvious.

If the processor QC program was originally established according to the recommended guidelines, the processor QC chart will most likely be back in control, although the first points plotted will reflect the effects of fresh chemicals.

If the processor QC program was originally established some months ago or longer, consider reestablishing the aims and even choosing different step numbers from the values from the sensitometric strips processed on five consecutive days. The longer the period of time, the more likely different individuals with different levels of training have been involved in daily processor QC activities, raising the likelihood that minor or even major mistakes in math and procedure have occurred.

## #12 Contact the Appropriate Individual(s) If the Problem Recurs

If the problem recurs after restarting the processor, contact the QC technologist, the processor service provider, the film representative, and/or the medical physicist. A bigger solution is required for the underlying problem.

To avoid the same problem again in the future, assess your situation in terms of whether or not it is ideal. Start following the manufacturer's recommendations.

### #13 Document All Actions on the Processor QC Chart

It is critically important to document all actions on the processor QC chart. Proper documentation is the key to resolving problems and maintaining proper processor functioning. Such documentation is also necessary to pass the scrutiny of federal and state inspectors and of the American College of Radiology (ACR).

Out-of-control points on the chart should be circled and replotted after the problem has been corrected.

It is also important that the solution(s) to the problem be documented.

## General Troubleshooting Guide

Refer to Table 7-10 for an example of a general processor QC troubleshooting guide that is not film specific. While such guides are helpful, they do not reflect all situations for all films.

## Fixer Retention

Medical imaging facilities should perform regular testing of the processor for fixer (hypo) retention. The American College of Radiology requires the performance of this test in mammography facilities on a quarterly basis. Refer to Chapter five for information on the performance of this test.

A chemical test solution and a test tool, called a hypo estimator, are used to evaluate the results of the fixer retention test. The test spot that is evaluated against the colors printed on a hypo estimator strip must not be darker than the third patch on the strip. This indicates an estimate of 0.05 grams of retained hypo (thiosulfate ion) per square meter of film.

Meeting this standard means that clinical images should be stable for a long period of time (approximately 100 years).

If a test spot is darker than the third patch, the test should be repeated. Use a freshly processed film or one that has been processed no more than two weeks previously.

Consider replacing the test solution first after a failed hypo test if the test solution is older than two years (or its age is unknown), if the conditions under which it has been stored are unknown, or if it is possible that the cap on the bottle may not always have been tightly closed. When purchasing a new bottle of test solution, write the date it is first opened on the bottle.

Other causes for failing the fixer retention test include the following:

- The water was not turned on; the film was not washed at all.
- The level of water was low; the film was not washed as long as it would have been if the water had been at the proper level.
- The water flow was impaired, possibly due to a clogged filter; the film was not adequately washed because water flow was restricted.

**Table 7-10** General Troubleshooting Guide

| *Trends in Graph/ Appearance of Images* | *Possible Causes* | *Corrective Action(s)* |
|---|---|---|
| MD (speed): ↑<br>DD (contrast): ↑<br>Base Plus Fog (fog): ↑<br>or<br>Images too dark | Developer temperature too high | Measure developer temperature<br>Lower thermostat to correct temperature<br>Check:<br>• Thermostat and recirculation pump for malfunction<br>• Water flow and temperature |
| | Developer replenisher improperly mixed, underdiluted | Check:<br>• Amount of water added<br>• Excessive evaporation<br>• Amount of developer starter<br>• Developer replenishment rate |
| MD (speed): ↓<br>DD (contrast): ↓<br>Base Plus Fog (fog): ↓<br>or<br>Images too light | Developer temperature too low | Measure developer temperature<br>Raise thermostat to correct temperature<br>Check thermostat for malfunction |
| | Developer replenisher rate too low | Check:<br>• Developer replenishment rate<br>• Developer replenishment line for kinks |
| | Developer replenisher improperly mixed, overdiluted | Check amount of water added |
| | Developing time too short | Check level of developer in tank and raise to correct height<br>Check replenishment pump for malfunction<br>Check the time in solution using a time in solution test tool<br>Check for:<br>• Leaks from valves, tank, or fittings |

*Continued on next page*

**Table 7-10** ***Continued***

| *Trends in Graph/ Appearance of Images* | *Possible Causes* | *Corrective Action(s)* |
|---|---|---|
| MD (speed): → <br>DD (contrast): ↓ <br>Base Plus Fog (fog): ↑ <br>or <br>Loss of contrast and increased fog level | Contamination of developer with fixer | Call service to clean the processor <br>Use splash guards <br>Remove and insert the fixer rack carefully <br>Process film with edges in contact with film feed tray guides |
| | Fixer replenishment rate too low | Check: <br>• Fixer replenishment rate <br>• Fixer replenishment line for kinks |
| MD (speed): ↑ <br>DD (contrast): ↓ <br>Base Plus Fog (fog): ↑ <br>or <br>Images too dark, loss of contrast, and increased fog level | Developer replenisher improperly mixed, underdiluted | Check amount of water added |
| | Fog-producing conditions in the darkroom | Evaluate the darkroom for light leaks <br>Evaluate the safelights <br>Perform the darkroom fog test |
| MD (speed): <br>DD (contrast): <br>Base Plus Fog (fog): <br>or <br>Images too light, loss of contrast, and increased fog level | Complete oxidation of developer | Check developer replenishment rate <br>Call service to check processor |
| | Contamination of developer with fixer | Call service to clean the processor <br>Use splash guards <br>Remove and insert the fixer rack carefully <br>Process film with edges in contact with film feed tray guides |

*Continued on next page*

**Table 7-10 *Continued***

| *Trends in Graph/ Appearance of Images* | *Possible Causes* | *Corrective Action(s)* |
|---|---|---|
| MD (speed): ↓<br>DD (contrast): ↓<br>Base Plus Fog (fog): →<br>or<br>Images too light and loss of contrast | Developer replenisher improperly mixed, overdiluted | Check amount of water added |
| | Under-replenishment of developer | Check:<br>• Developer replenishment rate and raise<br>• Developer replenishment line for kinks |
| Variable trends | Intermittent replenishment | Check replenishment pump<br>Check entrance detector crossover assembly microswitches |
| | Chemical activity not adequately maintained due to low film volume or infrequent use of processor | Use flooded replenishment |
| | Developer and fixer improperly mixed | Implement quality control for consistent chemicals |
| | Poor quality chemicals | Use higher-quality chemicals |
| | Intermittent loss of temperature control | Check thermostat for malfunction |
| | Chemicals exposed to extreme temperature | Store and use chemicals at acceptable temperature |
| Images appear cloudy or milky | Fixer solution depleted | Replace with fresh fixer<br>Check fixer replenishment rate<br>Check fixer temperature |

- The water was colder than the recommended temperature range; if the temperature of the fixer was affected by the cold water, impaired fixer activity may have occurred.
- The film was processed at a faster cycle than recommended for the film type; i.e., a very rapid processing cycle was used instead of standard or extended cycle.
- The circulation of the fixer was impaired.
- The effectiveness of the fixer was impaired by missing or reduced key components, a low fixer replenishment rate, a kink in the fixer replenisher line, cold fixer, overdiluted fixer, or an excess amount of developer entering the fixer (inadequate squeegee action at the developer-to-fix crossover assembly).

It may be possible to preserve clinical information on films that are known to have an excess amount of retained fixer by rewashing the films. In order to do this, remove the top of the processor, the evaporation covers, and the developer-to-fix crossover assembly. Insert the film into the fixer rack so it is reprocessed through the fixer, wash, and dryer sections of the processor. Note that the film should not be reprocessed through the developer section of the processor; contamination of the developer with fixer could result. A small magnet may be needed to operate the processor with the lid removed. If desired, a duplicate may be made of the radiograph before rewashing.

If a more quantitative evaluation of retained fixer is needed because of a fixer retention test failure (excessive amount of retained fixer) and the cause is not readily apparent, check with the film manufacturer regarding a methylene blue test. There may be a nominal charge for this test.

## Processing Artifacts

Chapter six covers processing artifacts.

To summarize, significant contributors to artifacts include:

- An improperly installed and/or vented processor.
- Lack of adequate preventive maintenance.
- Inadequate chemical quality.
- Incorrect replenishment rates.
- Processing film using a developer temperature higher than the temperature recommended by the manufacturer.
- The omission of the daily cleaning of the crossover assemblies.
- The use of abrasive materials to clean processor rollers.
- Incorrect processor replacement parts.
- Improperly trained service providers and operators.
- Service on the processor performed after image quality has been affected rather than with the first signs of roller wear.
- Defects in processor parts.

## Tools for Troubleshooting

Some simple but very effective tools are available for troubleshooting the processor and film. They include:

- Cleanup film.
- Reusing seasoned developer.
- The time-in-solution test tool.
- Emulsion number log.
- The split phantom test.

### Cleanup Film

Cleanup film, sometimes called roller transport cleanup film, is a specialized film designed to be used in conjunction with the processing environment. Each 14 x 17-inch (35 x 43-centimeter) sheet of film features a tacky, non-light-sensitive coating on both sides of the film base.

Cleanup film picks up lint, dirt, and other deposits and helps carry them out of the processor. For best results, one or two sheets should be processed.

Cleanup film is particularly useful to control a processing artifact called delay streaks. Cleanup film may be used in all processors, except those with area replenishment, which cannot sense this clear-based film.

Fogged or expired single- or double-emulsion film that has not previously been processed may also be used as cleanup film. Note that mixing films from different manufacturers in the same environment should be avoided, as undesired sensitometric changes may occur.

Any film used as cleanup film should be discarded after one use to avoid contaminating the developer solution with fixer and redepositing lint or dirt back onto the rollers in the processor.

### Reusing Seasoned Developer

As discussed in Chapter two, the developer solution in the developer tank of the processor changes from a fresh to a seasoned state as films are processed and replenishment occurs. At approximately one-month intervals or as experience indicates, the processor should be cleaned as recommended by the processor manufacturer. This cleaning is usually known as a preventive maintenance procedure (PM).

During a PM, the chemicals (seasoned developer and fixer) are typically drained from their respective tanks and discarded. After the tanks, racks, and other processor components are cleaned, checked, and adjusted, fresh chemicals are added to the processor.

The processor QC chart generally reflects the change in chemicals from a PM. A spike in the points plotted for speed, either up or down, occurs immediately after a complete chemical change. If the processor is well maintained with adequate replenishment rates, the spike for speed should rarely exceed the ±0.15 control limits. As films are processed and seasoning again occurs, speed usually returns to its normal pre-PM level.

Some facilities may wish to avoid this spike on the processor QC chart by retaining the seasoned developer and reusing it after the PM has been completed. The procedure for doing this is described below.

This procedure may be used indefinitely, provided the processor is in control at the time of the cleaning (within the ±0.15 control limits for speed) and contamination of the developer with fixer does not occur.

In addition, note the following important points:

1. Low film volume processors set up on flooded replenishment, as discussed in Chapter two, never reach a seasoned state; therefore, the chemicals should be completely changed at the appropriate intervals. (Developer starter should be added, as specified, to the developer replenisher holding tank or automixer; no developer starter should be added to the developer tank in the processor.)
2. Any processor whose processor QC chart shows a consistent downward trend over the previous month is not well controlled; the chemicals should be completely changed at the appropriate intervals. Increasing the developer replenishment rate may be necessary to prevent the downward trend.

Follow these steps to reuse seasoned developer after a processor PM:

1. Expose and process a sensitometric strip, as discussed in Chapter five. Plot and evaluate the results. Proceed if the processor is in control.
2. In processors with separate developer and fixer drains, carefully drain the seasoned developer from the processor developer tank into a clean receptacle. If an open receptacle is used, it should have a floating lid to prevent oxidation.

   In processors with common developer and fixer drains, the seasoned developer may either be left in the developer tank during the PM or it may be carefully pumped or ladled from the processor developer tank into a clean receptacle. If the seasoned developer is left in the tank, a normal PM with a complete chemical change should be completed at intervals of no greater than three months.

   Drain the fixer, according to the usual procedure recommended by the processor manufacturer, and discard.
3. Carefully remove the developer and fixer racks from their respective tanks. Care should be taken to avoid intermixing of solutions.
4. After the processor and racks are thoroughly cleaned, change the developer filter.
5. Pour or pump fresh fixer solution into the fixer tank of the processor. Make sure no fixer is splashed into the developer tank. The use of the splash guard is recommended.
6. Pour or pump the retained seasoned developer from the holding receptacle back into the processor developer tank. Do not add any starter.
7. Replace the developer and fixer racks. Check that the level of both solutions is at the required level.
8. Expose and process another sensitometric strip to check that the processor is still in control. Speed should be within 0.05 of the speed value from the first

strip processed prior to the PM. The processor must also be within the established control limits.

9. If the strips match, the processor is ready to resume operation; if the strips do not match, drain the developer solution, clean the rack and tank, and refill as after a normal PM.

## The Time-in-Solution Test Tool

Films used for medical imaging should be processed in the processing cycle or cycles recommended for the particular film type.

The temperature of the developer solution and the length of time a film spends in the developer solution are important considerations in determining whether the film will be properly processed (optimal optical density and contrast).

The length of time the film spends in the developer is known as developer time, development time, or developer immersion time. It is defined as the amount of time from the leading edge of the film into the developer to the leading edge of the film into the fixer.

Table 7-11 shows the developer time and temperature required for several mammography films in standard- or extended-processing cycles.

A time-in-solution (TIS) test tool (Figure 7-2), which consists of a strip of clear blue film base from which all film emulsion components have been chemically removed and two white tape strips, may be used to verify the correct developer immersion time as specified by film and processor manufacturers. Note that motor speeds will vary slightly from processor to processor of the same type; the TIS tool should be used to check for times significantly longer or shorter than the times listed in Table 7-11.

The procedure is as follows:

1. Remove the lid of the processor. A small magnet placed near the microswitch may be needed to allow the processor to operate without the lid in place.
2. Refer to Figure 7-3. Locate the approximately 1/4-inch (6.3-millimeter) gap between the entrance detector crossover and the guide shoe. Also locate the gap between the developer-to-fix crossover assembly and the guide shoe.
3. Feed the test tool into the processor, placing the tool either along the film feed tray guide or in the center of the film feed tray, with the bottom of the taped T feeding first.
4. With a stopwatch in hand, get ready to begin timing when the black line drawn across the tape passes through the space in the entrance detector crossover.
5. Begin timing when the cross of the T reaches that space. Either the top or bottom of the strip of tape may be used.
6. Stop timing when the same part of the cross of the T reaches the space in the developer-to-fix crossover assembly.
7. Repeat the timing sequence three times, and take an average to determine the developer time.

**Table 7-11** Processing Conditions for Kodak Mammography Films.[1]

| *Film and Processing Cycle* | *M35 M35A M35A-M M20* | *M6B M6-N M6A-N M6AW* | *M7 M7A M7B* | *M8* | *270 RA*[2] *3000 RA* | *480 RA*[3] *5000 RA* |
|---|---|---|---|---|---|---|
| Min-R H, Min-R M, or Min-R 2000, Standard Processing Cycle | | | | | | |
| Processing Time(s)[4] | 140 | 90 | 122[6] | 90 | 100 | 88 |
| Developer Time(s)[5] | 32 | 24 | 27 | 21.5 | 26.6 | 23.8 |
| Developer Temperature | 92°F (33.5°C) | 95°F (35°C) | 94°F (34.4°C) | 96°F (35.6°C) | 94°F (34.4°C) | 95°F (35°C) |
| Min-R E, Extended Processing Cycle | | | | | | |
| Processing Time(s)[4] | 203 | 172 | 188 | NA | 209 | 172 |
| Developer Time(s)[5] | 49.5 | 47 | 43 | | 52 | 47 |
| Developer Temperature | 95°F (35°C) | 95°F (35°C) | 96°F (35.6°C) | | 94°F (34.4°C) | 95°F (35°C) |

[1]Kodak RP X-Omat chemicals recommended.
[2]Listed processing times are associated with a front film exit. Top film exit processing times are 214 seconds for extended cycle processing and 107 seconds for standard cycle processing.
[3]Some variation may occur in the listed times due to adjustable turnarounds.
[4]Processing time is based on 24 centimeters of film travel, leading edge entering the processor to trailing edge exiting the processor.
[5]Developer time is based on the leading edge of the film into the developer to the leading edge of the film into the fixer.
[6]Some M7 and M7A processors may have a longer standard processing time (140 seconds rather than 122 seconds). The temperature recommended for standard processing of all Kodak mammorgraphy films, in this case is 92°F (33.5°C).
NA-not applicable.
(Reprinted with permission from Eastman Kodak Company.)

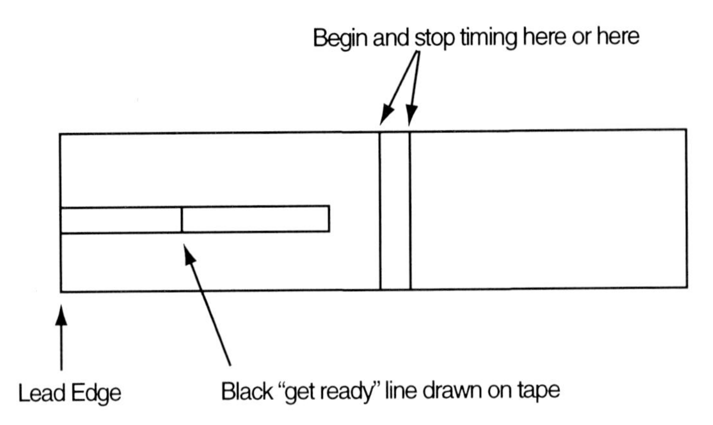


**Figure 7-2** Time-in-solution (TIS) tool

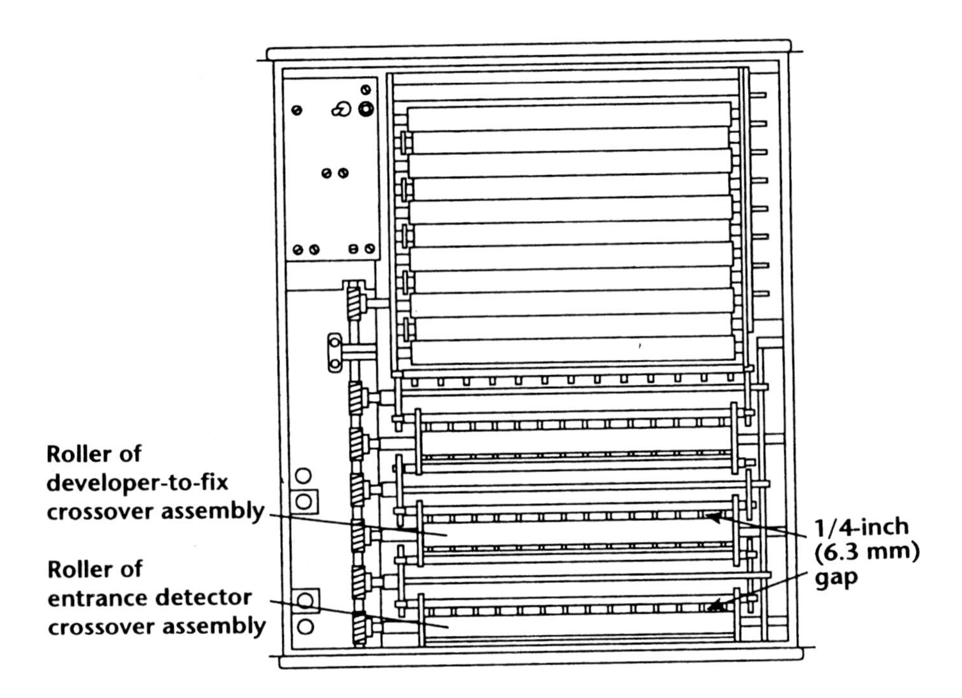


**Figure 7-3** Bird's eye view of a typical processor with the gaps identified for the TIS timing procedure

8. The total processing time should also be checked and verified by measuring the amount of time either a mammography film (18 x 24 centimeters, with 24 centimeters of film travel) or a general radiology film (35 x 43 centimeters, with 35 or 43 centimeters of film travel, depending on the width of the processor) takes to transport through the processor. Begin timing when the leading edge of the film enters the processor and stop when the trailing edge exits the processor. Repeat the timing sequence three times and take an average to determine the total processing time.

## Emulsion Number Log

Film manufacturers designate each box of x-ray film with a multiple-digit emulsion number. This number provides important information such as which particular emulsion batch was used as well as which roll it came from, the specific part (slit) of the roll, and which variation of emulsion was used (variation code).

Keeping track of the emulsion number of the film used for processor quality control is generally done in every processor quality control program. It is also extremely advantageous to keep track of the complete emulsion numbers of all film used clinically, especially for mammography. Doing so allows film manufacturers to make comparisons in speed, contrast, D-max, etc., between current clinical images and images taken one or more years previously. Such comparisons may assist in troubleshooting image quality concerns.

The easiest way to retain this information is to start an emulsion number log. The log should contain the following information:

- Type of film.
- Date box of film was opened.
- Expiration date.
- Film size.
- Complete emulsion number, including variation code.
- Any other comments, such as which emulsions were used for processor QC or phantom images.

Table 7-12 shows an example of a emulsion number log. Every time a box of film with a different emulsion number is opened, a new entry should be made in the log.

## The Split Phantom Test

A split phantom test should be performed to determine relative speed differences between a new film and the film in current use either clinically or for processor quality control. Speed comparisons made using a sensitometer may not accurately

**Table 7-12** Emulsion Number Log

*Film A*

| *Date Opened* | *Expiration Date* | *Film Size* | *Emulsion Number* | *QC* | *Comments* |
|---|---|---|---|---|---|
| 4/30/97 | 8/98 | 18 x 24 cm | 141 016 14 | | Phantom image |
| | 8/98 | 24 x 30 cm | 141 011 16 | | |
| 5/1/97 | 7/98 | 18 x 24 cm | 139 019 12 | ✓ | |
| | 8/98 | 18 x 24 cm | 141 016 14 | | |
| 5/5/97 | 8/98 | 18 x 24 cm | 141 023 25 | | |
| 5/6/97 | 8/98 | 24 x 30 cm | 141 011 15 | | |
| 5/12/97 | 9/98 | 18 x 24 cm | 143 018 11 HD | | Reestablished QC |

reflect the differences in speed between two films exposed by light from an intensifying screen.

The procedure is as follows:

1. Assemble the tools that are needed for the test:
   - A phantom used for mammography quality control testing.
   - The 18 x 24 centimeter mammography cassette normally used for the phantom test.
   - A piece of cardboard from the film box cut in half to use as a guide.
   - A pair of scissors.
   - A lead pencil.
   - The mammography x-ray unit and the processor will also be used for this test.
2. In the darkroom, in total darkness to reduce any additional density added to the films from long safelight exposure, cut a sheet of film from the current or normal box in half by using the cardboard as a guide. (This can be done by lining up the 18-centimeter edges of the cardboard and film, with the film closest to the countertop and the cardboard half on top. Use care in cutting the film in the dark.)
3. Place the film emulsion side up in the cover of the opened cassette with the film on the right side and the cut edge toward the right edge of the cassette; use a lead pencil to mark the corner "N" for normal.
4. Cut a sheet of film from the suspect box in half by using the cardboard as a guide.
5. Place the film emulsion side up in the cover of the opened cassette with the film on the left side and the cut edge toward the left edge of the cassette; use the lead pencil to mark the corner "S" for suspect.
6. Make sure the film edges in the center of the cassette are directly adjacent to one another and not overlapping before closing the cassette (Figure 7-4).
7. Place the cassette with the two film halves in the grid of the mammography x-ray unit.
8. Place the phantom on top of the grid in the standard location used for mammography quality control testing.
9. Position the photocell beneath the center of the phantom (standard location), assuming the phantom exposure is always made using the phototimer.
10. Select the same technique factors usually employed when imaging the phantom (same kVp, etc.).
11. Make the exposure and process the two film halves immediately in the same manner (e.g., emulsion side up and on the right side of the processor).
12. Use a densitometer to take two optical density readings in the center of the phantom, just to the right and left of the cut edges (one on the normal and one on the suspect film).
13. Calculate the density difference by subtracting the optical density value of the suspect film from the optical density value of the normal film. If the density difference is a negative value and the suspect film is darker than the normal film, the suspect film is faster. If the density difference is a positive

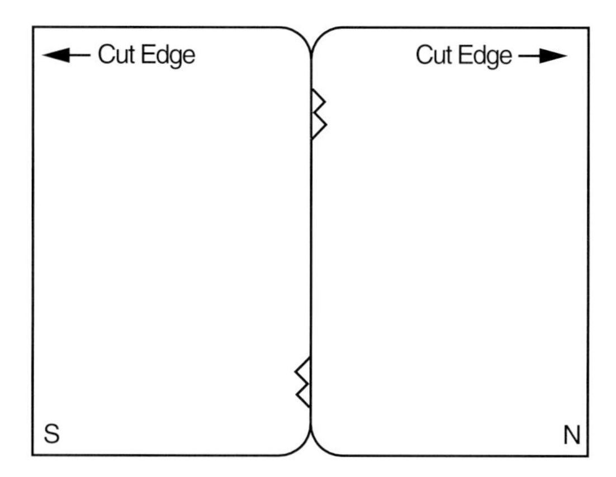


**Figure 7-4** Film placement in the cover of a mammography cassette for the split phantom test

value and the suspect film is lighter than the normal film, the suspect film is slower.

According to the ACR in *Recommended Specifications for New Mammography Equipment:*

- "A density difference of 0.30 between any two films of the same type from the same manufacturer, exposed and processed together, is a reasonable maximum to be expected from manufacturing variability for films of roughly the same age and storage conditions."
- "If the difference between the two film densities exceeds 0.30 at a density of approximately 1.25, then the film supplier should be contacted to determine the source of the problem."

## Information Needed by Manufacturers for Troubleshooting

Manufacturers of film, chemicals, and processors frequently assist in troubleshooting efforts.

As much of the following information should be provided as possible in order to maximize troubleshooting efforts:

- Make and model of processor.
- Mix of film types.
- Volume of films (number of sheets per typical eight-hour day).
- Brand of chemicals (developer and fixer).
- Type of chemicals (e.g., premixed, automixed, or facility mixed).
- Replenishment rates (fully defined in terms of number and size of films and length of film travel [Chapter two]).
- Dates last and previous PM performed.
- Developer temperature.
- Current and previous emulsion numbers.
- Date change noticed.
- Description of change (what was observed and who noticed it).

## References

1. American College of Radiology (ACR) Mammography quality control manual for radiologists, medical physicists and technologists. Reston, VA: American College of Radiology, 1994.
2. American College of Radiology (ACR) Recommended specifications for new mammography equipment: Screen-film x-ray systems, image receptors, and film processors. Reston, VA: American College of Radiology, 1995.
3. Retain developer during mammography processor preventive maintenance, Service Bulletin no. 205, Eastman Kodak Company publication no. N-919, 1994.

# A

# Reciprocity Law Failure and Latent Image Fading

- Reciprocity Law Failure
- Latent Image Fading

## Reciprocity Law Failure

The effect of reciprocity law failure can be important in medical imaging when long exposure times are used—especially for screen-film mammography—because additional exposure may be required due to reduced film speed. The definition of exposure (Exposure = Intensity x Time) states that the response of the film to radiation of a given quantity will be unchanged if the product of intensity and time remains the same. It is implied that this relationship remains constant, regardless of whether long or short exposure times are used, provided that time changes are compensated for by a proportional change in intensity. This relationship, also known as the reciprocity law, holds true for direct exposure of film by x-rays; however, for exposure by screen-produced light, the law fails.

Reciprocity failure can be either high or low intensity. High-intensity failure occurs when a large number of electrons are produced in a short period of time.This can result in the formation of a greater number of initial latent image specks than would normally occur. Each individual center would be smaller than if fewer centers were formed. These centers (greater in number but smaller in size) are less stable, and the probability of their growing to a size suitable for development is reduced. The result is the formation of less density as a function of exposure than might otherwise occur.

Low-intensity failure is a result of too small a number of light photons over an exposure time period. If the time period between two subsequently absorbed light photons is too long in the earliest stages of latent image formation, the resulting metallic silver speck is of subcritical size and can regress back to a smaller than optimum size for growth and amplification by the developer. Again, less density is achieved than would normally occur as a function of exposure. Consideration of reciprocity law failure becomes important because techniques in screen-film mammography may necessitate the use of long exposure times due to (1) use of grids, (2) use of small focal spots for conventional and magnification techniques (low mA settings), and (3) use of low-powered x-ray units with limited mA output settings. In mammography, reciprocity law failure may affect film density when long exposure times (approximately 1.0 second or longer) are used. When reciprocity law failure effects occur, additional exposure may be required in order to provide the proper optical density on the mammogram (Table A-1).

Some mammographic units have built-in features that compensate for loss of film speed caused by reciprocity law failure using data like that in Table A-1. There is very little change in film contrast caused by film reciprocity law failure.

## Latent Image Fading

If an exposure has been made on film and processing is postponed for a relatively long time, the optical density obtained may be lower than if processing had immediately followed the exposure. This effect is called "latent image fading" (sometimes also referred to as latent image keeping [LIK]) (Table A-2). It is caused by an instability of the latent image.

In screen-film mammography, film speed loss due to latent image fading can occur if the time between exposure and processing is delayed because of (1) transporting the film from a van or mammography facility to a central location for film processing, or (2) batch processing films at the end of the day. In order to minimize latent image fading in the clinical environment, it is important that the time interval between

**Table A-1** Example of Reciprocity Law Failure Data for a Medical X-Ray Film*

| Exposure time (seconds) | 0.001 | 0.01 | 0.1 | 1 | 5 | 10 |
|---|---|---|---|---|---|---|
| Percent film speed loss | 6 | 0 | 0 | 12 | 30 | 38 |
| Percent contrast change (average gradient) | 0 | 0 | 0 | 2 | 3 | 4 |

*Speed: determined at a density of 1.00 above base plus fog.

Average gradient: determined from the slope of the characteristic curve between densities of 0.25 and 2.00 above base plus fog.

**Table A-2** Example of Latent Image Fading Data for a Medical X-Ray Film*

| Time delay exposure for film processing (hours) | 0 | 4 | 8 | 24 | 48 |
|---|---|---|---|---|---|
| Percent film speed loss | 0 | 10 | 12 | 18 | 23 |
| Optical density difference | 0 | 0.12 | 0.15 | 0.21 | 0.27 |
| Percent contrast change (average gradient) | 0 | 2 | 3 | 3 | 5 |

*Speed: determined at a density of 1.00 above base plus fog.

Average gradient: determined from the slope of the characteristic curve between densities of 0.25 and 2.00 above base plus fog.

exposure and processing be as consistent as possible from day to day. Ideally, films should be consistently processed as soon as possible after exposure. To minimize time interval differences, films should be processed in the order in which they are exposed. If films with slightly greater speed losses (due to latent image fading) are created and the time between exposure and processing is relatively long, the exposure technique can be adjusted on a one-time basis to obtain and maintain the appropriate density. If it is known in advance that processing will be delayed (e.g., by more than eight hours), it may be advisable to increase exposure time (automatic exposure control [AEC] setting) in order to obtain proper optical density in the mammogram. There is very little change in contrast caused by latent image fading.

For film processor quality control, it is recommended that sensitometric strips be processed immediately after exposure by the sensitometer to minimize the effects of latent image fading.

A recent study has shown that latent image fading is not accompanied by clinical impairment of mammographic interpretation if films are processed at the end of the work day. Therefore, latent image fading should not deter the use of batch processing.

## References

1. Arnold BA, Eisenberg H, Bjarngard BE. Measurement of reciprocity law failure in green-sensitive x-ray films. Radiology 126: 493–498, 1978.
2. Haus AG. Screen-film image receptors and film processing. In: Haus AG, Yaffe MJ, eds. A categorical course in physics: Technical aspects of breast imaging syllabus. Oak Brook, IL: Radiological Society of North America, 1994: 85–101.
3. James TH, ed. The theory of the photographic process, 4th Edition. New York: Macmillan, 1977.
4. Sickles EA. Latent image fading in screen-film mammography: Lack of clinical relevance for batch-processed films. Radiology 194: 389–392, 1995.

# B

# Manual Method of Time-Temperature Processing



Time-temperature processing of radiographic film is an effective means of converting the invisible latent image into a diagnostically useful radiograph. This technique is important at facilities that are without automatic processing capabilities, even temporarily.

X-ray processing solutions are most effective when used within a comparatively narrow range of temperatures. At temperatures below those recommended, some of the chemicals are definitely sluggish in action, which may cause under-development and inadequate fixation. At temperatures much above those recommended, activity is too high for manual processing control. Such high temperatures may soften the emulsion to such an extent that it is easily damaged. In fact, processing temperatures should not be above the film manufacturer's recommendations.

The processing temperatures prescribed by the film manufacturer are generally recommended for several reasons. First, good sensitometric performance of the film is obtained; that is, the contrast and speed of the film are satisfactory and fog is kept to an acceptably low level. Second, the processing time is practical; and third, with modern solution heating and cooling, the temperature is usually conveniently maintained.

For these reasons, every effort should be made to keep the solutions at the recommended temperature during use. By doing so, the user can obtain the best sensitometric characteristics and will also have the advantage of a standardized time of development, fixation, and washing.

Whenever it is necessary to work with solutions at other than the recommended temperature but within a range acceptable to the film manufacturer, an adjustment must be made in processing procedure. This adjustment consists of increasing or decreasing the time of development (depending on the temperature) and ensuring adequate fixation in a relatively fresh solution. As developer temperature increases, development time decreases and vice versa. This procedure of adjusting time to suit the temperature is known as time-temperature processing.

Time-temperature processing is preferable to processing by visual inspection, which actually requires more attention and demands greater skill and judgment. Variations in eye accommodation, the low level of illumination in the darkroom, and the opacity of the uncleared film all make processing by inspection difficult and subject to error. Such "sight development" should be avoided. When time and temperature are carefully correlated, as recommended by the film manufacturer, any lack of density in the radiographs can be attributed to under-exposure, not to under-development; and excessive density can be charged to over-exposure rather than to over-development. This is important in determining adjustments in exposure. Problems have been detected where excessive radiation exposure has been combined with under-developing to produce an image that appeared acceptable using sight development. The increased patient dose and loss of image quality from this error in technique can be avoided by using a constant time-temperature method.

## Replenishment for Manual Development

The activity of an under-replenished developer gradually diminishes by exhaustion; that is, as it is used, its developing power decreases, partly because of the consumption of the developing agent in changing the exposed silver halide to metallic silver and partly because of the restraining effect of the accumulated reaction products on the development process. The extent of this decrease in activity will depend on the number of films processed and their average density. Even when the developer is not used, the activity may decrease slowly because of aerial oxidation of the developing agent. This exhaustion, unless counteracted, will gradually result in under-development and will affect contrast and speed adversely. In addition, some of the developer is physically carried out of the tank with each film. These conditions must be offset in some way if uniform radiographic results are to be obtained. The best way to compensate for these losses is to use a replenisher system in which the activity and volume of the solution are maintained by suitable chemical replenishment.

A replenisher system is efficient and simple; it merely requires adding a solution to the original developer to compensate for loss of activity, thereby permitting a constant developing time. The replenisher performs the double function of maintaining both the liquid level in the tank and the activity of the solution.

With this method, films should be removed from the developer quickly to minimize the amount of excess developer that can drain back into the tank. Normally, this will carry out the proper amount of solution to permit correct replenishment. The level of the developer in the tank should be kept at a fairly constant point, the amount lost being restored by adding replenisher. After a little experience, withdrawing the correct amount of solution on the film will become automatic. When too much of the used developer remains, however, it can be removed to allow the correct amount of replenisher to be added. Replenisher should be added in amounts and at intervals recommended by the film manufacturer. Adding replenisher in relatively small quantities and stirring immediately after each addition reduces fluctuations in solution activity.

It is not practical to continue replenishment indefinitely. The best results will be obtained if the instructions given for the various types of developer are followed. In any case, the developer solution should be discarded at the end of three months because of air oxidation and accumulation of gelatin, sludge, and impurities that find their way into the solution.

## Procedure

Some films that have been designed for automatic processing are not suitable for manual processing and should be manually processed only in an emergency. With respect to the procedures followed in manual processing, consult the film manufacturer's recommendations, as these vary from product to product. Personnel performing the manual processing procedure should follow all manufacturers' safety recommendations. Processing sites must also be in compliance with all applicable local, state, and federal regulations regarding the introduction of waste material into the environment.

For the following procedure, see Figure B-1. First, using separate paddles, stir the solutions thoroughly to equalize the temperature and chemical activity throughout each tank. Next, determine solution temperatures. To avoid contamination, rinse the thermometer after contact with each solution. Set the timer for the recommended period of development based on the temperature of the developer solution. Attach the film carefully to the correct size hanger. Start the timer and immediately immerse the film in the developer. Tap the hanger against the side of the tank to remove air bubbles from the film surface. If the film manufacturer recommends agitation, follow instructions as to how frequently and vigorously to agitate. When the timer rings, remove the film hanger immediately and drain the film for a moment into the space between the tanks.

## Rinsing

After a film has been developed, it should be immersed in a rinse bath of clean, circulating water, or better still, a stop bath solution and, unless otherwise recommended, agitated continuously. The minimum rinsing or stop bath time is usually about 30 seconds. Temperatures of all solutions in the processing cycle, including the rinse bath, should be maintained within a few degrees of each other. After rinsing, the film should again be well drained so that the least possible amount of liquid is transferred to the fixer.

**1 Stir Solutions** Stir developer and fixer solutions to equalize their temperature. (Use separate paddle for each to avoid possible contamination.)

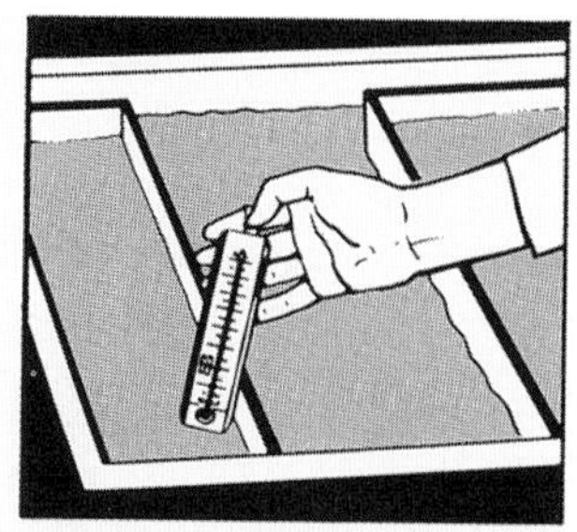

**2 Check Temperature** Check temperature of solutions with accurate thermometer. Rinse thermometer after each measurement to avoid contaminating next solution. Adjust to recommended temperature.

**DEVELOPMENT TIME**

| TEMP. | | MIN. |
|---|---|---|
| 60 F | 15.5 C | 8 1/2 |
| 65 F | 18.5 C | 6 |
| 68 F | 20 C | 5 |
| 70 F | 21 C | 4 1/2 |
| 75 F | 24 C | 3 1/4 |

**3 Set Timer** Set timer for recommended period of development based on temperature of developer solution.

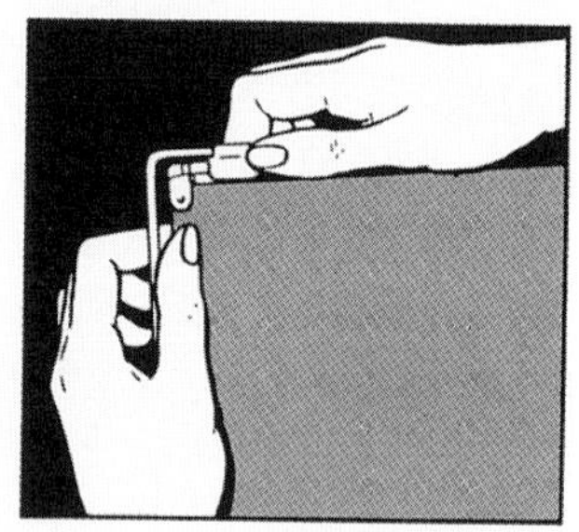

**4 Load Film on Hanger** Attach film carefully to correct size hanger. (Attach at lower corners first.) Avoid finger marks, scratches, or bending.

**5 Immerse Film in Developer** Start timer. Completely immerse film into the developer. Do it smoothly and without pause to avoid streaking. Rap film against side of tank to remove film surface air bubbles.

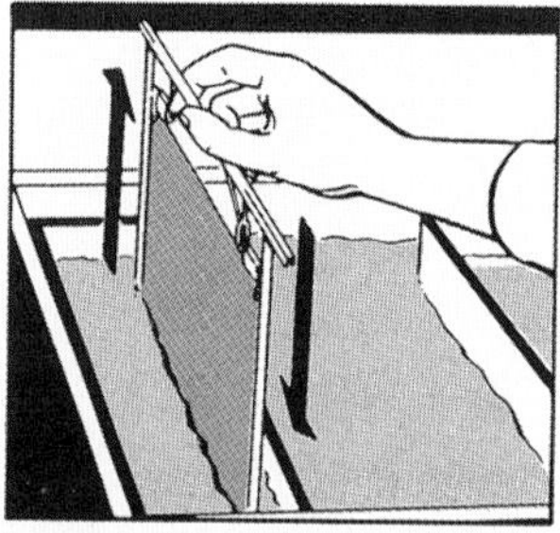

**6 Agitate Film, If Recommended** Follow the film manufacturer's instructions as to vigor and frequency and whether the film should be agitated within the tank or raised and lowered.

**Figure B-1** Basic steps for manual processing of x-ray film

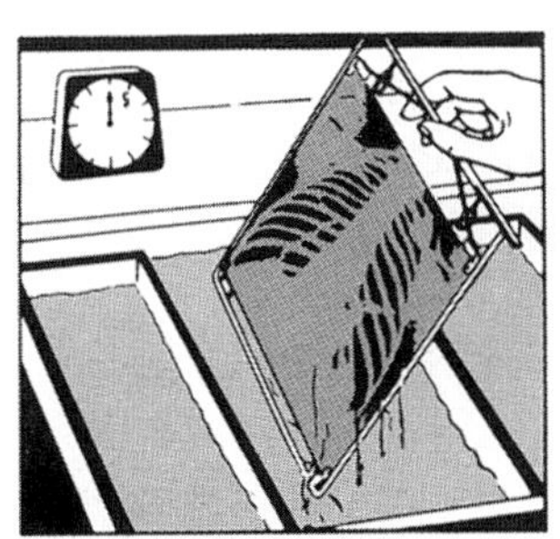

**7 Drain Outside Developer Tank** When timer rings, lift hanger out immediately. Then drain film for a moment into space between tanks. For fast drainage, tilt hanger.

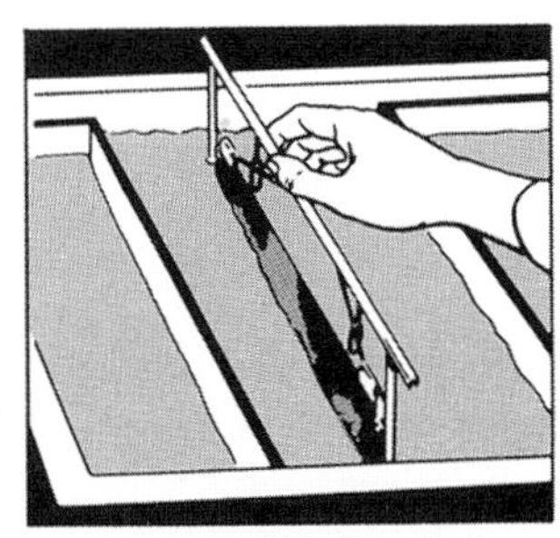

**8 Rinse Thoroughly** Place film in acid rinse bath or running water. Agitate hanger continuously. Rinse film about 30 seconds. Lift from rinse bath. Drain well.

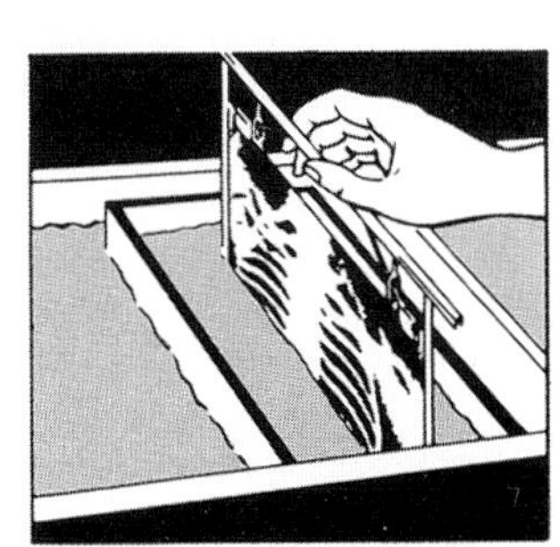

**9 Fix Adequately** Immerse film. Agitate hanger vigorously at start. Follow the film manufacturer's recommendations for time and temperature–at least twice the time required to "clear" film (when its milky look has disappeared).

**10 Wash Completely** Remove film to tank of running water (flow rate of about 8 complete changes per hour). Keep ample space between hangers (water must flow over their tops). Allow adequate time for thorough washing–usually from 5 to 30 minutes, depending on film.

**11 Use Final Rinse** If facilities permit, use a final rinse in a solution containing a wetting agent to speed drying and prevent water marks. Immerse film for 30 seconds, and drain for several seconds.

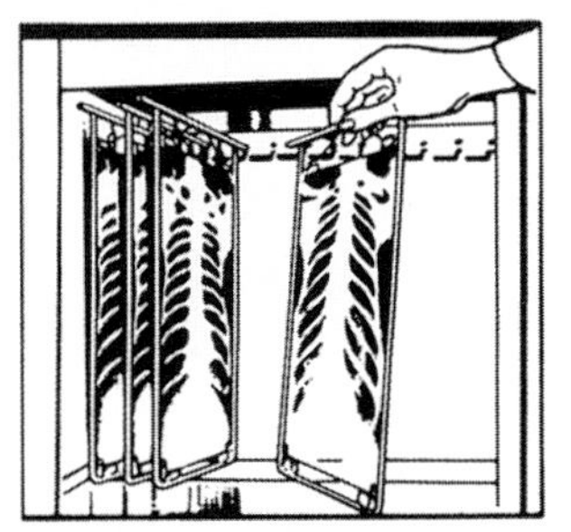

**12 Dry** Dry in dust-free area at room temperature or in suitable drying cabinet. Keep films well-separated. When dry, remove films from hangers and trim corners to remove clip marks. Insert in identified envelopes.

## Fixation

As in the case of the developer solution, the activity of the fixer is influenced by its freshness and temperature. If the fixer solution is allowed to become warmer than recommended, abnormal emulsion swelling and slow drying of the film may occur.

Fixation should be properly timed. Films must not be taken out before fixation is completed, nor left in for prolonged periods. The total fixing time for a film should be about twice the clearing time, which is the time required for the complete disappearance of the original milky opaqueness of the film. It is during this time that the fixer is dissolving the undeveloped silver halide. However, an equal amount of time is required for the dissolved silver salts to diffuse out of the emulsion and for the gelatin to be hardened adequately.

If fixing time is longer than developing time, it may be convenient to have a fixer tank that is larger than the developer tank. When many films are being processed, the number of films in the fixer tank will build up relative to those in the developer as processing continues.

## Fixation Procedure

Put the film into the solution and agitate it vigorously by moving it up and down several times; repeat periodically as recommended. This agitation will prevent stagnation of the solution in contact with the film so that fixation proceeds at a uniform rate. Follow the film manufacturer's recommendations for temperature and the time the film should be left in the fixer. This is usually at least twice the clearing time. Because the silver halide content of direct-exposure films is usually greater, their fixing time tends to be longer than for screen-type films. Keep in mind that the total fixing time is approximately twice the clearing time and that films are sensitive to light until after fixation.

## Fixer Replenishment

Unless the fixer solution is replenished, its activity diminishes with use. This results in longer times to fix films properly. Fixer replenishment maintains a minimum fixing time, thereby substantially increasing the fixing capacity of the processing system.

When using a fixer-replenishment system, remove the necessary amount of partially exhausted solution before adding replenisher. Because films and hangers carry about the same amount of liquid into the fixer tank as they carry out of it, the level of the fixer solution (unlike the developer) remains almost constant if little evaporation takes place.

## Washing

Films must be properly washed to remove the processing chemicals from the emulsion, or the image will eventually discolor and fade. Proper washing requires an adequate supply of clean, running water. It should flow so that both surfaces of each film will receive fresh water continuously. The hangers should be well-spaced in the tank and be completely immersed, including the top bar, so that the chemicals will also be removed from the tops of the frames.

The time required for adequate washing depends on water temperature, water quality, rate of water flow and turbulence, type of film, and, somewhat, on the type of fixer. The temperature of the water should be close to that of the other solutions.

The water flow in the wash tank should be rapid enough to provide approximately eight complete changes of water per hour. Follow the film manufacturer's recommended washing times, which usually range from 5 to 30 minutes. Because its emulsions tend to be thicker, direct-exposure film requires longer washing. Washed and partially washed films absorb fixer chemicals from the contaminated wash water, as well as release them to the fresh water. Therefore, washing time should begin when the last film is immersed. To minimize contamination, put the films most recently introduced into the wash water from the fixer into the section of the tank nearest the outflow. Then, as more films are added, the older films, which contain less chemicals, can be moved upstream in a countercurrent direction—moving progressively closer to the inlet. This allows the films that are about to be removed for drying to be washed in the freshest water.

A washing method that uses a cascade system in which water flows through two or more small tanks is more efficient than one large tank. Preferably, water enters at the bottom of the second tank, flows past the film and out the top into the first tank. There it flows past the film most recently removed from the fixer and out through the drain in the bottom. The film should be allowed to drain for two to three seconds after being lifted from the wash water.

## Biological Growth

A common problem with wash and processing solution tanks is the growth of slime deposits, particularly during seasons when the incoming water supply is warm. These slippery coatings are produced by bacteria, fungi, and, to some extent, algae. Sources of contaminants originate from the air, personnel, or the water supply. If not controlled, the deposits can cause both corrosion of metal surfaces and artifacts on radiographs. Control of these deposits requires continual housekeeping. Draining and scrubbing tanks with liquid household detergent at regular intervals is essential.

The use of filters in water supply lines and microbicides may also be helpful. Before using a microbicide, seek technical advice to determine that it is: (1) effective for the kind of growth present, (2) compatible with the process, and (3) in conformance with antipollution laws.

Information about biological growth control units that can be installed in a water supply line is available from manufacturers.

## Hypo Clearing Agent

If the supply of wash water is limited, the capacity of the wash tanks insufficient, or there is not enough time to wash the film properly, use a hypo clearing agent or a hypo eliminator between the fixing and washing procedures. Treating the film with a hypo clearing agent permits a reduction in time and the quantity of water necessary for adequate washing.

## Wetting Agent

To help prevent water spots and drying marks on the radiograph and to reduce drying time, give the film a 30-second rinse in a wetting agent after removal from the wash water. The wetting agent reduces the surface tension of the water on the film, thereby preventing the formation of water droplets that create marks as they dry.

## Drying

This is the simplest step in processing, yet it is an important one. Improper drying may result in water marks or damage to the gelatin from excessive temperature. Drying temperatures should not exceed those recommended by the film manufacturer.

Wet, processed films are usually dried in cabinets fitted with heaters and fans to circulate the warmed air. Such dryers should be vented to the outside to prevent an excessive rise in temperature and humidity in the room.

Another type of dryer consists of a cabinet in which the moisture is withdrawn from the air by chemicals and the dehumidified air recirculated over the film.

When dry, films should be removed promptly from the dryer to prevent them from becoming brittle. Corners should be trimmed to remove hanger clip marks. Films should be well separated. If they come in contact with each other while drying, they will be damaged by drying marks or may stick together.

# C

# Handling and Processing of Mammography Film



Mammography film has special handling and processing requirements for both clinical images and quality control (QC) films. This appendix covers:

- Handling mammography film.
  - Handling artifacts.
  - How to open a box of film.
- Evaluating the processor prior to establishing the processor QC program.
  - Processing emulsion side up or down.
- Processing QC films.
- Processing clinical images.
- Posting film feeding protocols.
- Room light processing systems.

## Handling Mammography Film

Many artifacts may occur because reasonable care is not taken while handling mammography film.

Mammography film should generally be held near a corner and allowed to dangle freely to avoid creasing the film (Figure C-1).

Fingerprints can occur if hands are not clean, dry, and free of lotions and other contaminants. Minus-density (light) fingerprints result from moisture on the hands transferred to the film prior to exposure. Plus-density (dark) fingerprints result from moisture on the film after exposure and prior to processing.

Plus-density static marks may occur if the relative humidity in the darkroom is low (i.e., below 30 percent). A humidifier may be used to raise humidity to between 30 and 50 percent. Antistatic rubber mats for the floor may also be helpful. In addition, personnel should wear natural fibers and use antistatic laundry products to reduce static in the darkroom. (Static marks may also be caused by an improperly grounded processor, malfunctioning component, or low solution levels.)

Pressure marks may occur if cases of film are stacked flat one on top of another (most likely to be seen on the film where the box top meets the body of the box), if exposed film is carelessly unloaded from cassettes or if fingernails press against the film emulsion.

Scratches, both minus- and plus-density, are more likely to occur from handling than processing. Scratches may occur from the following:

- Routinely slamming the film bin without having the box wedged in place. Instead, place the box in the film bin, making sure it is sufficiently immobilized to prevent shifting back and forth as the film bin is opened and closed. Shifting can cause chucking abrasions on the film—multidirectional scratches, usually grouped together in approximately 1 to 2 millimeter diameter clusters.
- Carelessly removing film from the box or film bin so the film scrapes across the box, envelope, or part of the film bin.

**Figure C-1** Mammography film should generally be held near a corner of the film and allowed to dangle freely to avoid creases. Hands should be clean, dry, and free of lotions and other contaminants. (Reprinted with permission from Eastman Kodak Company.)

- Carelessly loading film into cassettes.
- Sliding film across a dirty film feed tray, especially if processing film with the emulsion side down.
- Improperly opening boxes of film. Open as follows:
  - Check the outside of the box for correct film type and size, emulsion number, expiration date, etc. Record the emulsion number and other pertinent information in the emulsion number log (Chapter seven).
  - Open the box by pulling the tear strip and removing the top.
  - If the film is packaged in a sealed, foil-lined paper envelope, use a pair of scissors to cut straight across the envelope, close to the top. Tearing the envelope will create a rough edge that could potentially cause handling artifacts as the film is removed from the box/envelope. It also creates free paper fibers along the edge that can contribute to minus-density artifacts, called shadow images, on the processed film (Chapter six).
  - Remove the cardboard stiffeners used to protect the film during shipping from the box and from the darkroom because fibers from the cardboard are another source of shadow images.
  - Do not remove the film from the envelope or store loose film in the box, film bin, or film safe at any time.
  - If the film will be stored in a film bin, the film in its foil-lined paper envelope is better protected if left in the box. Before placing the box in the bin, one side of the top of the cardboard box should be broken at the perforations and the flap removed, to minimize dirt in the film bin. Pull the envelope forward so film does not scrape across either the envelope or any portion of the box as it is removed. Stabilize the box in the bin.
  - If storing the film flat on a shelf in a film safe, discard the cardboard box and top outside the darkroom to reduce dust and fibers.
  - If storing film in the box on top of the darkroom countertop, do not break the box at the perforations near the top.
  - Film should generally be removed from the center of the box and drawn straight up from the box so it does not drag across any part of the film bin or cassette as it is being loaded.
  - Once film has been opened, avoid fog from accidental white light exposure.

## Evaluating the Processor Prior to Establishing the Processor QC Program

Each processor used for mammography film (manually fed) should be evaluated before establishing the processor QC program to determine whether single-emulsion mammography film should be processed emulsion side up or emulsion side down. The decision should be based on which orientation provides the best uniformity and the fewest processor-induced artifacts. There are no regulations that require a specific film orientation. Room light processing systems are excluded from this evaluation because film orientation during processing is fixed.

Chapter six describes a similar procedure that should be used prior to the one that follows to assess whether any artifacts are being caused by the x-ray equipment, especially the grid(s). Both the 18 x 24 and 24 x 30 centimeter grids should be evaluated.

Use the following procedure:

1. Select a 24 x 30 centimeter mammography cassette that is known to have good screen-film contact.
2. In the dark, load a sheet of film into the cassette so that the film notch is positioned at the lower right corner or the upper left corner. The emulsion is upward facing if the notch or notches are located as described.
3. Place the cassette in the nongrid cassette holder or on top of the grid. A uniform sheet of acrylic (1-inch [2.54 centimeters]) thick may be placed on top of the cassette.
4. Select an exposure technique that will provide an optical density of 1.10 to 1.50 on the processed film.
5. In the darkroom under safelight illumination, remove the exposed film from the cassette. Lay the film on the film feed tray so that the following apply :
   - The widest dimension of the film is the leading edge.
   - The emulsion side is up.
   - The film edge is along the guide on the right side of the film feed tray.
   - Using a pencil, mark "↑UR" on a corner of the film nearest the film guide immediately before processing. The leading or trailing film edge corner may be used, as desired, as long as consistency is employed. The "↑" indicates the direction of film travel, "U" indicates that the emulsion side is up, and "R" indicates that the right-hand side of the processor was used.
6. Repeat the steps above for three additional films, all processed with the widest dimension of the film as the leading edge, and as follows:
   - Process film #2 emulsion side up on the left side of the film feed tray; mark the film "↑UL."
   - Process film #3 emulsion side down on the right side of the film feed tray; mark the film "↑DR."
   - Process film #4 emulsion side down on the left side of the film feed tray; mark the film "↑DL."
7. Place the four films on a viewbox and evaluate. Careful analysis will indicate whether emulsion up or down gives the best overall processing results. All clinical and QC films should then be processed in that orientation.

It may also be possible to determine if one side of the processor is significantly different from the other in terms of artifacts. If so, it may signal a need to have one or more rollers replaced. It is important to check whether any film feeding practices (e.g., all single films habitually fed on the right side) may be accelerating roller wear on one side of the processor versus the other. Clinical films should be processed on both sides of the processor to prolong roller life. All QC films, however, should be processed consistently on only one side.

Note that while the above evaluation should take place in all processors used for mammography film, many processor manufacturers do recommend a particular orientation. Information on specific processors may be obtained by consulting the processor manufacturer.

Also, processing mammography films with the emulsion side up will generally provide the best overall results.

## Processing QC Films

Consistency is important in generating and processing QC films, both sensitometric strips and phantom images.

For properly exposed sensitometric strips, single-emulsion film must be inserted into the simulated light sensitometer so the emulsion side of the film is toward the light source (down) (Figure C-2). The location of the film notch(es) should also be defined so all personnel follow the same procedure. The notches of the film shown in Figure C-2 are located away from the body of the sensitometer, toward the operator, and on the left. The placement of the film in this orientation consistently as the sensitometric strip is exposed helps ensure that the film is properly exposed (emulsion toward the light source). (Note that if the film were rotated 180 degrees, it would also be correctly exposed.)

How QC films (sensitometric strips and phantom images) are placed on the film feed tray when processing manually should be fully defined. Consistency is important, as there may be artifactual and slight sensitometric differences in films processed emulsion up or down, or on the right or left sides of the processor. The orientation

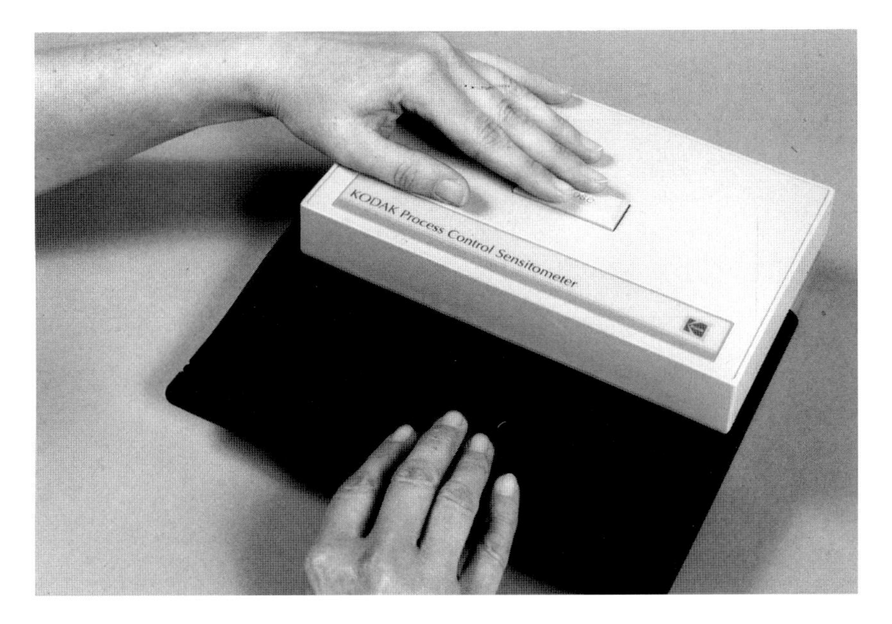


**Figure C-2** A single-emulsion mammography film should be inserted into the simulated light sensitometer so the emulsion side of the film is toward the light source (down). The location of the film notch(es) should also be defined so all personnel follow the same procedure.

that was previously determined to give the best overall results, up or down, is part of the equation.

In addition, the American College of Radiology (ACR) requires that sensitometric strips be processed so that the least exposed density step is fed into the processor first. If the film exposed in Figure C-2 is placed on the film feed tray as shown in Figure C-3 (notched edge as the leading edge, emulsion side up, and along the guide on the right-hand side of the processor), the least exposed density step will be fed first. If films are to be fed emulsion side down, the film from Figure C-2 will also be processed with the least exposed density step first if placed with the notched edge as the leading edge, emulsion side down, and on the left-hand side of the processor.

Using either of the two protocols described above has the added benefit of positioning the exposed portion of the sensitometric strip toward the center of the processor and away from the edges of the rollers, usually the first part of the roller to show signs of wear.

Phantom images should be processed similarly to the sensitometric strips, using the same emulsion orientation (up or down), same notch location, and same side of the processor (right or left).

## Processing Clinical Images

Mammography film is available in 18 x 24 and 24 x 30 centimeter sizes.

**Figure C-3** Sensitometric strips should be processed in a consistent manner (e.g., notched edge as the leading edge, emulsion side up, and along the guide on the right-hand side of the processor) and with the least exposed density step fed into the processor first.

In most processors, the film feed tray will accommodate two 18 x 24 centimeter films, side-by-side, with the short dimension edge of both films as the leading edge.

When single sheets of 18 x 24 and 24 x 30 centimeter mammography film are fed, it has been common practice to rotate them 90 degrees so that the long dimension edge (24 and 30 centimeters, respectively) feeds into the processor as the leading edge, to reduce chemical replenishment and costs.

All mammography films that are manually processed should be placed on the film feed tray so that the short dimension edge is the leading edge. There are two reasons this is important:

1. Small plus-density guide shoe marks (Chapter six) may appear at regularly spaced intervals on the leading and trailing edges of all processed films. When mammographic films are fed with the long dimension edge as the leading edge, guide shoe marks may appear along the chest wall. While these marks do not interfere with diagnostic interpretation, they can be eliminated from this area by merely feeding all film with the short dimension as the leading edge (Figure C-4). As there is a minimal amount of breast tissue imaged along the short dimension edges, any marks positioned here are less obtrusive.
2. All personnel processing films in the same way helps to stabilize replenishment rates. The proper amount of replenishment is important for image quality and to reduce costs and environmental impact.

## Posting Film Feeding Protocols

After all decisions about processing have been made, the protocols for clinical and QC films should be posted in the darkroom, so that all films are processed consistently. In addition, they should be included in the policy and procedures manual, and all personnel should be trained on those protocols.

## Room Light Processing Systems

Room light processing systems are excluded from much of the information in this appendix because film handling is accomplished mechanically and film orientation during processing is fixed. A typical orientation is emulsion side down with a long dimension edge of the film as a leading edge. As the leading edge is the nonchest wall edge, guide shoe marks may be visible along it. However, no breast tissue is imaged in this area, so marks should not be a concern.

Care must still be taken when loading film into a film magazine, so the emulsion side is correctly oriented as film is loaded into cassettes. Additionally, room light processing systems used for mammography should be located in clean areas, as clean as required for a typical mammography darkroom. The area should be vacuumed and mopped frequently. The ceiling above a room light processor must not shed dirt or dust that could be drawn inside the unit.

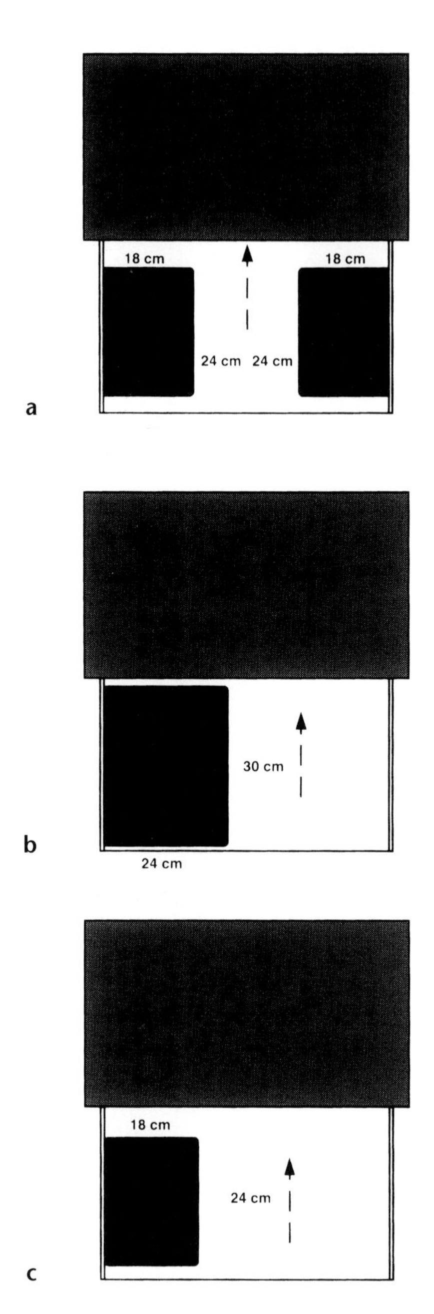


**Figure C-4** Suggested ways to position mammography film on the film feed tray to eliminate guide shoe marks along the chest wall and to help provide consistent replenishment. (a) Two 18 x 24 centimeter films should be placed side-by-side if the processor film feed tray will accommodate them. Note that some model processors with narrower film feed trays (14 inches [35 centimeters]) recommend that only single films be processed. (b) A 24 x 30 centimeter film should be processed with the 24-centimeter edge as the leading edge. Both sides of the processor should be used. The length of film travel is 30 centimeters. (c) An 18 x 24 centimeter film processed alone should be processed with the 18-centimeter edge as the leading edge. Both sides of the processor should be used. The length of film travel is 24 centimeters. (Reprinted with permission from Eastman Kodak Company.)

QC films can be processed by either inserting film into the room light processor through a back door, if the unit is installed through the darkroom wall, or turning off the overhead lights, if the room in which the unit is installed has appropriate safe-lighting. QC films may also need to be loaded into cassettes for processing. Note that any procedures requiring films to be processed with a 90-degree change in orientation may not be possible in room light processing systems because of mechanical limitations.

# D

# The Mammographic Darkroom

- Assessing the Darkroom
- Darkroom Fog Test
- Troubleshooting the Darkroom for Dirt and Dust
- Troubleshooting the Darkroom for Fog or Light Leaks

Deficiencies in the mammographic darkroom are responsible for several mammographic image quality problems, notably shadow images caused by dirt and dust (Chapter six) and increased fog.

This appendix covers the following topics:

- Assessing the darkroom.
- Performing the darkroom fog test.
- Troubleshooting the darkroom for dirt and dust.
- Troubleshooting the darkroom for fog or light leaks.

Considerations for the layout of the darkroom and film processing area are discussed in Chapter three.

## Assessing the Darkroom

An assessment of the darkroom should include the following key areas:

- Cleanliness.
  - Single-emulsion films used for mammography are more sensitive to shadow images caused by dirt and dust.

- Clean or filtered incoming air.
  - Clean fresh air provides a better environment for personnel and reduces shadow images.
- Lighttightness.
  - No white light should be seen from any part of the darkroom after standing with all lights and safelights off for five minutes.
- Number of safelights.
  - Too many safelights and/or safelights placed too closely together result in increased safelight intensity, which may potentially increase film fog and decrease contrast.
- Wattage of lightbulbs and distance from safelights to countertops.
  - The rule of thumb is that there should be a minimum of 4 feet from the bottom of a safelight to a countertop where mammography film is handled in order to use a 15-watt frosted incandescent lightbulb.
- Safelight filters.
  - The filter typically used with orthochromatic mammography films is the KODAK GBX-2 Safelight Filter, or equivalent.
  - If the filter has been correctly installed, printed information on the face of the filter may be read.
  - If the filter has been incorrectly installed (gel coating toward the heat of the lightbulb), the printed information will be backwards.
- Temperature.
  - Fifty to 70°F (10 to 21°C).
- Relative humidity.
  - Thirty to 50 percent.
- Ventilation.
  - A minimum of 10 complete changes of air per hour inside the darkroom and two per hour outside the darkroom are recommended.
  - Air exchanges are necessary for high-quality processing and safe handling and storage of film and chemicals.
  - A service person from a heating and air-conditioning firm should be able to measure this.
- Safelight test.
  - Provides information on the amount of time unexposed, or "green," film and exposed film with a latent image may be safely handled under safelight illumination without causing unacceptable levels of fog.
  - Safelight test materials and instructions are commercially available.
- Darkroom fog test.
  - The level of increased fog on mammography film may not exceed 0.05.
  - Generally, a loss of contrast is the first sign of image quality degradation when the level of safelighting is excessive.

## Darkroom Fog Test

The darkroom fog test should be performed before establishing the processor quality control (QC) program (Chapter five) because if conditions that produce fog exist in the darkroom, the sensitometric characteristics of the QC film will be affected. Thereafter, it should be performed at least semiannually, as recommended by the American College of Radiology (ACR) and any time changes affecting lighting are made in the darkroom.

Follow this procedure:

1. Assemble the tools required for this test, including the following items:
   - A mammography phantom (Figure D-1).
   - A fresh box of mammography film.
   - A piece of cardboard, cut so it will cover half of the film.
   - A watch or timer.
   - A spot-reading densitometer.
2. Load a film from a freshly opened box into a mammography cassette in total darkness.
   - The film should not be close to the expiration date.
   - If the film is close to the expiration date, review how film is handled within the facility and take steps to ensure stock rotation on a first-in, first-out basis.
3. Place the cassette in the grid tunnel of the mammography unit.
4. Place the phantom and photocell in the positions normally used when taking a phantom image.

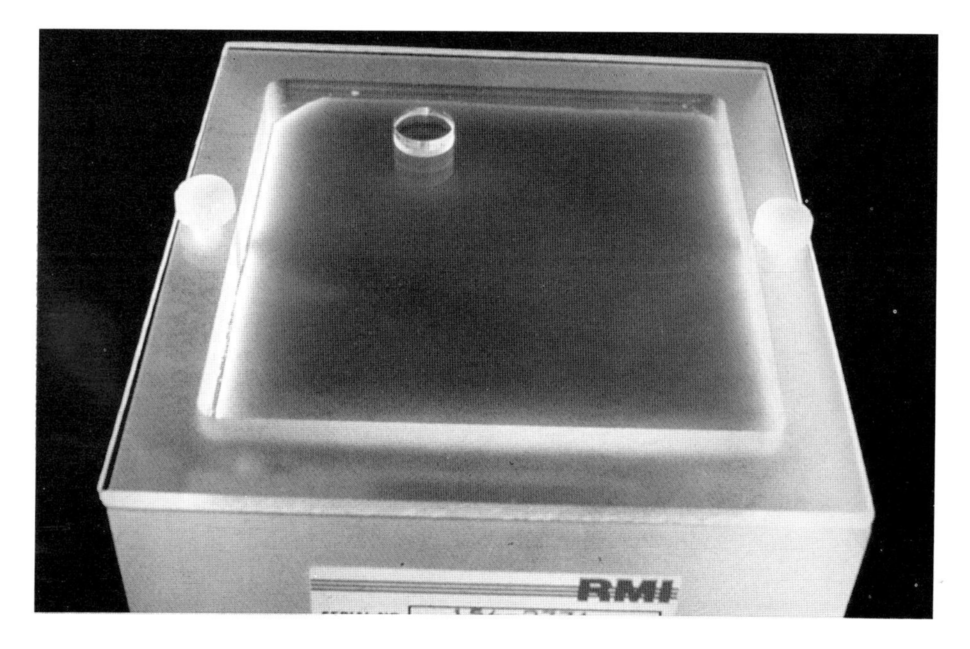


**Figure D-1** A mammography phantom is used for the darkroom fog test. (Printed with permission from Radiation Measurements, Inc.)

5. Expose the phantom so that the optical density on the processed radiograph is between 1.20 and 1.50, as measured in the center of the radiograph.
6. Back in the darkroom, remove the film with the latent image of the phantom from the cassette, and place the film emulsion side up on the darkroom countertop (Figure D-2).
7. Cover the film with the cardboard so that the latent phantom image is divided in half (Figure D-2).
8. Turn the safelights on for two minutes, and then process the film, placing the film on the film feed tray in a consistent manner each time (e.g., emulsion side up, notched edge as the leading edge, and with the film edge along the guide on the right-hand side of the processor) (Appendix C).
9. Use the densitometer to measure the optical density near the center of the radiograph on either side of the location of the edge of the cardboard that covered the film (Figure D-3).
10. Calculate the difference in optical density from the covered and uncovered halves.
11. If the density difference does not exceed 0.05, the level of fog in the darkroom is acceptable.
12. If the density difference exceeds 0.05, repeat the test, checking that correct procedure was followed. If the results still indicate a failed test, begin troubleshooting the darkroom for fog or light leaks, discussed later in this appendix.

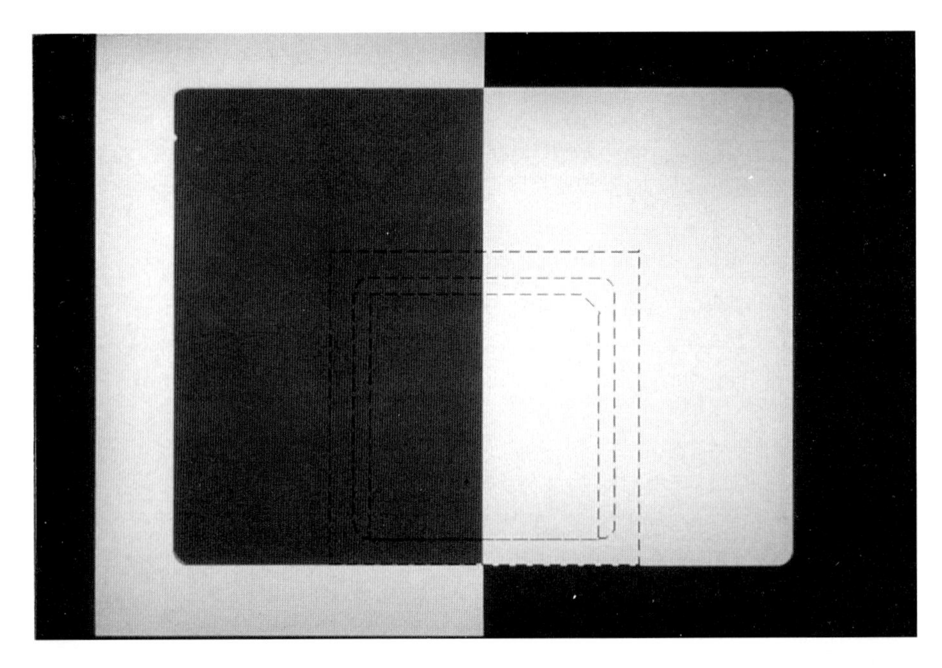

**Figure D-2** The mammography film with the latent image of the phantom, shown by the dotted lines, should be placed emulsion side up on the darkroom countertop. Cover half of the film, shown by the shaded area, with a piece of cardboard. (Printed with permission from Eastman Kodak Company.)

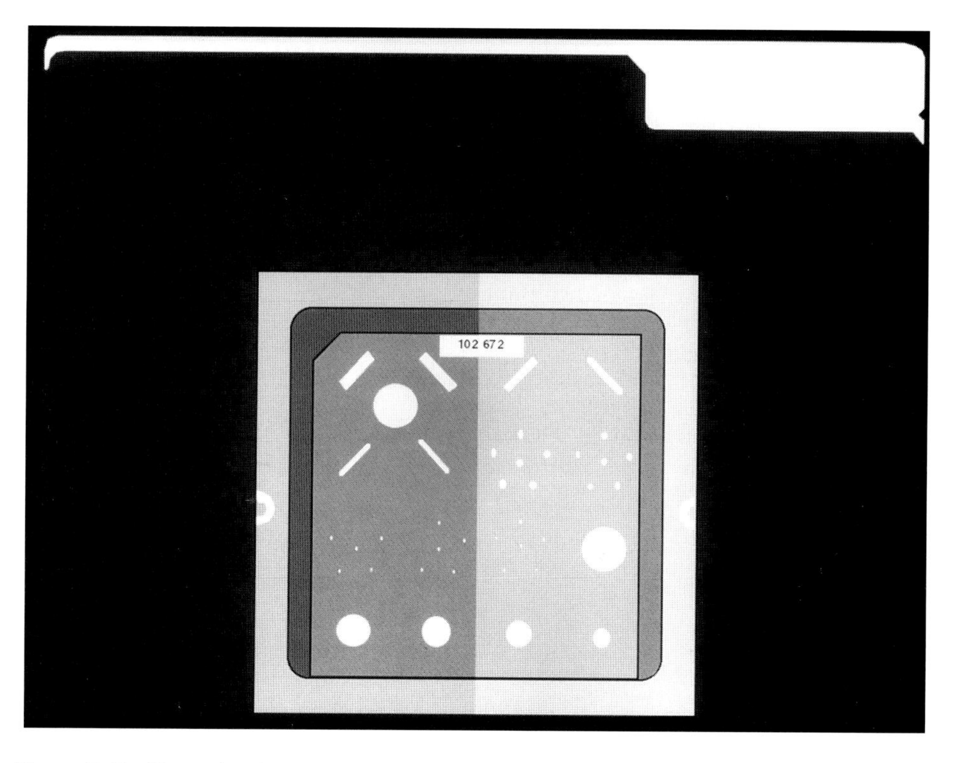


**Figure D-3** Use a densitometer to measure the optical density near the center of the radiograph on either side of the location of the edge of the cardboard that covered the film. The density difference should not exceed 0.05. (Reprinted with permission from RSNA Categorical Course in Physics 1996; pp. 49–66.)

## Troubleshooting the Darkroom for Dirt and Dust

The following suggestions are intended to help reduce airborne dust, dirt, and lint from the environment and make the darkroom easier to keep clean:

- Use an ultraviolet, or black, light to help identify sources of dust and dirt in the darkroom.
  - Note that all dust particles do not fluoresce under ultraviolet illumination.
  - Limit exposure prudently and observe appropriate safety precautions for eyes, face, and exposed skin; do not look directly at the light.
  - Do not expose film to ultraviolet light.
  - Use the light again after cleaning.
- Maintain relative humidity between 30 and 50 percent to minimize the attraction of particles onto films and intensifying screens.
- Do not routinely place mammography cassettes on the floor.
  - Dirt may be carried into the darkroom along with the cassettes.
- If passboxes are used, check the insides for paint particles, metal flakes, etc.
  - Clean periodically.
- Filter air coming into the darkroom, either at the system level or by placing furnace-type filters in the air vents feeding the darkroom.

  - Change air filters according to the frequency recommendations of the manufacturer.
  - Electrostatic air cleaning devices may be helpful if a significant dust problem exists in the darkroom.
- Locate ventilation louvers in darkroom doors near eye level, not near the floor, and vacuum or wipe periodically.
  - Use a vacuum cleaner in the darkroom only if the dust particles that could be stirred up from the vacuum exhaust have enough time to settle (e.g., overnight).
  - Clean the countertops, floors, and film feed tray after using a vacuum cleaner.
- Contain dust from any new construction in or near the darkroom.
  - Avoid dry wall dust accumulation in vent pipes during construction, or long-term dust/shadow image problems may result.
  - Thoroughly clean vents before access is limited.
  - Avoid installing any new equipment, such as a room light processor, in an area still under construction.
- Darkroom walls should be smoothly finished and painted using a light-colored semigloss paint.
  - Walls painted with semigloss paint are easier to keep clean.
  - The use of a light-colored paint maximizes safelight illumination and makes it easier to accomplish all tasks performed in the darkroom.
  - Avoid flat paint because dust can cling more easily to walls, and it makes them difficult to wash.
  - Repaint walls with peeling paint.
- Replace any open shelving.
  - No open shelving of any kind should be used, especially above darkroom countertops.
- Install cabinets with doors to reduce areas that may accumulate dirt and dust.
  - Use cabinets that reach all the way to the ceiling for upper storage.
- If possible, darkroom ceilings should be solid, smoothly finished, and painted.
  - Suspended ceilings, especially in darkrooms with a single door entrance, may be a significant source of dirt and dust.
  - Slamming the darkroom door can cause suspended ceiling tiles to move and dust to sift down from the tiles and dead space above the ceiling.
  - It may be helpful to seal tiles to the suspension frame to prevent ceiling tile movement when the darkroom door is closed (slammed); this should be done only if there is no periodic need for access to the area above the ceiling.
  - It may also be helpful to seal the surface and edges of the ceiling tiles themselves to decrease/eliminate airborne particles from the tiles; check with the ceiling tile manufacturer for an appropriate sealant (e.g., semigloss paint or polyurethane) and with local authorities to ensure that sealing the tiles in any manner does not violate local fire codes.

  - If access to the area above the suspended ceiling is required, securely tape a large continuous sheet of plastic over the ceiling to contain the dust; it may be easily removed when access is necessary.
  - Install a revolving (barrel-type) door to eliminate door slamming/dust sifting down from the ceiling and to improve traffic flow into and out of the darkroom.
  - Check the felt installed as part of revolving doors, as it may be an additional source of lint.
  - If the darkroom is old (i.e., constructed many years ago), check all materials used in the ceiling for brittleness and replace if necessary.
- Wipe darkroom walls, the fronts of cabinets, safelights, vent surfaces, exposed pipes, etc., with a damp lint-free wipe periodically to eliminate clinging dust.
- Vacuum or use a damp lint-free wipe to remove dirt and dust from the inside of the film bin occasionally (e.g., once a year, or as needed).
- Minimize storage in the darkroom.
- Open cases of film and other corrugated cardboard boxes outside the darkroom, and carry the contents inside to reduce the introduction of additional cardboard fibers into the darkroom.
- Open new boxes of film slowly and carefully to reduce the number of cardboard fibers that become airborne.
- Remove the flap and box top from the darkroom after placing the film (in its envelope and box) in the film bin.
- Remove the stiffener cardboard from film boxes and the darkroom.
- If the film is packaged in a sealed, foiled-lined paper envelope, use a pair of scissors to cut the envelope to reduce airborne fibers.
  - Do not remove the film from the envelope or store loose film in the box, film bin, or film safe at any time.
- Avoid replacing unexposed film that has been inside a cassette back into the film supply in the film bin.
  - This could introduce additional dust into the film bin and into the film supply.
- Switch from an identification (ID) printer or camera located in the darkroom to a camera located outside or to an ID-capable mammography unit to eliminate fibers from the thin paper cards inserted into the printer or camera in the darkroom.
- Eliminate nonessential items that contribute to dust and paper fiber.
  - These include newspapers, magazines, facial tissue, notebooks, paper pads, etc.
- Require all darkroom personnel and technologists to wear lint-free clothing or to wear smocks or lab coats over clothing.
- Avoid hanging or storing articles of clothing in the darkroom.
- Avoid bringing materials that shed fibers into the darkroom (e.g., knitting).
- Avoid washing machines and dryers in the darkroom.
- Clean screens and cassettes at least weekly (Appendix E).
- Clean screens and cassettes more frequently (e.g., daily) if necessary.

- Thoroughly clean the entire darkroom if it is necessary to clean screens and cassettes more frequently than once a day.
- Clean darkroom countertops daily, using a lint-free wipe and appropriate nonresidual cleaner.
- Damp mop darkroom floors daily to remove any dust that settled overnight.
- Clean the processor film feed tray last.
  - Moisten a lint-free wipe with a small amount of the antistatic cleaning solution used for intensifying screens.
  - If the feed tray needs to be cleaned more frequently, use the ultraviolet light to check the cleanliness of the incoming air.

## Troubleshooting the Darkroom for Fog or Light Leaks

Fog caused by safelights and light leaks in the darkroom may result in low-contrast mammographic images because the toe area of the characteristic curve (Chapter four) may be altered (i.e., filled in), even without a corresponding increase in base plus fog on the processor QC chart.

The following suggestions are intended to help improve the lighttightness and level of fog in the darkroom:

- Always turn on all lights in adjacent rooms when troubleshooting the darkroom for light leaks and before performing a darkroom fog test.
  - Light may be leaking into the darkroom through an area cut out in the ceiling for the passage of pipes or ventilation tubing.
  - If increased fog is an intermittent problem, check adjacent areas that may normally have subdued lighting, but where brighter illumination is used occasionally.
- Check for the correct type of filter on all safelights.
  - Use KODAK GBX-2 Safelight Filter or equivalent.
- Check that all filters have been properly installed.
  - If the filter has been correctly installed, printed information on the face of the filter may be read.
  - If the filter has been incorrectly installed (gel coating toward the heat of the lightbulb), the printed information will be backwards, possibly causing the filter to deteriorate and crack.
- Check the age of the filter(s).
  - Write the installation date with a permanent black marker on the face of a newly installed filter.
  - Check with the filter manufacturer for information on expected filter life.
  - For a darkroom in which the safelights are on approximately 10 hours per day, the typical specification for filter life is four months.
  - If the results of the darkroom fog test are acceptable, it is not necessary to replace filters older than four months.
- Check for cracked filters or safelight housings.
  - View the filtered light exiting from the bottom of safelights.
  - Closely inspect housings from all other angles.

- Check for an appropriate number of safelights for the size of the darkroom.
  - The darkroom should be bright enough only to perform darkroom functions safely.
- Check the distance from safelights to countertops and to the processor film feed tray.
  - The distance should be at least 4 feet in order to use a 15-watt frosted incandescent lightbulb.
  - Try using a 7.5-watt frosted incandescent lightbulb if the distance is less than 4 feet.
- Check/replace the 15-watt frosted incandescent lightbulb.
  - Lightbulbs are not classified as medical devices, and light intensity from lightbulb to lightbulb of the same wattage, even from the same manufacturer, differs widely. A 15-watt frosted incandescent lightbulb may be too bright.
  - Replacing the lightbulb may quickly, easily, and economically solve the problem of a failed darkroom fog test if all other aspects of the darkroom check out as appropriate.
- Check direct versus indirect safelighting.
  - If all safelights are ceiling or wall mounted and point downward, and the level of fog in the darkroom is too high, change some of the safelights so that the light is aimed at the ceiling; then repeat the darkroom fog test.
- Check for devices in the darkroom with indicator lights that illuminate when the device is in use, such as a telephone.
  - Fog is usually an intermittent problem with such devices.
  - Even if the device has been modified so the indicators are red, it may still be causing film fog.
- Check for white light leaks.
  - Turn on all lights in adjacent rooms.
  - Stand in the darkroom for five minutes with all of the lights turned off to allow eyes time to adjust.
  - Check all areas and levels of the darkroom, especially around processors, passboxes, darkroom doors, processor exhaust tubing, ceiling tiles, ceiling vents, etc.
  - Use extreme care when moving around a totally darkened darkroom.
  - Mark any areas where a light leak is seen in the darkroom with masking tape.
- Check that the lid on the processor (installed through the darkroom wall) is always properly replaced.
  - This may cause an intermittent problem.
- Check for an electrical problem in the processor.
  - May be an intermittent problem.
- If the darkroom has a single door and film is occasionally fogged by opening the door before the film has fed far enough into the processor, consider the installation of a lighttight cover for the processor film feed tray.

- Even with passing the darkroom fog test and no significant increase in base plus fog on the processor QC chart, the contrast of clinical images may be reduced due to the darkroom environment. Perform this quick test:
  - In a totally dark darkroom, open a fresh box of film, expose, and immediately process a sensitometric strip (Chapter five).
  - Expose two additional films in total darkness, using film from the freshly opened box, and place them emulsion side up on the darkroom countertop.
  - Turn on all safelights.
  - Process one of the films after one minute; process the second film after two minutes.
  - Measure base plus fog and average gradient values on all three films using an automatic scanning densitometer.
  - If there is a decrease in average gradient from the first to the second to the third film, darkroom fog may be contributing to the decreased contrast of clinical images.

# E

●●●●

# Cleaning Intensifying Screens

Intensifying screens should always be cleaned following the recommendations of the screen manufacturer. Generally, for effective cleaning and prolonged screen life, only the screen cleaning solution offered by the specific manufacturer and any other products recommended should be used.

Intensifying screens used for medical imaging other than mammography should be cleaned periodically.

Mammography screens, which are used as a single screen with single-emulsion mammography film, must be cleaned at least once a week, or more frequently as needed, to reduce shadow images caused by dirt and dust. Shadow images are discussed in Chapter six.

The following procedure, recommended by one manufacturer of intensifying screens, is provided as an example of the steps required to clean mammography screens:

1. Choose a clean location to clean screens and cassettes.
   - If working on a countertop in the darkroom used for processing mammography film, wipe the outside of the cassettes and clean the countertop prior to cleaning the screens.
   - Manage cassettes so they are empty at the end of the day, the best time to clean mammography screens.
   - Avoid replacing film previously loaded into cassettes back into the film box in the film bin, since this could introduce dust into the film supply.
2. Moisten a lint-free wipe with a small amount of 70 percent isopropyl alcohol, and gently rub the wipe across the intensifying screen.
   - Check with the manufacturer of the specific screens used regarding the use of 70 percent isopropyl alcohol.
   - Seventy percent isopropyl alcohol, if appropriate for the type of screens used, should be used as the first cleaning solution on all new cassettes after

removal from the packaging, on all cassettes prior to performing screen-film contact testing, and any time a thorough cleaning is desired.
- Seventy percent isopropyl alcohol, if recommended, may be used on a daily basis without causing damage to mammography intensifying screens.
- Any time 70 percent isopropyl alcohol is used on the screen, intensifying screen cleaner and antistatic solution must be used afterwards.
- Clean around any labels on the screen (e.g., those used to number cassettes individually).

3. Dampen a second lint-free wipe with a small amount of intensifying screen cleaner and antistatic solution (Figure E-1).
   - Clean and dry the screen.
   - Avoid excessive pressure or rubbing on the screen surface or edges, abrasive wipes such as surgical gauze pads, pouring of any solutions directly onto the screens or into the cassettes, or using an excessive amount of any solution.
   - A mild soap and water solution may also be used if screen cleaner is not available.
   - Soaps or detergents containing brightening agents should not be used.
4. Clean the inside plastic cover of the cassette (tube side panel) using a lint-free wipe that has been dampened with a mild soap and water solution or 70 percent isopropyl alcohol.
   - Dry the cover.

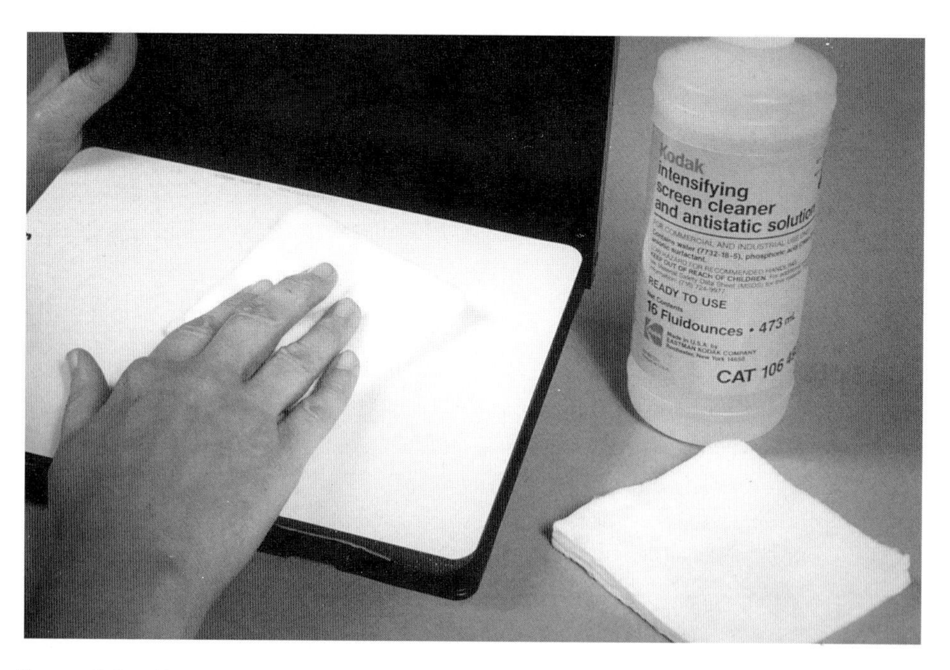


**Figure E-1** Clean intensifying screens in mammography cassettes, using the cleaning solution(s) and procedures recommended by the manufacturer. (Printed with permission from Eastman Kodak Company.)

5. Stand the cassettes on edge to dry.
   - Note that the procedure of applying a thin, even coating of intensifying screen cleaner and antistatic solution to the screen after cleaning and allowing the screen to dry is no longer recommended.
6. After all screens and cassettes have been cleaned and allowed to dry, use an antistatic brush with the bristles fully extended to remove any dust accumulated in the corners of the cassette cover and along the hinge.
   - Avoid touching the bristles of the brush to prevent the transfer of body oils.
   - The antistatic element may have an expiration date, after which it loses its effectiveness to remove dust and should be replaced.
7. Retract the brush approximately halfway and, holding the cassette suspended in midair, carefully stroke the antistatic brush across the screen surface with the antistatic element toward the screen (Figure E-2).
   - Use care to avoid scratching the screen.
8. Inspect the screen and cassette cover for any stray particles of dust.
   - An ultraviolet, or black, light is helpful in determining if screens and cassettes are clean.
   - Limit exposure to the ultraviolet light prudently and observe appropriate safety precautions for eyes, face, and exposed skin; do not look directly at the light.

**Figure E-2** After all screens and cassettes have been cleaned and allowed to dry, use an antistatic brush with the bristles fully extended to remove any dust accumulated in the corners of the cassette cover and along the hinge. Then retract the brush approximately halfway and carefully stroke the antistatic brush across the screen surface with the antistatic element toward the screen. (Printed with permission from Eastman Kodak Company.)

9. Reload the cassette with film.
   - Wait at least 15 minutes after loading before using.
10. Clean the countertop, if the cleaning procedure was done in the darkroom.
    - Between weekly cleanings, the brush may be used to remove dust particles from the cassette and screen surface, or use screen cleaner followed by the antistatic brush. Care should be taken to protect the screen surface from scratches, which will degrade image quality.

# F

# Mobile Van Film Processing

- Recommendations for Film Processors Installed in a Mobile Environment
- Chemicals and Replenishment
- Air Exhaust
- Room Temperature
- Service Access
- Effluent Storage
- Movement of Processors
- Maintenance of the Crossovers and Racks
- Start-up of Processors at a New Site
- Recommendations for Batch Processing

For medical imaging procedures (e.g., mammography, MRI, CT, chest, etc.) performed in mobile vans, there are two options for film processing: (1) an automatic film processor can be installed in the mobile van, or (2) exposed films can be accumulated and transported to a facility equipped with an automatic film processor (sometimes referred to as batch processing). In either case, the processor must be properly set up and serviced according to the manufacturer's recommendations, and appropriate quality control (QC) procedures must be implemented.

## Recommendations for Film Processors Installed in a Mobile Environment

Consult the processor's site specifications for processor installation requirements. Items that are usually covered include: (1) supply voltage, (2) chemicals (developer and fixer), (3) replenishment, (4) water supply, (5) air exhaust, (6) room temperature (operational and storage) and relative humidity, and (7) service access.

Items unique to mobile installations that may not be covered in the site specifications include: (1) effluent storage, (2) movement of processors, (3) maintenance of the racks, and (4) startup of processors at a new site.

Each of these factors must be considered when installing a processor in a mobile environment. The KODAK M35A-M X-OMAT Processor will be used as an example throughout this appendix.

Mobile vans typically use a generator to meet the power needs of the x-ray unit and processor. Alternatively, they may "plug into" a power outlet at the site being visited. In either case, the processor's site specifications provide the voltage, current, and frequency requirements for the processor, as shown in Table F-1.

Fluctuations in the power supply are tolerable as long as they are within the limits specified for the processor. If the voltage of a generator exceeds the tolerance zone of the processor, a malfunction may occur. Fluctuations that are tolerable for a processor may be intolerable for the x-ray unit. Total power supply for the entire mobile van must be consistent.

## Chemicals and Replenishment

Processors require a source of fresh chemicals for replenishment as films are fed. Separate tanks for the mixing and storage of developer and fixer solutions are necessary. Tank size depends on three factors: (1) the typical number of films that will be processed, (2) the length of time the mobile van is on the road before returning to home base, and (3) whether fresh chemicals can be mixed or added to the storage tanks while on the road. Typical mix batch size for developer and fixer is five or ten gallons. Five-gallon mix will minimize the space required for storage, but will limit the number of films that can be processed. The tanks also must be located where it will be easy to add more chemicals.

The location of the chemical supply must also be free from extremes in temperature. The processor must maintain a relatively constant developer temperature to provide optimum film processing and consistent image quality, and the fixer temperature must be maintained at a minimum temperature to ensure adequate fixing. It is recommended that the replenishment solutions be maintained at the suggested room temperature limits listed in the site specifications, i.e., between 59 and 86°F (15–30°C) for the M35A-M and most other processors. If the temperature of the replenishment solutions falls below these limits, the processor may not be able to maintain the developer and fixer temperatures at the required setpoints. If the temperatures go above these limits, the developer may exceed the recommended temperatures. Elevated fixer temperature is not a significant problem. Exact specifications

**Table F-1** Processor Site Specifications

| | |
|---|---|
| Voltage: | 120 volt +6% -13% or 104-127 volt |
| Current: | 15 amps maximum |
| Frequency: | 50 or 60 Hertz |

cannot be given for replenishment solution temperatures because the temperature control depends on many factors, e.g., dryer temperature, number of films fed in rapid succession, etc. Therefore, the recommendations given here are only suggested limits.

The processor's site specifications usually suggest a flow rate for the water entering the processor for film washing and developer temperature control. A mobile van installation does not typically include access to the large volumes of running water needed and must, therefore, achieve adequate film washing by other means. This is normally done by converting the processor to a batch water system. In this mode, a volume of water is recirculated through the processor for washing the film. The amount of water depends on the volume of film to be fed. Secondarily, the amount of water is dictated by the need to maintain developer temperature control.

To determine whether adequate washing is achieved, tests must be run with the film type that will be used in the mobile application and a specified amount of water. This is determined by performing a fixer retention test and ensuring that the amount of hypo retained by the film is acceptable. Table F-2 lists recommendations for the M35A-M processor for different films and different keeping requirements.

The amount of water required and the size of the water tank vary greatly, depending on the film type and desired keeping time. To ensure that adequate washing is maintained, conduct testing with the specific film type and processor to be used.

The second issue affecting the size of the wash tank is developer temperature control. A typical processor uses water to control the developer temperature, which normally rises due to the heat from the dryer. The water ensures that the dryer heat does not cause the developer temperature to exceed the control limit. However, it is not easy to estimate the size of the water tank because many factors must be considered, e.g., the number of films fed, the ambient temperature of the air around the water tank, the dryer temperature, etc. The site specifications for the processor generally give the upper and lower limits for the water temperature (temperature range). For the M35A-M processor, the range is 40–85°F (4–30°C). If the water supply tank is positioned to maintain the water temperature within this range, no problems should occur with developer temperature. Larger tank sizes are more stable and will have more cooling capacity. If the water temperature falls below the lower limit, the developer temperature may never reach operating temperature, may take a long time to reach operating temperature, or may fluctuate when films are fed. If the water

**Table F-2** KODAK M35A-M X-OMAT Processor Recommendations

| | *Keeping Limits* | |
|---|---|---|
| *Film Type* | *Long Term* | *Archival* |
| KODAK T-MAT RA Film | 20-35 x 43 centimeter sheets/gallon | 7-35 x 43 centimeter sheets/gallon |
| KODAK MIN-R M Film | 90-18 x 24 centimeter sheets/gallon | 30-18 x 24 centimeter sheets/gallon |

temperature exceeds the upper limit, the developer will likely exceed the temperature control limit. Fluctuations in developer temperature caused by the temperature of the water supply will result in variable image quality.

## Air Exhaust

A processor does generate some heat (generally from the dryer) and some fumes (from the heated developer and fixer). The heat and fumes are removed from the processor through the exhaust air flow. In a mobile van, the exhaust from the processor should be vented to the outside of the van. The processor's site specifications describe the amount of air that is exhausted. The exhaust vent on the van should also be configured to prevent exterior air from entering the van.

## Room Temperature

Room temperature is important for proper operation of the processor. The processor's site specifications identify the upper and lower limits for the ambient room temperature. Exceeding these limits may affect the performance of the processor. Proper room temperature and relative humidity must be maintained at all times on the mobile van when film and chemicals are being stored there.

## Service Access

Because space is limited on a mobile van, a processor of small size is generally desirable. Service access should also be considered when designing the layout of the van. The site specifications generally suggest service access for either side of the processor. Every effort should be made to provide adequate access to the feed tray, receiving bin, and the side with the highest suggested service access requirement.

## Effluent Storage

A processor will generate effluent consisting of developer and fixer. The amount of effluent depends on the usage rate, number, and size of films processed. Table F-3 summarizes typical usage rates based on 35 x 43 centimeter films for a processor. Consult the manufacturer of your processor and film for actual rates.

Based on this data, processing 50 sheets of film in one day would yield 4,000 milliliters (1.1 gallons) of developer and 5,000 milliliters (1.3 gallons) of fixer. It is

**Table F-3** Typical Usage Rates

| *Usage Rate* | *< 25 films/day* | *25-75 films/day* | *> 75 films/day* |
|---|---|---|---|
| Developer Effluent | 100 milliliters/sheet | 80 milliliters/sheet | 60 milliliters/sheet |
| Fixer Effluent | 120 milliliters/sheet | 100 milliliters/sheet | 85 milliliters/sheet |

suggested that the effluent tanks and supply tanks be of equal size so that the effluent tanks will accommodate all of the replenishment supply volume. The effluent tanks should also be located below the processor to allow for chemical draining.

## Movement of Processors

If a mobile van is moved while the processor is filled with chemicals, the processor solutions will spill, causing contamination. The processor should, therefore, be drained of all solutions prior to moving the van. These solutions are still usable and should not be drained into the effluent tanks. They should instead be drained into separate storage containers for reuse at the next site. Storage tank volumes must be slightly greater than the processor solution volumes. This information may not be part of the processor's site specifications, but it can be obtained from the manufacturer. The storage tanks should also be sealed so that no spillage occurs in transport.

## Maintenance of the Crossovers and Racks

Because the tanks in the processor must be emptied for transport, the crossovers and racks require special attention. If these components are left in the drained processor, residual chemicals may dry on the rollers, producing artifacts and potential transport problems when the processor is restarted at the next site. It is, therefore, suggested that they be removed from the processor and stored in separate water-filled storage tanks. This will keep the rollers wet and prevent chemicals from drying on them. The tanks must be covered to prevent spillage during transport. The water in the tanks should be changed whenever the wash water is changed. Each crossover and rack must be stored in its own tank to prevent contamination.

## Start-up of Processors at a New Site

On arrival of the mobile van at a new location, the processor must be leveled to ensure proper processing and to avoid contamination. Then all processor components should be installed, the developer, fixer, and wash water solutions refilled, and the processor allowed to warm up. Once the processor has reached the proper operating temperature, several cleanup films should be processed first to help prevent artifacts. The processor should be evaluated sensitometrically to ensure that it is performing properly. Processor QC charts should be used to monitor the processor and help maintain film contrast, speed, base plus fog, and developer temperature within required limits.

## Recommendations for Batch Processing

Any time exposed films are transported from the mobile van to another location for processing—or when exposed films are held for processing—there is a delay between the time the film is exposed and when it is processed. If this delay is long enough, the optical density may be lower than if the film had been processed immediately after exposure. This effect, called latent image fading (sometimes also referred to as latent image keeping [LIK]), is a normal characteristic of all medical x-ray films. It is caused by instability of the latent image. Different films have different levels of speed loss; contrast changes may also occur.

Table F-4 shows the percentage of film speed loss and film contrast changes when the time between exposure of a mammographic x-ray film and processing is 0, 4, 8, 24, and 48 hours. To minimize latent image fading effects in the clinical environment, it is important that the time interval between exposure and processing be as short as possible and consistent each day. Ideally, films should be processed immediately after exposure. To minimize time interval differences, films should be processed in the order in which they are exposed. If the time between exposure and processing is relatively long and consistent, the exposure technique can be adjusted on a one-time basis to compensate for latent image fading effects and to obtain the appropriate film density. For the film shown in Table F-4, if it is known in advance that processing will be delayed for more than eight hours, it may be advisable to increase exposure time (by adjusting the automatic exposure control [AEC] setting) to obtain the proper optical density in the clinical images. Also note in Table F-4 that there is very little change in film contrast due to latent image fading.

In summary, follow these two suggestions to minimize the effects of latent image fading:

- *Film processor quality control.* Process film strips immediately after exposure with the sensitometer.
- *Batch processing of films (in both mobile and stationary environments).* (1) Be as consistent as possible on a day-to-day basis in the time interval between exposure and processing. Ideally, films should be processed immediately after exposure. To minimize time interval differences, process films in the order that they are exposed. (2) If the time between exposure and processing is relatively long but consistent, the exposure technique can be adjusted, on a one-time basis, to compensate for the effects of latent image fading and to obtain and maintain the appropriate film density.

**Table F-4** Example of Latent Image Fading Data for a Mammographic X-Ray Film

| Time delay between exposure and film processing (hours) | 0 | 4 | 8 | 24 | 48 |
|---|---|---|---|---|---|
| Percent film speed loss | 0 | 10 | 12 | 18 | 23 |
| Optical density difference | 0 | 0.12 | 0.15 | 0.21 | 0.27 |
| Percent contrast change (average gradient) | 0 | 2 | 3 | 3 | 5 |

Note: Speed determined at an optical density of 1.00 above base plus fog. Average gradient determined from the slope of the characteristic curve between optical densities of 0.25 and 2.00 above base plus fog.

(Reprinted with permission from Eastman Kodak Company. EKC publication no. N-325.)

## References

1. Haus AG, ed. Film processing in medical imaging. Madison, WI: Medical Physics Publishing, 1993.
2. Haus AG. State-of-the-art screen-film mammography: A technical overview. In: Barnes GT, Frey DG, eds. Screen-film mammography: Imaging considerations and medical physics responsibilities. Madison, WI: Medical Physics Publishing, 1990.
3. Haus AG, Batz TA, Dickerson RE, Lillie RF, Oemcke KW, Lanphear JD. Automatic film processing in medical imaging: System design considerations. In: Seibert JA, Barnes GT, Gould RG, eds. Specification acceptance testing and quality control of diagnostic x-ray imaging equipment. New York: American Institute of Physics, 1994.
4. Sickles EA. Latent image fading in screen-film mammography: Lack of clinical relevance for batch-processed films. Radiology 194: 389–392, 1995.

# G

# The Sensitometric Technique for the Evaluation of Processing (STEP) Test

Whether or not facilities involved in medical imaging are appropriately processing film in automatic film processors (i.e., following the manufacturer's recommendations) may be an area of regulatory concern. Under-processing (reduced film speed) is the major concern because unnecessary increased patient radiation exposure may occur as a result.

In the United States, during surveys conducted by the Food and Drug Administration (FDA) under two programs (the Nationwide Evaluation of X-Ray Trends [NEXT] since 1984 and the Mammography Quality Standards Act [MQSA] since 1995), processing speed is evaluated by an empirical test, called the Sensitometric Technique for the Evaluation of Processing (STEP) test. This test measures only processing speed. Furthermore, it is designed to identify film-chemistry-processor systems deviating significantly (±20 percent, or log relative exposure ±0.08) from film manufacturers' recommendations for a standard processing cycle. These action limits correspond to a nominal ±4°F temperature difference. This wide a range (±4°F) was selected to encompass: (1) control film emulsion variability over the useful life of the film, (2) variations attributable to the calibration of the sensitometers and densitometers, and (3) the normal daily fluctuations associated with a properly operating automatic film processor. Note that under MQSA, an extended processing cycle may also be evaluated.

The STEP test essentially compares the density of a control film processed in a facility's processor with the density of the same control film processed according to the manufacturer's recommendations. The sensitometers and densitometers used are calibrated, i.e., they are matched to reference sensitometers and densitometers maintained by the FDA. In addition, control film that is of the same type and from the same emulsion batch as the film used to calibrate the sensitometers is used. The control film used for MQSA surveys is known to be sensitive to the differences between standard- and extended-cycle processing and also has adequate sensitivity in detecting changes in chemical solutions from different manufacturers.

When the STEP test method is used on a processor using a standard processing cycle, if the densities of the film processed are equivalent to the densities of the film processed according to the manufacturer's recommendations, the processing speed is 100. Standard-cycle processors with a STEP test number lower than 80 are currently cited for noncompliance. The standard-cycle value of 100 is also used as the benchmark for processors using an extended processing cycle for mammography, although a higher number (e.g., 130 to 140) is expected. Extended-cycle processors with a STEP test number lower than 100 are also cited for noncompliance. (Note that processors are evaluated according to the processing cycle stated by facility personnel. As discussed in Chapter three, some processors with longer processing times deliver standard-cycle performance. If such a processor is evaluated against extended-cycle criteria, as indicated to the MQSA inspector, the processor may be cited for noncompliance.) Higher than expected STEP test numbers for either standard- or extended-cycle processing receive no citation.

When a mammography facility receives a noncompliance citation under MQSA, it is required (as of this writing) to respond to the FDA within 30 days as to the actions it has or will take to correct the processing deficiency. Generally, three variables influence the speed resulting with a particular processor: (1) the particular chemicals used, (2) developer temperature, and (3) development time. Check the following in a processor cited for noncompliance:

1. Quality of the chemicals.
2. Method used for mixing chemicals.
3. Contaminated or outdated chemicals.
4. Replenishment rates.
5. Accuracy of developer temperature using an accurate thermometer.
6. The manufacturer's recommendation for developer temperature for the model processor and processing cycle in comparison with the current developer temperature.
7. The accuracy of the development time (refer to Chapter seven).

## References

1. Suleiman OH, Rueter FG, Antonsen RG, Conway BJ, Slayton RJ. Automatic film processing: Analysis of 9 years of observations. Radiology 185: 25–28, 1992.
2. Suleiman OH, Belella SL, Fuller RE, Spelic DC. How do I evaluate my automatic film processor for compliance with MQSA? Scientific Exhibit, 27th National Conference on Breast Cancer, 1996.
3. Suleiman OH, Conway BJ, Rueter FG, Slayton RJ. The sensitometric technique for the evaluation of processing (STEP). Radiation Protection Dosimetry 49: 105–106, 1993.

# Glossary

**absorption, x-ray:** describes the reduction in fluence as x-rays traverse body tissues and emerge to expose an x-ray film.

**aerial image:** the pattern of x-ray fluence emerging from the body, incident on the image receptor; also referred to as the image in space.

**afterglow:** the tendency of a phosphor to continue to emit light after the x-ray exposure has stopped; also called phosphorescence.

**American National Standards Institute (ANSI):** an organization that deals with standards; now known as the National Institute of Standards and Technology (NIST).

**antifoggant:** a chemical added to film during emulsion manufacturing to keep the fog level to a minimum.

**antistatic solution:** a liquid that, when applied to the surface of the intensifying screen in a cassette, will reduce artifacts caused by static electricity.

**archiving:** long-term storage of images, documents, or digital data for subsequent retrieval.

**artifact:** any unwanted irregularity on a radiograph due to lint, dust, static electricity, film processing, or improper storage or handling.

**artifact, minus density:** any unwanted artifact on a radiograph characterized by a white mark or one that is lighter than the density of the surrounding image.

**artifact, plus density:** any unwanted artifact on a radiograph characterized by a black mark, or one that is darker than the density of the surrounding image.

**artifact, processor:** any artifact arising from the automatic film processor.

**attenuation, x-ray:** see absorption, x-ray.

**automatic exposure control (AEC):** a system, often referred to as a phototimer, that automatically terminates the x-ray exposure, producing a predetermined level of exposure in the image.

**automatic film processing:** a processing method in which temperature, recirculation, and replenishment are automatically controlled as film is transported through processing solutions.

**average gradient:** the slope of the straight line joining corresponding points of specified densities on the characteristic curve; used to indicate the contrast characteristics of a film.

**background radiation:** radiation that occurs naturally in the environment.

**base plus fog:** the optical density of the film base plus any additional density where no exposure has occurred.

**biochemical oxygen demand (BOD):** a measurement of the amount of dissolved oxygen that an effluent consumes. It is usually measured over a five-day period and is referred to as BOD5.

**blur:** the lateral spreading of the structural boundary of the image; also referred to as unsharpness.

**cassette:** a lighttight case, usually made of thin, low x-ray absorption plastic for holding x-ray film. One or two intensifying screens for the conversion of x-rays to visible light photons are mounted inside the cassette in close contact with the film.

**characteristic curve:** a type of input-output response curve. In radiography, this curve expresses the change in optical density (output) with the change in exposure of x-ray film (input); also called an H&D curve, D log E curve, or sensitometric curve.

**chemical oxygen demand (COD):** an analytical method for measuring the oxygen demand of an effluent.

**contrast, film:** a measure of a film's property to convert input changes in exposure (light or x-radiation) to output changes in optical density (film blackening).

**contrast, radiographic:** the difference in optical density (film blackening) between areas of interest in a radiograph. The combination of subject contrast and film contrast determines radiographic contrast.

**contrast, subject:** the ratio of x-ray photon fluences in the aerial image corresponding to different regions in the object being radiographed due to the effects of absorption.

**control limit:** the maximum acceptable deviation or limit of film contrast, speed, and fog as plotted on a processor quality control (QC) chart, used to monitor processor performance.

**crossover:** in dual-emulsion film, light from the front screen that penetrates the front emulsion layer and base, causing exposure and increased blur of the back emulsion.

**crossover assembly:** part of an automatic film processor that is designed to move x-ray film from one section of a processor to another.

**crossover procedure:** the procedure done during quality control that compares the film characteristics of film from a particular emulsion number with film from a different emulsion number so that the aims on a processor quality control chart can be adjusted appropriately.

**D log E curve:** see characteristic curve.

**densitometer:** a device consisting of a light source, an aperture, and a light sensor, used to measure optical density.

**density difference (DD):** the difference in optical density between two predetermined steps on a sensitometric strip; also referred to as contrast or contrast index.

**density, optical:** the degree of blackening of film after exposure and processing.

**density, physical:** the mass of a substance per unit volume (e.g., grams per cubic centimeter).

**detail:** a quality criterion corresponding to the ability of an imaging system to resolve small objects.

**developer:** a chemical solution that converts the latent image on film to a visible image.

**digital radiography:** an image acquisition process that produces radiographic images in digital form.

**D-max:** see maximum density.

**effluent:** used or exhausted chemical overflows and wash waters.

**electrolytic silver recovery:** a method of recovering silver by applying a direct current across two electrodes in a silver-rich solution.

**emulsion cracking:** an artifact on film caused by improper storage and handling of film.

**emulsion, film:** a mixture of gelatin and silver halide crystals (in suspension) where latent image formation takes place.

**exposure:** a measure of the amount of ionization produced in air by an x-ray beam, measured in milliroentgens (mR) or Roentgens (R).

**exposure rate:** Roentgens per minute measured at a point of interest.

**film, direct exposure:** silver halide film primarily sensitive to the direct action of x-rays but having low sensitivity to light from screen fluorescence.

**film speed:** the sensitivity of film emulsion to x-ray or light exposure.

**film, x-ray:** silver halide film that is primarily sensitized to the spectrum of light emitted by fluorescent intensifying screens.

**fixer:** a chemical solution that both removes the unexposed and undeveloped silver halide crystals from the coated film emulsion and hardens the gelatin.

**flooded replenishment:** a technique used in low-volume processing that adjusts replenishment pumps in an automatic film processor to automatically add a predetermined amount of developer and fixer replenisher on a timed cycle to maintain the chemical activity of the processor; also called flood replenishment.

**fluence:** in electromagnetic imaging, the number of photons per unit area.

**fluorescence:** the property of a phosphor to emit light in the visible region of the electromagnetic spectrum as a result of absorbing higher energy radiation, such as x-rays.

**focal-film distance (FFD):** the distance from the anode target (focus) of an x-ray tube to the radiographic film; also referred to as the target-film distance (TFD) or the source-image receptor distance (SID).

**fog:** an unwanted exposure to the film emulsion from light, radiation, heat, or chemicals; also refers to the optical density on a processed radiograph arising from this exposure.

**gamma:** the slope of the straight-line portion of the characteristic curve that is related to film contrast.

**gelatin:** a material made from animal tissue in which silver halide crystals are suspended to make an emulsion.

**gradient:** the slope of the tangent to the characteristic curve at a given density.

**graininess:** the appearance of optical density fluctuations on film, caused by the random inhomogeneous distribution of silver grains as they are unevenly coated in the emulsion.

**granularity:** the microdensitometric measurement of the graininess fluctuations in an emulsion.

**gray scale:** the range of optical densities present in an image.

**grid:** a device composed of parallel or slanted lead strips and radiolucent spacer material, used to reduce scattered radiation.

**grid, moving:** a device comprising a stationary grid and a mechanism for moving it that blurs the image of the grid lines, making them indistinguishable on the radiographic film; also referred to as a Potter-Bucky diaphragm or a Bucky mechanism.

**H&D curve:** see characteristic curve.

**halation:** a double image or "halo" effect that occurs when some of the exposing light penetrates the film emulsion and is reflected back by the film support.

**heavy metals:** metals with a specific gravity greater than 5.0; this includes some metals found in photographic effluents such as iron, silver, and chromium.

**hypo:** the agent in the fixer solution that dissolves the unexposed, undeveloped silver halide crystals, leaving the black metallic silver in the exposed and developed areas of the film more readily discernible; also called clearing agent.

**image in space:** see aerial image.

**image, latent:** the invisible changes that occur in an unprocessed image receptor and which arise from x-ray exposure.

**image processing:** any operation performed to enhance or clarify an image as an aid in diagnosis.

**image quality, radiographic:** a description or quantitative measure of the excellence of an image with respect to contrast, blur, noise, and artifacts.

**image receptor:** the x-ray detection material (for example, intensifying screen[s] with radiographic film) on which a latent image corresponding to the aerial image is formed.

**intensifying screen:** fluorescent crystals coated on a plastic base that convert x-radiation into light.

**intrinsic efficiency:** the ratio of light energy generated in a phosphor crystal to the x-ray energy absorbed by the crystal; also called phosphor conversion efficiency.

**inverse square law:** the relationship between distance and radiation fluence in which the exposure varies inversely as the square of the distance from the source.

**latitude, exposure:** a measure of the tolerance allowed in the choice of exposure values required to produce the desired density range.

**latitude, film:** the capability of a film emulsion to record a wide range of exposures, resulting in the film having a lower inherent contrast. Wide-latitude film is often used for chest radiography.

**line-spread function (LSF):** spatial distribution of optical density in the slit image of an x-ray beam.

**maximum density:** the highest optical density that would be obtained if all the silver halide grains on a film were developed; also referred to as saturation density or D-max.

**metallic replacement:** a method of recovering silver from silver-rich solutions by an oxidation-reduction reaction with elemental iron and silver thiosulfate to produce ferrous iron and metallic silver. The device used is commonly called a metallic replacement cartridge (MRC).

**mid-density (MD):** the optical density for a given step on a sensitometric strip whose value is approximately 1.0 plus base plus fog; also referred to as medium density, speed or speed index.

**minus-density artifact:** any unwanted artifact on a radiograph characterized by a white mark; usually occurs from improper handling, dust (i.e., shadow images), etc., before a latent image is placed on the film, but may also occur during processing (e.g., pick-off).

**modulation transfer function (MTF):** a graphic representation of the resolution capability of an imaging system or component.

**mottle, quantum:** a type of radiographic mottle due to the statistical variation in the number of photons incident on any given area of the image receptor.

**mottle, screen:** an image imperfection resulting from the structural composition of the fluorescent screen; also known as structure mottle.

**National Electrical Manufacturers' Association (NEMA):** an organization of manufacturers active in standardization.

**National Institute of Standards and Technology (NIST):** an organization that deals with standards; formerly known as the American National Standards Institute (ANSI).

**noise:** any undesired variation of input or output signal that interferes with the detection of the desired signal.

**overcoat:** a very thin layer of clear gelatin in film, usually coated on top of the emulsion layer to protect it from abrasion and provide important surface properties.

**parallax effect:** the impression of a change in position of an object as seen on dual-emulsion film when a tube-angled technique is used.

**pH:** a numeric value from 0 to 14 that describes the acidity or alkalinity of a solution with 7 representing neutrality; numbers less than 7 represent increasing acidity; and numbers greater than 7 represent increasing alkalinity.

**phantom:** a test object that simulates the x-ray absorption of various structures within the body.

**phosphor conversion efficiency:** see intrinsic efficiency.

**phosphorescence:** the ability of phosphor to emit light as a result of the previous absorption of higher energy radiation, such as x-rays; see afterglow.

**photon:** a bundle of electromagnetic energy.

**phototimer:** an electronic timing device that automatically measures and terminates the x-ray exposure to achieve a preselected density on a radiograph; also called an automatic exposure control (AEC) device.

**plus-density artifact:** any unwanted artifact on film characterized by a black mark; usually occurs from improper handling, etc., between the time a latent image is placed on the film and before the film is processed.

**Potter-Bucky diaphragm:** a device comprising a stationary grid and a mechanism for moving it that blurs the image of the grid lines and makes them indistinguishable on radiographic film; also referred to as a Bucky mechanism or a moving grid.

**processing cycle:** the amount of time required for the leading edge of the film to enter and the trailing edge to exit an automatic processor; also called drop time.

**processor, automatic film:** a machine that transports film through developing, fixing, washing, and drying sections to produce a finished radiograph.

**processor quality control (QC) chart:** a chart used to record and monitor performance of a film processor in a quality control program.

**publicly owned treatment works (POTW):** a wastewater-treatment plant owned by the public (municipality of service authority).

**quality assurance (QA):** a set of guidelines, tests, and procedures followed on a planned schedule to help ensure that optimum quality images are produced to facilitate diagnosis.

**quality control (QC):** the tests and procedures used in a quality assurance plan to routinely assess the imaging environment.

**radiograph:** an image captured on film (or other hard copy media) that is created as a result of differential absorption of radiation by objects in the x-ray beam.

**reciprocity law:** a law stating that the optical density of a radiograph will be the same if the milliampere-seconds (mAs) is constant, regardless of mA.

**replenishment:** manually or automatically adding fresh chemicals to developer and fixer solutions to maintain proper chemical activity, stability, and liquid levels.

**replenishment rate:** the amount of replenisher chemicals added to the developer and fixer solutions based on film type, film size, and usage rate (number of sheets of film processed in 24 hours).

**resolution:** the ability of an imaging system to distinguish closely spaced lines visually.

**Roentgen (R):** a unit of radiation exposure measured in air.

**safelight:** a device containing a low-wattage lightbulb and filter that produces visible light to which film is relatively insensitive, while still allowing one to see and function in the darkroom.

**saturation density:** see maximum density.

**screen:** fluorescent crystals coated on a plastic base that convert x-radiation into light.

**screen conversion efficiency:** the ratio of the light energy emitted by an intensifying screen to the x-ray energy absorbed by the screen.

**screen-film contact:** the close proximity of the intensifying screen to the film emulsion that is necessary to minimize image blur.

**screen speed:** a measure of the relative light output of various intensifying screens for a given x-ray exposure.

**sensitometer, simulated light:** a device that exposes film to a controlled, stepped set of light exposures. It is used to evaluate a film's response to exposure.

**SID:** source-image receptor distance.

**silver halide:** a compound produced by mixing a solution of silver nitrate with a solution of potassium halide. Used in photographic emulsions because of its sensitivity to light and other forms of radiation.

**silver estimating test paper:** a test paper coated with a chemical reagent that changes color according to the amount of silver in the solution.

**silver recovery:** any method in which a recovery unit removes silver from the fixer effluent.

**speed:** a measure of the exposure needed to produce a given density on film.

**stepped-wedge:** a device with graduated thicknesses that is used to demonstrate or test various degrees of x-ray penetration; also known as a step-wedge.

**subtraction:** a radiographic, photographic, or digital electronic technique in which unwanted information can be subtracted from images to enhance the visibility of the desired information. Subtraction is especially useful in angiography.

**time-temperature processing:** a technique used in manual processing that involves adjusting development time and developer temperature to produce optimum optical density and image quality for a given exposure.

**viewbox:** a device that provides a uniform field of white light for viewing the radiographic image.

**viewing station:** an electronic device for the display of text and medical images in digital format.

# Index

Page numbers followed by f or t indicate a figure or table, respectively.

## A



## B



## C

# D



# E



# F

# G



# H

# Q



# R



# S



# T

## U



## V



## W